Zooarchaeology and Field Ecology

Zooarchaeology and Field Ecology

A Photographic Atlas

Jack M. Broughton and Shawn D. Miller

Foreword by Frank E. Bayham

The University of Utah Press
Salt Lake City

The Defiance House Man colophon is a registered trademark of the University of Utah Press. It is based on a four-foot-tall Ancient Puebloan pictograph (late PIII) near Glen Canyon, Utah.

LIBRARY OF CONGRESS CATALOGING-IN-PUBLICATION DATA

Names: Broughton, Jack M. | Miller, Shawn (Shawn D.)
Title: Zooarchaeology and field ecology : a photographic atlas / Jack M. Broughton and Shawn D. Miller ; foreword by Frank E. Bayham.
Description: Salt Lake City : The University of Utah Press, 2016. | Includes bibliographical references and index.
Identifiers: LCCN 2015045226| ISBN 9781607814856 (paperback) | ISBN 9781607814863 (ebook)
Subjects: LCSH: Animal remains (Archaeology)—West (U.S.)—Pictorial works. | Animal remains (Archaeology)—Canada, Western—Pictorial works. | Natural history—West (U.S.)—Pictorial works. | Natural history—Canada, Western—Pictorial works. | Animal ecology—West (U.S.)—Pictorial works. | Animal ecology—Canada, Western—Pictorial works. | Animal remains (Archaeology)—West (U.S.)—Laboratory manuals. | Animal remains (Archaeology)—Canada, Western—Laboratory manuals. | Natural history—West (U.S.)—Laboratory manuals. | Natural history—Canada, Western—Laboratory manuals. | BISAC: SOCIAL SCIENCE / Anthropology / General.
Classification: LCC CC79.5.A5 B76 2016 | DDC 930.1/0285—dc23
LC record available at http://lccn.loc.gov/2015045226

All photographs are by the authors unless otherwise noted in the caption. Photos not credited to the authors are used with permission or are in the public domain.

Printed and bound by Sheridan Books, Inc., Ann Arbor, Michigan.

Contents

Illustrations

Figures

Color Plates (following page 126)

Tables

Foreword

On a cold spring morning in 1986, I hiked up a volcanic, boulder-strewn ridge to get a view of Eagle Lake and the surrounding mountains and valleys. It was my introductory visit to the Eagle Lake Field Station of Chico State University in northeastern California, and I sat quietly absorbing the beauty. I was captivated by the majesty of the setting and the diversity of the environment. A small herd of deer crept out of a mountain mahogany grove in the distance and tentatively approached the water's edge. As I watched them for a good half hour, a vision and resolve developed to find a way to teach zooarchaeology in this isolated, world-class field setting.

Over the next year in dialogue with biologist Raymond J. Bogiatto, this nascent idea evolved into a course, *Zooarchaeology and Field Ecology*, which we launched in the summer of 1988. We invited an energetic, irrepressible Chico State graduate student, Jack Broughton, to serve as the teaching assistant that year. Jack embraced and nurtured the vision of teaching zooarchaeology with its interdisciplinary nature in a field setting. That vision has evolved into a rich, cohesive treatise, strongly rooted in the academic traditions of vertebrate paleontology and evolutionary ecology.

Zooarchaeology is the study of animal remains from archaeological sites in order to understand the complex and changing relationships of people and animals. At its core, zooarchaeology involves the identification of what are quite often fragmentary skeletal remains of fishes, amphibians, reptiles, birds, and mammals. The careful examination of these bone remains, comparison with known osteological specimens, and written and graphic documentation of taxonomic skeletal variation has long-standing roots in vertebrate paleontology. Paleontologists J. A. Shotwell, who strongly influenced eminent zooarchaeologist Donald K. Grayson, and Stanley J. Olsen, who authored many osteological manuals for vertebrate identification in zooarchaeology, exemplified and put a premium on the integrity of bone identifications. The quality of graphic skeletal images presented in a comparative fashion in this atlas reflects Shawn Miller's extensive experience in human dissection and publication of human anatomy references and hearkens to an important tradition in paleontology: sound interpretations and inferences must be predicated on accurate, replicable bone identifications. The clarity and descriptive detail of the osteological images in this volume with attention to diagnostic criteria across an array of vertebrate classes is unique yet very traditional.

Because fragmentary bone data, individual or aggregated, are fundamentally zoological, it is important to understand the life history and ecology of the animals being identified. The tradition of appreciating the position of an organism in its environment and evolutionary context has its roots in evolutionary ecology. Robert MacArthur was an eminent theoretical ecologist whose breadth of knowledge of animal communities and population ecology influenced the development of evolutionary ecology and generations of ecologists. Sievert Rohwer and Stephen Fretwell were influenced by MacArthur and in turn directly or indirectly affected this work. Broughton worked with Rohwer at the Burke Museum–University of Washington and accompanied him on field excursions, including a collecting expedition to Siberia. Steve Fretwell taught me animal ecology and served on my doctoral committee; his ideas and approach to science have subtended much of my teaching and research. While it may not be manifest directly in this study, this tradition of evolutionary ecology has infused an integrated vitality into virtually all of Broughton's archaeofaunal research. Certainly, Broughton and Miller's treatment of animal systematics, life histories, and ecology across a spectrum of taxa from fishes to mammals in this volume is a testament to the strength of this tradition.

While the rich integration of detailed bone identifications, comparative osteology, and animal ecology could perhaps be done in any academic classroom, the original decision to teach such disparate knowledge realms in a relatively isolated setting invoked another important tradition from biology and archaeology: the field school. In the biological sciences and in archaeology, fieldwork and the field school are important traditions that bring like-minded individuals and researchers together. In that setting, personal interactions and relationships, both formal and informal, can give moment and significance to the mundane, leaving an indelible imprint and enhancing the learning experience.

Modern zooarchaeological interpretations are informed by a variety of theoretical perspectives that may well span the anthropological spectrum. This diversity is wonderful and will lead to innovative interpretations and discoveries in the future. Yet the foundations of our intellectual heritage should be acknowledged and remembered. This volume by Broughton and Miller, *Zooarchaeology and Field Ecology*, brings together several important intellectual traditions that can benefit students and researchers alike as they embark in their study of the dynamic field of zooarchaeology.

— Frank E. Bayham
California State University, Chico

Preface

Zooarchaeology is an interdisciplinary subfield of archaeology. It focuses on the analysis of animal remains from archaeological contexts in order to understand the long-standing and complex relationships between past people and animals. The focus is often on reconstructing and explaining variation in ancient human subsistence behavior, but considerable emphasis is also placed on what archaeological vertebrate remains can tell us about climate change and past environments. Such analyses are also increasingly providing information relevant to the conservation and management of living animals.

The field of zooarchaeology thus transcends the traditional academic boundaries between zoology, anthropology, paleontology, and anatomy. Making taxonomic identifications by comparing ancient bone fragments with known modern comparative skeletons is the core of zooarchaeological lab work. This task requires a substantial background in vertebrate osteology, of course, but bone identifications are only the starting point. Much of what can be learned from them derives from knowledge about the behavior, ecology, and natural history of the animals they represent. Without this knowledge, the analytical and inferential potential residing in an archaeofaunal assemblage cannot be fully realized. As a result, teaching zooarchaeology, in our view, is best carried out in a combined field and laboratory setting where students can directly experience animal ecology and behavior in the field while they also master vertebrate bone identification and analysis in the lab. Integrating wildlife ecology with osteology, along with instruction in relevant zooarchaeological method and theory (quantification, foraging theory, taphonomy, research design, etc.), has been the focus of our *Zooarchaeology and Field Ecology* field course that we have taught at the remote Eagle Lake Field Station for decades.

Students of *Zooarchaeology and Field Ecology* must thus engage both vertebrate osteology and natural history, but unfortunately the currently available guidebooks and atlases only deal with such subjects as independent fields. That is, separate guides for: 1) the bones—often focused on specific kinds of animals and specific parts (e.g., mammal skulls)—and, 2) the taxonomy, natural history, and ecology of animals—also focused typically on specific classes such as birds, mammals, or fishes. This separation has long been a source of frustration when teaching our zooarchaeology courses and has required the compilation of various technical osteology identification guides along with separate books and articles that cover the taxonomy and natural history of a wide range of vertebrate groups.

We designed this atlas explicitly as a laboratory manual for zooarchaeology courses and integrate the two realms here in a single volume. We present the osteological details and natural history information of all the vertebrate classes and, with few exceptions, all of the orders that are native to inland North America west of the Rocky Mountains and north of Mexico. Because these represented orders have much broader ranges that encompass much of North and Central America, the atlas will be useful for students working across these broader regions. The atlas should also be a useful reference for zoologists, forensic and wildlife biologists, vertebrate paleontologists, or anyone with a general interest in identifying animal bones and understanding the natural history and ecology of the species they represent.

We begin the atlas with an introductory chapter that provides important background information essential for zooarchaeology students (taxonomic classification, bone biology, etc.). The next five chapters then present the detailed osteology and the ecology and natural history of the vertebrate classes native to the west. We begin each class-level treatment with *General Osteology* sections that provide background information for each group followed by a series of labeled photos that clearly depict the skeletal elements and features of representative animals of the different classes. Using a variety of magnifications and views, the photos allow easy recognition of each labeled structure. Legends corresponding to the labeled photos occur below the photo on the same page to allow self-testing and easy review.

For each vertebrate class, we then present photos with labeled features to illustrate the distinguishing osteological characteristics that allow the identification of selected bony elements to lower taxonomic levels such as orders, families and, in some cases, genera. These *Taxonomy and Osteological Variation* sections present up-to-date taxonomic nomenclature for vertebrates occurring in the region. And, again, because zooarchaeological analysis depends on an intimate knowledge of the living animals, these presentations are annotated with information on the habitat affinities and natural history features of the various taxa, accompanied by photographic images of animals in their characteristic habitats or other useful contexts.

A few technical matters should be mentioned here. Our approach to literature citation differs from many standard academic works in order to avoid interrupting and cluttering the text with repetitive lists of citations. We do, however, provide at the close of each chapter a *Notes* section that provides the references we used with notes on how we used them and relevant topics. In addition, and because the primary intended audience for the atlas is students of zooarchaeology, we provide an extended sampling of additional references related to the chapter topics that we feel may provide a springboard to additional reading on a group's biology, natural history or osteology. These works are listed by chapter in "Suggestions for Further Reading," which follows the "References" section at the end of the Atlas.

As noted, our taxonomic focus covers interior western North America and fauna occurring during Holocene times, exclusive of marine taxa. We thus do not systematically cover historically introduced species or those that succumbed to extinction toward the end of the Pleistocene (e.g., mammoths, horses, camels, sloths, etc.).

Regarding capitalization of animal common names, accepted styles vary considerably. The American Ornithologists' Union and the American Society of Ichthyologists and Herpetologists require that the first letters of the common names of birds, amphibians, reptiles, and fishes be capitalized in their peer-reviewed journals (*The Auk* and *Copeia*). The American Society of Mammalogists, on the other hand, advocates usage of lower case for the first letters of mammalian common names, unless they include a proper noun. To maintain consistency, we follow here the accepted convention for fishes, birds, and herps and capitalize the first characters of all common names.

Finally, taxonomic identification of fragmentary vertebrate bones and teeth is the foundation of zooarchaeological studies. All meaningful conclusions hinge on their accuracy. We view the material presented here, once mastered, as an essential foundation for the more detailed analyses that typically require the use of comparative osteological collections. For example, the atlas provides features that are useful for distinguishing the skulls of rabbits from squirrels, and the femora of hawks from falcons, but no guidance on how to distinguish the different species of squirrel, rabbit, hawk, or falcon. Species-level identifications must be determined, in most cases, by actual comparisons with known skeletal samples of all potential species that may have occurred in a region prehistorically. As helpful as guides such as this one may be, there remains no substitute for direct comparison with known skeletal reference specimens.

Acknowledgments

We are grateful and indebted to Frank E. Bayham and Raymond Jay Bogiatto who conceived the *Zooarchaeology and Field Ecology* field course from which this work was inspired. We also thank them and R. Lee Lyman and Virginia Butler for reviewing the manuscript and providing many helpful suggestions that significantly improved the final product. We would also like to thank all of our zooarchaeology students who have made comments and caught various errors in earlier versions of the atlas.

CHAPTER 1

Introduction

Vertebrate faunal materials derived from a range of depositional contexts can inform our understanding of past human subsistence, lifeways, and the nature of ancient environments. In many cases our work involves the analysis of animal remains from archaeological contexts where clear evidence of human involvement is indicated. Bones recovered inside ancient ruins in direct stratigraphic association with stone tools, ceramics, or human burials would be an example of such an archaeological fauna, or archaeofauna. In other cases, we may be interested in analyzing faunal material from deep cave deposits in order to reconstruct trends in past climate and environments where owls and carnivores introduced the bones and past people played no role in the accumulation. This would represent a paleontological deposit. In many settings, especially cave and rockshelter contexts, animal bone materials owe their presence to a mix of agents that include past people, raptorial birds, and carnivores.

Determining the taxonomic identity of the recovered vertebrate bones and teeth thus lies at the heart of zooarchaeological data collection. The effort requires a range of skills, including familiarity with a number of classification systems and terminology relating to taxonomic nomenclature, skeletal biology, and anatomy. Before we can systematically present the osteological details of western vertebrates, which will provide a basis for making zooarchaeological identifications, it is important to discuss this relevant background information.

Identification, Classification, and Taxonomy

Since our focus is on identifying vertebrate bones and teeth, we begin with several key terms. A fitting start is thus the term "identification" itself, which refers to the assignment of something to an existing category or class. Proper identification of a bone fragment must utilize two distinct classification systems: one pertaining to the different named parts of the skeletal system (e.g., skull, femur) and the second to an established biological class (e.g., deer, dog, mouse). We will examine skeletal nomenclature in detail, but first we address biological classification.

Biological classification involves the process of establishing groups or categories of organisms. As many might recall from high school biology, it has its roots in the work of Carolus Linnaeus, an eighteenth-century Swedish naturalist. It was Linnaeus, in fact, who coined the term taxonomy (Gr., taxis = arrangement, nomia = method) to refer to the methods of naming and classifying organisms in an ordered system. While Linnaeus grouped species according to shared physical characteristics reflecting what he believed to mirror the links in God's "Great Chain of Being," modern biologists construct the groupings so as to reflect evolutionary relatedness following the Darwinian principle of common descent and ancestry. Traditionally, biologists estimated this degree of relatedness by comparing the anatomy (coloration, number of legs, body size, etc.), physiology, and behavior of organisms, but much recent work is based on molecular data (e.g., DNA) where relatedness is determined by similarities and differences in genetic makeup (see Systematics and Phylogenetics below).

The basic structure of the Linnaen system of zoological nomenclature, as it is called, is a series of divisions or ranks. Moving from top to bottom, divisions reflect ever-decreasing levels of inclusiveness and ever-increasing levels of evolutionary relatedness.

Kingdom
  Phylum
    Subphylum
      Class
        Order
          Family
            Genus
              Species

Starting at the bottom, groups of closely related species are placed in the same genus (plural: genera). In turn, groups of related genera are placed in the same family and so on. Species names follow the binomial system of nomenclature, so named because they are formed by two Latin words (bi = two, nomial = name) that are usually chosen to describe features of the organism. For example, the binomial or scientific name for the Great Horned Owl is *Bubo virginianus*, and the full taxonomic hierarchy of this bird is presented in Table 1.1.

TABLE 1.1. Taxonomy of the Great Horned Owl (*Bubo virginianus*)

Kingdom	Animalia
Phylum	Chordata
Subphylum	Vertebrata
Class	Aves
Order	Strigiformes
Family	Strigidae
Genus	*Bubo*
Species	*Bubo virginianus*

Note first of all in this table that the Great Horned Owl is in the subphylum Vertebrata, a group that contains all animals that possess a backbone or vertebral column and is the taxonomic focus of this atlas. Lower down the list we see that *Bubo* is the genus (or generic) name, which is Latin for owl, while *virginianus* is the specific epithet (or specific name) and refers to Virginia, the state from which the bird was first described. The genus and specific epithet together comprise the species, in this case *Bubo virginianus*. Other distinct but similar owls are grouped in the same genus, but are assigned to different species (i.e., they have different specific epithets) such as the Snowy Owl, *Bubo scandiacus*. Remember that the plural for genus is "genera" but "species" is both singular and plural. For example, *Bubo* is a genus of owl and *Bubo* and *Strix* represent two genera of owls; but, I saw three squirrel species yesterday but only one species today.

Lastly, geographic variation within a species is typically formalized by a subspecies designation, commonly referred to more casually as races. In Great Horned Owls, for example, the plumage color varies considerably and is correlated with regional humidity regimes, with the darkest populations occurring along the humid north Pacific coast (northern California to southern Alaska). The unique owls in this setting have thus been assigned to a unique subspecies, *Bubo virginianus saturatus*. Subspecific epithets (*saturatus*, in this case) are also italicized but not capitalized. Note that the genus name (*Bubo*) is capitalized and italicized while the specific epithets (*virginianus* and *scandiacus*) are italicized but presented in lower case. For all higher taxonomic ranks or categories, the first letter of the name is capitalized but italics are not used (e.g., Strigidae, the owl family; Strigiformes, the owl order). Note too that the specific epithet should never be written without the genus name (full or abbreviated) preceding it. These conventions are strictly observed in scientific writing.

A name for any of the various ranks may be referred to as a taxon (plural: taxa). Depending on the formality of the publication, a name may also be followed by the "authority" for that taxon. The authority is represented by the last name of the author who first published a valid description of the taxon. In our example, R. Ridgway first described *Bubo virginianus saturatus* in 1877, and thus represents the authority for this taxon—thus denoted as *Bubo virginianus saturatus*, Ridgway. These authority names are frequently abbreviated. For example, the abbreviation "L." is universally accepted for Linnaeus. It is also accepted convention that if any part of a taxon has been revised since the original description, the original authority's name is placed in parentheses, thus, *Bubo virginianus saturatus* (Ridgway).

Of all the different taxonomic ranks used in this system, only the species has a relatively unambiguous definition provided by the biological species concept: all the members of different populations that are capable of interbreeding and producing fertile offspring. This allows a biological test to determine species membership, at least among living, sexually reproducing organisms. Fossil species are usually defined by the morphological species concept, as an assemblage of fossils representing organisms that lived at the same time that have similar skeletons with consistent differences between them and all other organisms.

The membership of all other higher taxa (e.g., genera, families, orders) must be determined more subjectively. Not surprisingly, there is often considerable disagreement among taxonomists on how this should be done. Indeed, whether individual taxonomists have a tendency to make many separate taxonomic categories or a fewer number of larger groupings are referred to colloquially as "splitters" and "lumpers." Fortunately, within each of the major vertebrate classes, authoritative scientific councils have been established to revise taxonomic nomenclature as new information on evolutionary relationships becomes available. In our presentations of the various western vertebrates in this atlas, we provide the current taxonomies according to these authorities: for fishes, we follow the taxonomy of Froese and Pauly (2015); for amphibians and reptiles, Collins and Taggart (2015); for mammals, Wilson and Reeder (2015); and for birds, the American Ornithologist's Union (2015).

There are several commonly used conventions that are helpful to recognize when referring to these higher-level taxonomic names. The endings of higher-level taxa, for example, follow the pattern whereby order names (in fishes and birds) end in "-iformes," family names end in "-idae," and subfamily names end in "-inae." In standard parlance, one can refer to an individual within a given family by omitting the "ae." For example, we would say that a Red Fox (*Vulpes vulpes*), a species in the family Canidae, is a canid; similarly, all dogs, wolves and foxes are canids. Animals in the order Artiodactyla are referred to as artiodactyls; rabbits and hares in the order Lagomorpha are lagomorphs. Note only the formal order names are capitalized.

In the context of zooarchaeological identification, we strive to identify a given bone or tooth to the lowest taxonomic level possible—that is, to the species level—because a species determination carries the greatest amount of information relating to past environments and human foraging behavior. It is routinely the case, however, that some specimens will be too fragmented, or from a nondiagnostic element or portion (e.g., a rib shaft), to allow a species-level identification. When this occurs—common in zooarchaeology—the identification must be elevated to a higher taxonomic level. We may have criteria, for example, that allow us to identify a distal femur specimen to the kangaroo rat genus *Dipodomys*, but not enough information to determine which species it may represent. The identification of such a specimen should thus be left at the genus-level: *Dipodomys* sp., where sp. is an abbreviation for "species" that is unknown in this case. The abbreviation "spp." refers to the plural form of species and is used to indicate that several different species are or may be represented. There is some inconsistency in how these abbreviations are used. In some academic traditions, "sp." is used to indicate simply that species identification is impossible or has not been attempted but in other cases that a single, unknown species is represented. Regardless, sp. and spp. are not italicized.

There are some occasions in which a secure, genus-level identification is made for a specimen with strong reason to believe a particular species is represented; yet a lack of definitive osteological features precludes a secure species-level identification. In such cases of provisional identifications, the abbreviation "cf." (from the Latin, *confer* = bring together, compare) can precede a likely, but unconfirmed, species (e.g., *Dipodomys* cf. *microps*). It is advisable to discuss in the text when a cf. designation is used to provide a rationale for its use. To take one example with rabbits, the site locality may lie within the elevational and geographic range of only a single rabbit species, say *Sylvilagus nuttallii* (Mountain Cottontail), and several thousand secure, species-level identifications were made for cranial elements of this species. In this context, an author may wish to designate post-cranial materials from the site that compared closely with *Sylvilagus nuttallii* reference material, but lacked specific features to allow definitive identifications, as *Sylvilagus* cf. *nuttallii*.

It can often be the case that many specimens only permit secure identifications to higher-level taxonomic categories, that is, to the family, order, or even the class level. This may occur due to fragmentation or simply because the element involved is not diagnostic or distinctive below these higher levels. For example, the vertebrae of fishes belonging to the order Cypriniformes (minnows and suckers) are widely known not to differ consistently among the many species in the order, so when present in archaeological faunas, they can only be identified to the order level. Another example from many sites in western North America is the presence of specimens that can only be identified to the taxon "medium Artiodactyla." This is because the three widespread medium-sized artiodactyl species in the region—Pronghorn (*Antilocapra americana*), Mule Deer (*Odocoileus hemionus*), and Bighorn Sheep (*Ovis canadensis*)—may be present, but specimens are too fragmented or represent elements that are not distinctive (ribs, vertebrae) to allow secure species-level identifications (Collared Peccary [*Pecari tajacu*] is our only "small" artiodactyl, while bison [*Bison bison*], elk [*Cervus elaphus*] and moose [*Alces americanus*] are "large" artiodactyls). Depending on the nature of the research question, such higher-level (order or family) identifications may or may not be informative. If the research questions are geared toward paleoenvironmental concerns that typically require genus or species-level identification (because it is typically different *species* that exhibit unique tolerations to habitat and environmental conditions), such specimens would provide little, if any, relevant information. If, on the other hand, the interest is in trends through time in human diet and the energetics of foraging behavior, then changing frequencies of minnows and suckers relative to medium artiodactyls would be especially telling.

Systematics and Phylogenetics

The current taxonomic nomenclature that is used in presenting zooarchaeological data is based on our understanding of the evolutionary relationships between vertebrate taxa. The study of these evolutionary relationships in order to understand the diversification of life over time is known as systematics and as noted previously is carried out through detailed comparisons between living and fossil organisms. Evolutionary relationships of a set of organisms can be determined by comparing the similarity between structural, functional, behavioral, molecular, genetic, or other features. All can contribute to our understanding of evolutionary relationships, but our focus here is on osteological features. Corresponding osteological parts in different organisms may come to resemble each other, however, in several different ways, with not all of them related to common ancestry or origin. This latter form of similarity is referred to as homology. Similarity in the structures between two or more organisms can be considered homologous if the trait can be traced back through time to a common ancestor. The bones in a duck's wing and a monkey's arm are homologous structures that can be traced to a common reptilian ancestor living about 300 million years ago. Strong clues suggesting that different bones may be homologous often come from their relative position and connections to surrounding parts within the body. Similarity in overall shape can also be telling. For example, consider the bones of the forelimb in a human, a deer, and a bat. Although these bones have evolved over the last 60 million years to serve different functions, respectively, grasping, running, and flying, they are still recognizable homologies based on their shapes and spatial relationships in the body.

It is also the case that homologous structures may become dissimilar over time. One such example is the morphological changes evident in the evolution of the auditory ossicles (incus, malleus, stapes) in the mammalian ear, which were derived from the lower jawbones (quadrate, articular, hyomandibular) of fish ancestors. Similarities also need not represent common ancestry at all. Indeed, unrelated organisms may evolve very similar structures from different underlying anatomical sources if they enter the same habitats and

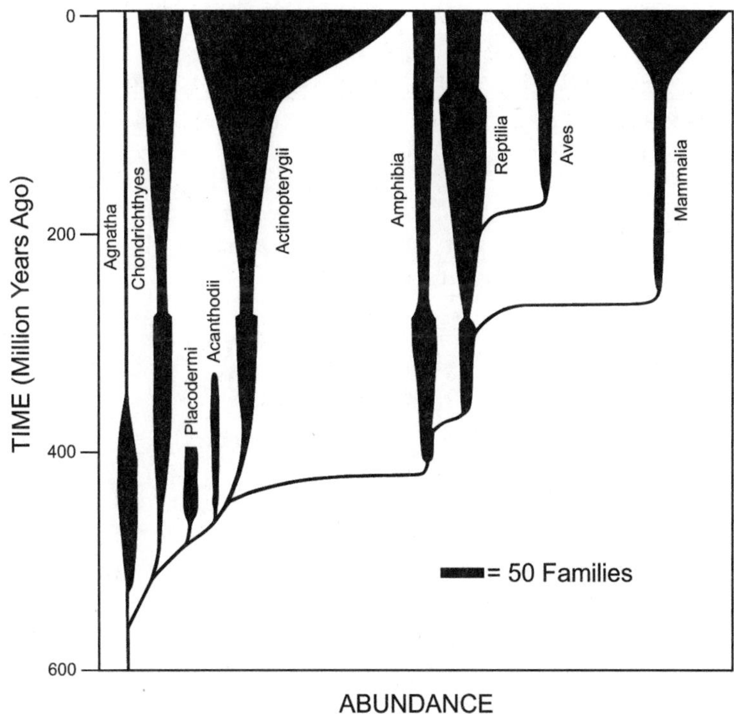

FIGURE 1.1. Spindle diagram of major vertebrate groups.

are subjected to comparable selective pressures. Similarities resulting from this process, known as convergent evolution, are said to be analogous (or analogies). Sharks and dolphins, for instance, share many superficial similarities: both groups have streamlined bodies and strong fins for efficient swimming and large, sharp teeth for securing prey. However, while sharks are clearly fishes, fossil evidence makes it clear that dolphins share a common ancestor with artiodactyls, the order of mammals that includes such species as deer, sheep, antelope, and hippos. The similarity in body form has resulted from convergence.

In establishing evolutionary relationships, contemporary systematic biologists are not concerned so much with general homologous similarities, as they are with identifying specific features that two groups share but are not found in any other group. Such a situation would imply a very close evolutionary relationship. In the parlance of systematics, they are referred to as shared derived characters or synapomorphies (Gr., syn = together, apo = away from, morpho = shape).

The product of systematic analyses is commonly depicted in phylogenetic trees (or phylogenies, dendrograms), diagrams showing the hypothetical evolutionary relationships of a group of organisms derived from a common ancestor. The ancestral form represents the tree "trunk" while forms that have arisen from it are put at the tips of the tree "branches." The distance from one group to another indicates the degree of relationship: closely related taxa are located on branches close to one another. Ideally, each higher taxon should contain all of the species derived from a common ancestor and, if so, are said to represent a clade or monophyletic group (Gr., monos = single, phyle = tribe). Paraphyletic groups (Gr., para = beside) by contrast do not contain all of the species derived from a common ancestor. A phylogenetic tree for the major vertebrate classes, including all the representatives in this atlas, is provided in Figure 1.1.

This particular form of phylogenetic tree is referred to as a spindle diagram: the horizontal dimension reflects the relative abundance of taxa within each class while the vertical axis represents time.

The result of systematic analyses that reconstruct the relationships of organisms is reflected in the taxonomic arrangement of different taxa. So while this spindle diagram shows the currently

TABLE 1.2. The Geological Time Scale

Eon	Era	Period	Epoch	Date at Beginning (mya)
Phanerozoic				
	Cenezoic			
		Quaternary		
			Holocene	0.01
			Pleistocene	2.6
		Tertiary		
			Pliocene	5
			Miocene	23
			Oligocene	35
			Eocene	56
			Paleocene	65
	Mesozoic			
		Cretaceous		146
		Jurassic		208
		Triassic		250
	Paleozoic			
		Permian		290
		Carboniferous		362
		Devonian		408
		Silurian		439
		Ordovician		510
		Cambrian		550
Precambrian				4560

understood relationships of major vertebrate classes, such relationships have also been established for lower taxa within each class. For example, within each class, some orders are known to be more ancient, while others are more recently derived. And within each family, some genera are known to have occurred before others in evolutionary time. These established sequences of occurrence within any taxon provide a standard convention for the listing of taxa in a wide range of contexts, from museum collections of skins and skeletons, to field guides, to zooarchaeological reports that provide the relative abundances of different taxa recovered at a site. We follow this convention here and use the established sequences for western vertebrates according to the currently accepted taxonomies noted earlier.

The Geological Time Scale

As we introduce and discuss the major vertebrate classes and the evolutionary relationships among them, we will frequently acknowledge current evidence as to the timing and origins of those groups using the Geological Time Scale. This scale provides a classification system within which earth history is divided into a hierarchical sequence of named time periods (Table 1.2). Three main eras are commonly recognized: Paleozoic, Mesozoic, and Cenozoic. The eras are further divided into epochs and periods. The vertebrate story begins with the jawless fishes (Agnatha) about 500 million years ago (mya) during the Paleozoic era and the Ordovician period. All of the extant vertebrate classes are present by the end of the Jurassic period of the Mesozoic, about 150 mya. Dramatic increases in vertebrate diversity occur for all major classes over the last 65 million years during the Cenozoic era.

Bone Biology, Development, and the Skeletal System

Given that soft tissues generally do not survive the ravages of time, our ability to reconstruct past environments and human-animal relationships must almost always be based on analyses of preserved bones and teeth. In this section, we briefly examine the structure of bones and teeth as connective tissues and introduce key aspects of the skeletal system including its primary functions and the ways in which the skeleton grows and develops. Although bone might appear to represent an inert, static, and silent residuum of vertebrate life, it is in fact a highly malleable living tissue that can represent a rich chronicle of both the evolutionary as well as the individual life histories of animals.

Skeletons are rigid structures that provide support and protection for living organisms. For many invertebrate animals skeletons are comprised of a hard protein known as chitin that surrounds and encloses the soft tissues and organs of the body. These are referred to as exoskeletons. In vertebrates, however, the skeleton is internal and thus called an endoskeleton. It serves three primary functions under normal conditions. First, the skeleton provides rigid attachment sites for muscles and so is integral to body movement and locomotion. Second, although the skeleton is internal and covered by muscle and skin, it also serves to support and protect many vital, soft tissues and organs. The cranium, of course, provides a solid wall of protection for the brain; the vertebral column encloses and protects the spinal cord; and the heart, lungs and major blood vessels are encased within the rib cage. Third, the skeleton serves a number of important physiological functions. Most notably, both red and white blood cells are produced from red and yellow marrow that lies within bones. Bones also provide an important reservoir of critical elements such as calcium and fat that can be readily mobilized in periods of nutritional stress.

Bones

As an organic, living connective tissue, bone is a composite material comprised of two primary constituents, one inorganic and one organic. The inorganic component is a crystalline calcium carbonate material known as hydroxyapatite. This mineral component represents about 65% of the dry weight of bone and gives bone its rigidity and compressive strength. The organic component is comprised of protein, primarily collagen, and makes up the remaining 35% of bone. Collagen forms long fibers that during bone growth are

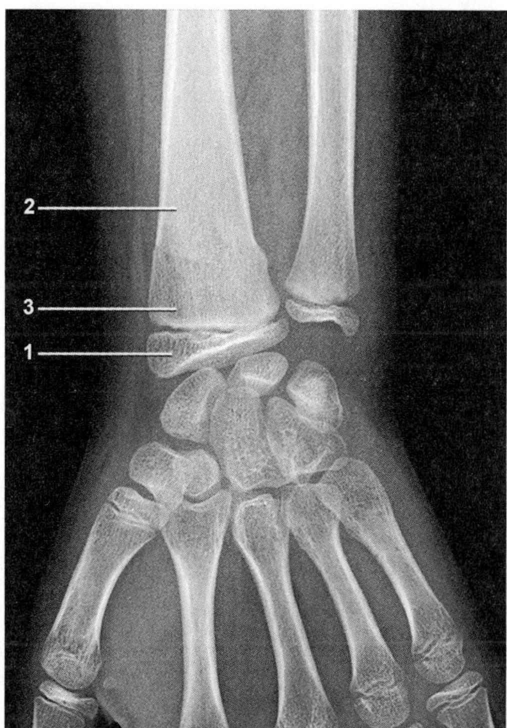

FIGURE 1.2. X-ray of a human distal (away from point of attachment) forelimb. 1, epiphysis; 2, diaphysis; 3, metaphysis.

infiltrated by hydroxyapatite to form a densely packed matrix. This organic component of bone gives a certain elasticity as well as added strength.

Bones are also typically classified according to their shapes. Cylindrical bones of the limbs are referred to as long bones; plate-like bones of the skull, girdles, and ribs are known as flat bones; and bones with other odd shapes are referred to as irregular bones. Short bones are those that are nearly as wide as they are tall, such as the wrist bones or carpals. Long bones also have distinct named portions: the main shaft of a long bone is known as the diaphysis, while the ends or caps are known as epiphyses (Figure 1.2). The region where they come together is known as the metaphysis. The hollow within bones is known as the medullary cavity. The medullary cavity is typically filled with red or yellow bone marrow—fat-rich substances that produce blood cells. During life, bones are covered externally by thin tissue called the periosteum, a tough, vascularized membrane that nourishes bone. The inner surfaces of bones are covered with an analogous membrane known as the endosteum.

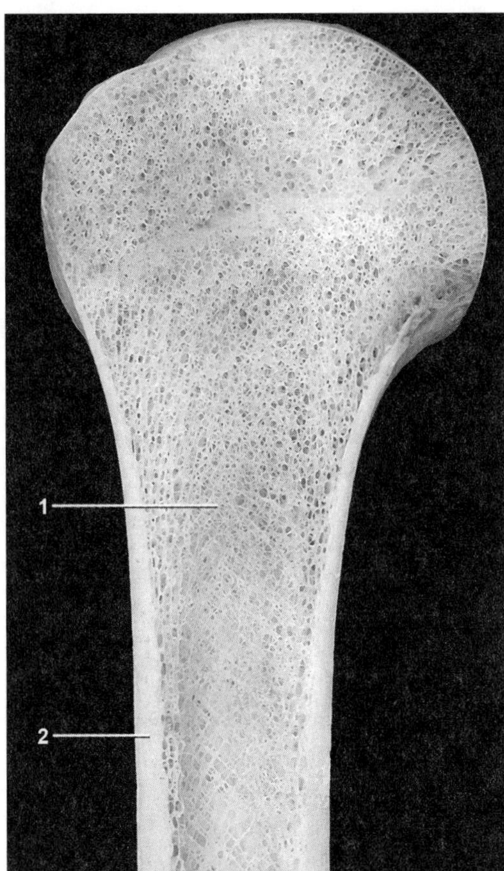

FIGURE 1.3. Frontal section of proximal (nearer point of attachment) human humerus. 1, spongy bone; 2, compact bone.

Different portions of bones also have unique structural properties with two primary kinds of mature bone being commonly recognized: compact bone and spongy bone (Figure 1.3). Compact bone is also called dense bone or cortical bone. Spongy bone is also called cancellous bone or trabecular bone. Compact bone, as the name implies, is solid and dense; spongy bone consists of a thin, air-filled lattice of bony fibers. Compact bone and spongy bone are found in different locations in the skeleton. Most of the diaphyses of long bones, for instance, are made up of compact bone, with a relatively small amount of spongy bone forming along the medullary cavity. The epiphyses of long bones, by contrast, consist primarily of spongy bone that is covered with a thin shell of compact bone. The flat bones of the skull are comprised of an interior layer of spongy bone that is sandwiched between two relatively thick layers of compact bone.

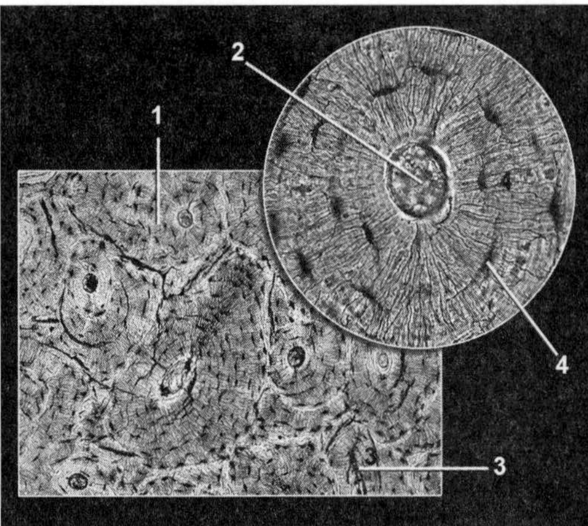

FIGURE 1.4. Micrograph of Haversian System. 1, secondary osteon; 2, Haversian canal; 3, Volkmann's canal; 4, lacuna housing osteocyte.

At the cellular level, compact bone is comprised of cylindrical structures called secondary osteons or Haversian systems that typically lie parallel to the long axis of the bone (Figure 1.4). A secondary osteon, which appears similar to a tree trunk, consists of concentric layers, or lamellae, of bone matrix that surrounds a central Haversian canal. The Haversian canal contains small nerves and blood vessels. Small canals that link different Haversian canals together and with the marrow cavity are referred to as Volkmann's canals. Situated within the lamellae of secondary osteons are gaps or lacunae containing living bone cells that are involved in both bone deposition and bone resorption, or the removal of bone tissue. There are three main types of bone cells. Osteoblasts are often concentrated at the surface just beneath the periosteum and actually produce bone-forming materials that eventually surround and enclose them; at that point, these mature bone cells are referred to as osteocytes. Osteoclasts, the third main cell type, are responsible for resorption. Both cell types play a critical role in the bone remodeling that occurs throughout the life of an individual in response to a variety of mechanical and physiological stresses.

In addition to the lamellae contained within a secondary osteon, remnant lamellae not associated with osteons are also evident in compact bone. These structures are referred to as interstitial lamellae and represent vestiges of old remodeled secondary osteons. Finally, the exterior and interior surfaces of long bones contain lamellae that extend the length of the shaft and are referred to as outer and inner circumferential lamellae, respectively.

Different bones also have different embryological origins that are characterized by different forms of growth. Bone originating internally within cartilage develops through a process known as endochondral ossification. Bones derived from the integument or connective tissues from the skin grow by dermal ossification. With endochondral ossification, bone is deposited gradually within a cartilaginous precursor or template. Bones formed this way are sometimes referred to as cartilage bones or replacement bones. In mammals, endochondral ossification proceeds through a process of fusion between initial or primary centers of ossification—typically within the diaphysis—and secondary centers that form the epiphyses. Until adulthood, the main shafts (primary centers) are separated from the epiphyses (secondary centers) by a cartilaginous growth plate. Ultimately, the plate is invaded by bone and epiphyseal fusion is thus said to be complete (Figure 1.5).

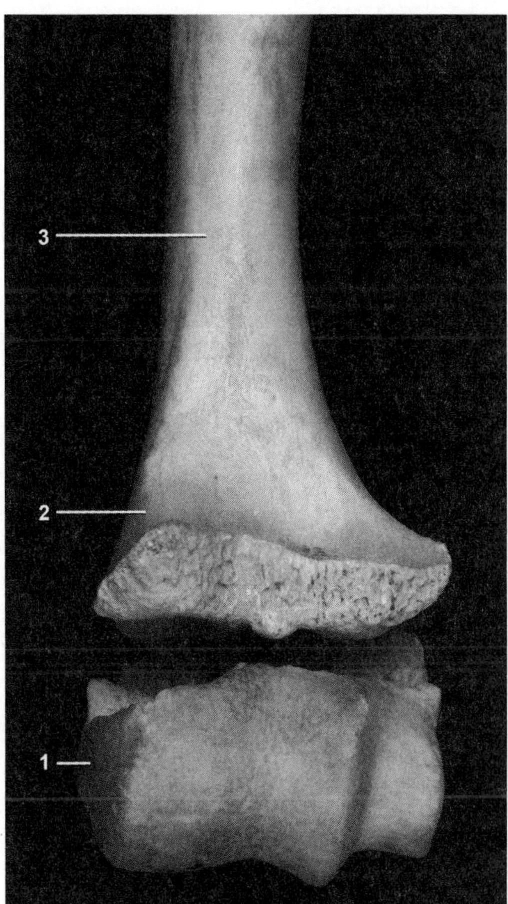

FIGURE 1.5. Distal humerus of immature Elk (*Cervus elaphus*). 1, detached distal epiphysis; 2, metaphysis; 3, diaphysis.

Alternatively, with dermal ossification, bone is formed directly within connective tissue, such as within the skin or dermis, without a cartilaginous template. With dermal ossification, bone growth starts within a central condensation of collagen fibers and osteoblasts and gradually expands outward at the margins. As bone grows, it thickens, forming a plate-like structure; when adjacent dermal bones contact, the junction, often forming a jagged, squiggly line, is known as a suture. Dermal bones tend to be flat and are commonly found on the skull, whereas endochondral bones take on a wide range of shapes and include most bones of the post-cranial skeleton. Single bones (e.g., the mammalian clavicle) may be derived from the fusion of both dermal and endochondral bone.

Teeth

Teeth are small, calcified structures found in the jaws of many vertebrates that are used to seize and masticate (chew) food materials. In some animals, especially carnivores and primates, teeth also serve defensive and communicative functions. Teeth appear to have evolved from dermal denticles, armor-like scales found in primitive fishes that contain both dentin and enamel, the two main hard components found in teeth. For several reasons, teeth have an importance in paleontology and archaeology disproportionate to their size in the body. First, they have greater preservation and survivorship potential than bones because they are much harder, denser structures. The fossil record is thus far richer for teeth than it is for other parts of the skeleton. Second, the morphology of teeth closely reflects an animal's feeding habits, so dietary adaptations can be readily reconstructed from them. Finally, because teeth exhibit great variation in structural detail and also tend to be evolutionarily stable, they are invaluable structures for tracking phylogenetic relationships between vertebrates.

Teeth are also among the most diagnostic structures in the vertebrate skeleton that can

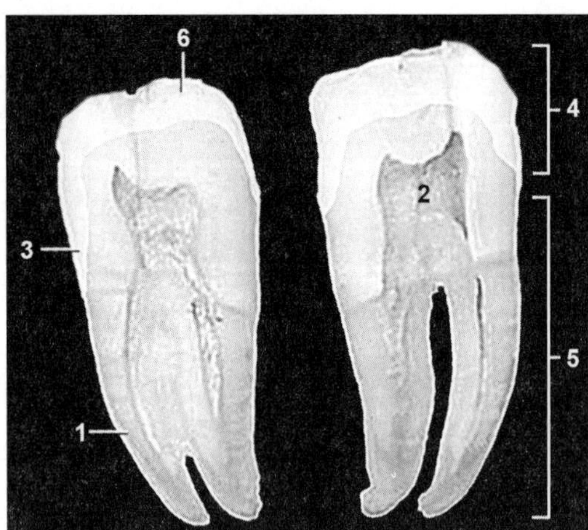

FIGURE 1.6. Longitudinal section of human molar. 1, dentin; 2, pulp chamber; 3, enamel; 4, crown; 5, root; 6, occlusal surface.

typically be identified to the species level. A single isolated tooth will almost always allow a species- or genus-level identification, whereas a rib shaft fragment, for instance, may only allow a family- or order-level identification. Teeth can also be studied to provide information on the age, sex, and health of individual animals.

Teeth are attached to the jaws of vertebrates in several ways. The teeth of cartilagenous fishes, such as sharks, are anchored only by collagenous fibers. The teeth of most vertebrates, however, are anchored more securely with bony attachments. Most mammals have what is referred to as thecodont attachment, where the teeth have roots that fit snugly into hollows or sockets referred to as alveoli (singular: alveolus) in the upper and lower jaw. Most reptiles and amphibians have pleurodont teeth, where they are anchored on a bony shelf. Acrodont dentition, found in most bony fishes, refers to teeth that are attached to bones by pedestals at their base.

Structurally, the living core of a tooth is a dense, yellowish, bone-like material called dentin (Figure 1.6). Through its base, dentin is penetrated by a pulp chamber that contains nerves and blood vessels. A thin layer of extremely dense, hard, and brittle material, called enamel, covers the dentin. Enamel is 96% mineral, mainly hydroxyapatite crystals. The part of the tooth evident above the gum line is referred to as the crown; the part that anchors the tooth to the jaw, below the gum line is the root. Tooth roots attach to the bone within the alveoli by a substance known as cementum. The chewing surface of a tooth is referred to as its occlusal surface, and when upper and lower teeth come to rest on one another they are said to be in occlusion. Unlike the continuous remodeling that takes place in the bony skeleton, a tooth's crown morphology can only be affected by attrition (wear), breakage, or demineralization—bones can heal, tooth damage is permanent.

Because teeth can be readily damaged or worn during feeding, a variety of approaches have evolved to maintain these structures that are so essential for survival. Most vertebrates, in fact, replace teeth continuously throughout their lives. Sharks, for instance, may grow a new set of teeth every several weeks to replace worn or lost ones. This is referred to as polyphyodont tooth replacement. On the other hand, most mammals have only two generations of teeth: an initial set of a milk or deciduous dentition that is followed by a second set of permanent or secondary teeth. This is referred to as diphyodont dentition. In rare instances, such as in toothed whales, only a single set of teeth forms, a pattern referred to as monophyodont dentition.

Tooth formation, eruption, and replacement sequences are patterned and occur within fairly narrow age ranges for different mammal species and thus can be used to determine the ontogenetic (growth) age at death of an animal. Additionally, the degree of tooth wear as measured by tooth crown height is a common method used to determine age at death. However, eruption sequences and crown height are less useful methods of determining age at death in vertebrates that have addressed the problem of excessive tooth wear by evolving ever-growing or ever-erupting teeth. Ever-growing teeth, also referred to as hypselodonty, occurs in many rodents and lagomorphs. Ever-erupting teeth occurs in Pronghorn, horses, and other ungulates. For these taxa, age at death determination can be evaluated by examining differential wear patterns on the occlusal surfaces of the cheek teeth.

Teeth come in a wide range of shapes and sizes among the vertebrates. In many groups, however, the overall form of individual teeth in the mouth does not vary much. All the teeth in the Bullfrog (*Lithobates catesbeianus*), for instance, are small, cylindrical, pointed structures of nearly identical size and serve the same function in grasping and

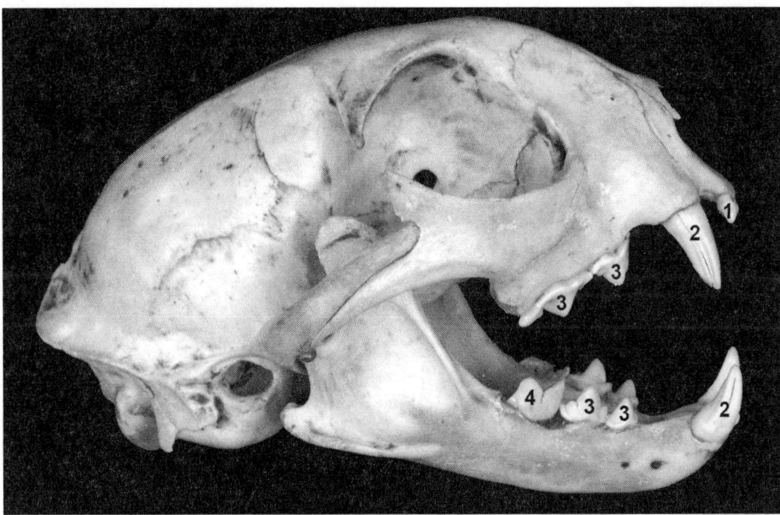

FIGURE 1.7. Cranium of Bobcat (*Lynx rufus*). 1, incisor; 2, canine; 3, premolar; 4, molar.

securing prey. This pattern is referred to as homodont dentition and is characteristic of most fishes, reptiles, and amphibians. Other vertebrates, especially mammals, have teeth differentiated with respect to function, or heterodont dentition.

For a mammal the typical pattern comprises four distinct types of teeth (Figure 1.7).

The anterior-most teeth are called incisors. They are usually chisel-shaped and adapted for securing food and sometimes also used for grooming. By convention, the numbers of different types of teeth are expressed for a given species by the number found in one side of the mouth. These are commonly depicted in a fraction with the count of teeth for the upper jaw being expressed as the numerator and the count for the lower teeth forming the denominator. The common pattern for mammalian incisors, for example, is 3/3.

Moving back in the mouth, the next set of teeth is the canines. They are usually long, strong, and spike-like and serve for piercing and holding prey as well as for fighting and defense. Unless secondarily lost, the canines number 1/1. The teeth immediately behind the canines are the premolars, and posterior to them are the molars. Molars are typically larger than premolars and have more roots and cusps, the latter being small projections on the occlusal surface of tooth crowns. The primitive counts for most mammalian premolars and molars are normally considered to be 4/4 and 3/3, respectively. Specialization can make it difficult to distinguish premolars from molars, so for convenience these teeth are often referred to collectively as cheekteeth. Patterns of cusps on the cheekteeth are varied and intricate, and a detailed terminology exists to identify them.

The overall shape of cheekteeth varies between major groups of mammals, reflecting broad types of feeding strategies. Classifications also exist to characterize these variations (Figure 1.8). Bunodont cheekteeth are moderately broad, with low cusps effective for crushing a variety of food materials and are common in omnivores. High-crowned, or hypsodont, teeth provide for long wear for herbivorous animals that feed on large quantities of coarse abrasive materials (e.g., grasses with silica) that cause rapid attrition. In hypsodont teeth, as the occlusal surface of the crown wears down, the roots gradually rise higher in the jaw, exposing more of the crown. Additional terminology applies to teeth with different cusp shapes. One type of hypsodont cheekteeth found in deer, camels, and cattle is called selenodont dentition. This literally translates as "moon tooth" and refers to the crescent moon shape characteristic of the cusps of their teeth. Another type of hypsodont cheekteeth is the lophodont pattern where the cusps form a series that are elongated transversely (medial-lateral) to the toothrow and are found in beavers as well as horses and their relatives.

Finally, the numbers and kinds of different teeth that characterize a species are expressed in a dental formula that consists of the complete set of fractions denoting the separate teeth mentioned above. For instance, the dental formula of the

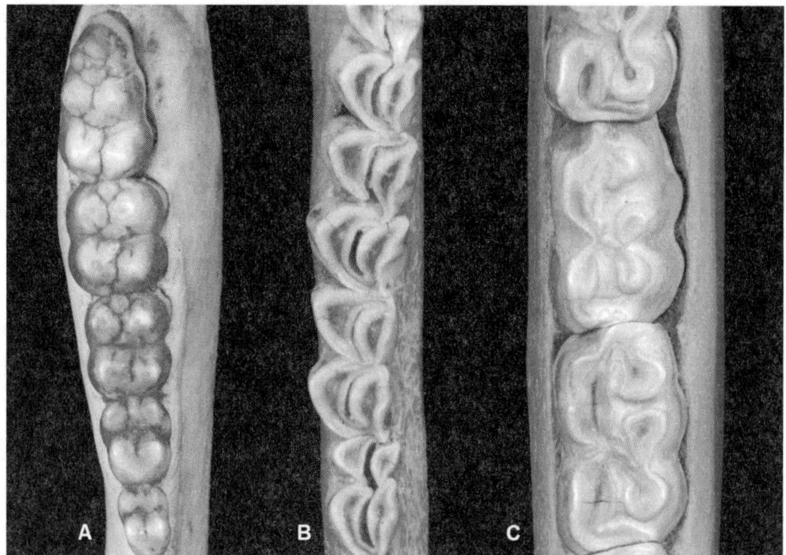

FIGURE 1.8. Mandibular tooth rows of Collared Peccary (*Pecari tajacu*) (A), Mule Deer (*Odocoileus hemionus*) (B), and Horse (*Equus*) (C), showing bunodont, selenodont, and lophodont tooth types, respectively.

Bighorn Sheep (*Ovis canadensis*) is 0/3, 0/1, 3/3, 3/3; for the California Kangaroo Rat (*Dipodomys californicus*) it is 1/1, 0/0, 1/1/, 3/3; and for humans it is 2/2, 1/1, 2/2, 3/3. Because premolars and molars may be morphologically indistinguishable, it is often necessary to consult a reference guide that provides dental formulae in order to identify a specific tooth as either a premolar or molar. Finally, individual teeth are denoted by the combination of the first letter of the tooth type (e.g., "I" for incisor, "C" for canine, "P" for premolar) and by their position in the tooth row moving from front to back (1, 2, 3, etc.). Teeth in the upper jaw (maxilla) and lower jaw (mandible) are denoted by superscript and subscript fonts, respectively. Thus, M^1 refers to the first molar in the upper tooth row while P_2 refers to the second premolar in the lower jaw.

Anatomical Terminology

Zooarchaeologists, paleontologists, and comparative anatomists throughout the world share a standard terminology to describe specific parts, portions, and orientations of the vertebrate skeleton. This terminology is essential for effective study, comparison, communication and finding your way around the vertebrate skeleton, just as knowing the terms right and left, and north, south, east, and west are critical to navigating from map directions. We will briefly describe the most essential anatomical terminology used in this atlas in reference to the vertebrate skeleton.

Directional Terms

These directional terms apply to quadrupedal vertebrates. Note that several differ from their usual application to bipeds, as encountered in human anatomy contexts (Figures 1.9–1.10). All of the terms are also useful only in relation to another specific landmark: most are also directly opposing, just as north is directly opposite south. The first five directional terms are specific to teeth.

Mesial: opposite of distal; a point nearer the midline of a dental arch where the central incisors contact. The incisors are mesial to the premolars.

Distal: opposite of mesial; a point farther from the midline of a dental arch where the central incisors contact. M^3 is distal to M^1.

Lingual: opposite of buccal or labial; towards the tongue side of the mouth or a tooth. The lingual tooth surfaces are not visible even when an angry dog shows its teeth.

Labial: opposite of lingual; a point in the mouth or on a tooth located toward the lips. Usually reserved for incisors and canines. The labial tooth surfaces are what you see when a person smiles.

Buccal: opposite of lingual or labial; a point located toward the cheek side of the mouth or a tooth. Usually reserved for premolars and molars.

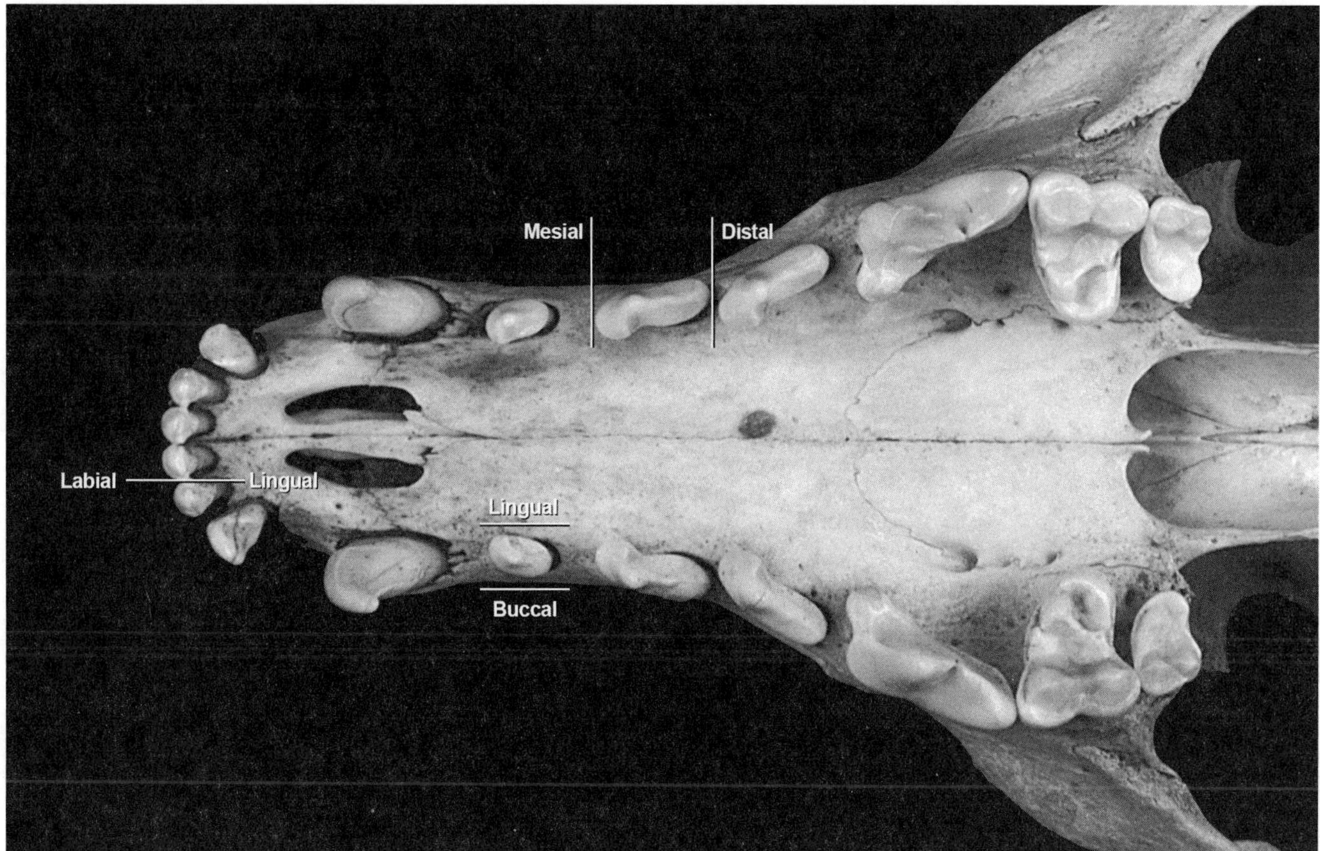

FIGURE 1.9. Cranium of Red Fox (*Vulpes vulpes*), ventral view, showing directional terms used for the dentition.

Medial: opposite of lateral; a point located toward the midline of the body. The sternum is medial to the pectoral girdle.

Lateral: opposite of medial; a point located away from the midline of the body.

Proximal: opposite of distal; a point located closer to the attached end of a limb. The upper arm bone, the humerus, is proximal to the hand.

Distal: opposite of proximal; a point located farther away from the attached end of a limb. The foot is distal to the shin.

Anterior (or cranial): opposite of posterior; toward the head or nose of the body. The skull is anterior to the vertebrae.

Posterior (or caudal): opposite of anterior; toward the tail end of the body. The vertebrae are posterior to the skull.

Ventral: opposite of dorsal; toward the underside or belly side of the body. The ribs project ventrally from the vertebrae.

Dorsal: opposite of ventral; toward the back of the body. The vertebrae are located dorsally to the ribs.

Endocranial: the inner surface of the skull. The brain fills the endocranial cavity.

Ectocranial: the outer surface of the skull. Unless broken, the ectocranial surface of the skull is what we usually see.

Parts of the Skeleton

In the sections that follow, we present the osteological details of the major vertebrate groups in a consistent sequence starting first with the axial skeleton, the parts of the skeleton that fall along the main axis of the body. The axial skeleton thus includes the skull or more formally, the cranium, and the bones of the trunk including the vertebrae, ribs and sternum. Reference to the post-cranial axial skeleton, refers to all of the axial skeleton posterior to, or not including, the cranium. The other primary components of the skeleton are the portions that support the appendages, and are thus referred to as parts of the appendicular skeleton. Finally, there are two paired sets of appendages, those comprising the front limbs, or forelimbs, which are supported by the pectoral

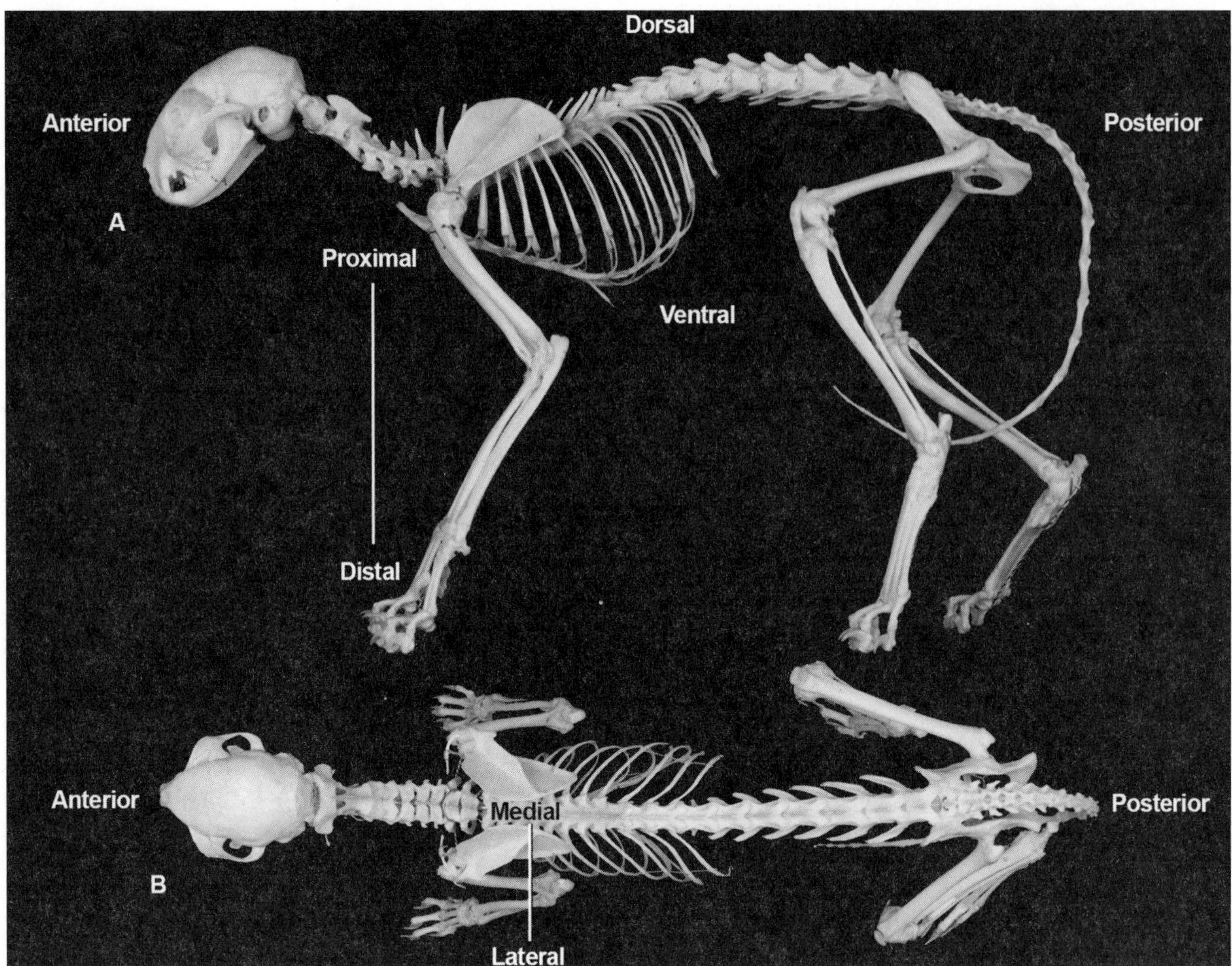

FIGURE 1.10. Skeleton of Domestic Cat (*Felis catus*), lateral (A) and dorsal (B) views, showing directional terms for the skeleton.

girdles, and the back limbs, or hindlimbs, which are supported by the pelvic girdles. For both the axial and appendicular skeletons, we present all of the separate bones, or elements, plus many of the more commonly used features on those bones.

Notes

Most introductory texts in general zoology, comparative vertebrate anatomy, and vertebrate paleontology provide detailed discussion on classification, systematics, phylogenetics, and taxonomy. Our discussion on these topics draws from Liem et al. (2001), Lederer (1984), Kardong (2004) and Hildebrand (1995). See Bengston (1988) for further discussion on the use of sp., cf., and other qualifiers in identification. For current taxonomic nomenclature we follow Froese and Pauly (2015), Collins and Taggart (2015), Wilson and Reeder (2015), and the American Ornithologist's Union (Alderfer 2014, 2015) for fishes, amphibians and reptiles, mammals, and birds, respectively. Many works address the geological time scale, which undergoes periodic changes, especially with the timing of the different units. Our treatment follows Benton (1997), with updates from Walker et al. (2012). White and Folkens (2005) provide a succinct treatment of bone biology, development, and the skeletal system; we adapt our discussion from their excellent work. Our presentation of tooth attachment types and tooth replacement draws on Hildebrand (1995), and Liem et al. (2001), respectively.

CHAPTER 2

Fishes

I. General Osteology of Fishes

Characteristics

Since the group "fishes" is polyphyletic and represented by five separate vertebrate classes, it is difficult to characterize. However, all fishes are completely aquatic—occurring in marine, freshwater, and estuarine environments—and use branchial (gills) respiration to extract oxygen from water. Fishes have ectothermic ("cold-blooded") metabolism and streamlined bodies that allow for efficient swimming. Most have a series of paired and unpaired fins, oviparous reproduction, and skin covered by keratinous dermal scales (the scales of birds and reptiles are of epidermal origin). They are also characterized by indeterminate growth, where individuals continue to grow in size throughout their lives. Currently, over 32,000 species of fishes are recognized, more than all other vertebrates combined.

Origins

Although current taxonomies are based on established or hypothesized phylogenetic relationships, broad groupings based on shared characteristics are still used in our everyday lives as well as in discussions of vertebrate evolution and diversity. The term tetrapod, for example, which literally means four-footed, includes animals that have four feet even though they come from a wide assortment of classes—amphibians, reptiles, birds, and mammals—including taxa within them that are understood to have substantially modified or secondarily lost legs such as snakes, legless lizards, and birds. Fishes are another such grouping that includes a diverse collection of separate vertebrate lineages and classes distinguished from the tetrapods by the lack of legs. Another broad pair of groups is distinguished by the absence or presence of jaws: hinged elements supporting the borders of the mouth. Crosscutting the fishes-tetrapod distinction, jawed vertebrates are referred to as gnathostomes, while those without jaws are known as agnathans (Gr., no jaws). Only a few groups of fossil and living fishes are agnathans, including the very first vertebrates. These first vertebrates had a mouth but lacked a biting apparatus and first appeared during the early Cambrian explosion, about half a billion years ago. Only two living classes, lamprey (Cephalaspidomorphi) and hagfishes (Myxini), retain this condition, but they are otherwise highly modified, having adapted to specialized niches.

Jawed fishes first appear in early Silurian (~450–500 mya) freshwater deposits and represent the ancestors of all subsequent gnathostome vertebrates. The evolution of the hinged jaw from anterior gill arch elements close to the mouth was one of the most significant developments in early vertebrate evolution. Two sets of paired fins are also present with the early gnathostomes. Jaws provided a means by which they could bite and crush larger prey, and the paired appendages allowed for more efficient balance and swimming. The combination enabled a major radiation. Aquatic habitats of the subsequent Devonian period were the first in Earth history to contain a diverse set of large, mobile, predatory vertebrates. Representatives of the other living fish classes emerged soon thereafter, including the cartilaginous sharks, rays and chimeras (Chondrichthyes) and two classes of bony fishes: the ray-finned fishes (Actinopterygii) and the fleshy-finned fishes (Sarcopterygii). Members of the latter group (e.g., Rhipidistia), which have paired fins that rest on the ends of short, bone-supported appendages gave rise to the first terrestrial vertebrates.

Osteology

The fish skeleton is primarily comprised of axial elements (Figures 2.1–2.14), to a degree surpassed only in some reptiles that have secondarily lost their appendages such as snakes and legless lizards. The skeleton thus consists mainly of cranial elements, vertebrae, and ribs that together make up over 90% of the bones in the fish skeleton. While fish vertebrae and ribs are repetitive simple structures and relatively easy to identify as elements, the fish cranial skeleton is a seriously complicated structure that varies tremendously in composition, shape, and morphology between different groups of fishes.

Before examining the fish skull in detail it is helpful to appreciate several major parts of the vertebrate cranial skeleton that have both distinct ontogenetic origins and phylogenetic representations: the chondrocranium, the splanchnocranium, and the dermatocranium.

The brain and sense organs of primitive vertebrates were protected by capsules and rods of cartilage that merged to form the chondrocranium. Vertebrates that lack bony skulls, such as the cartilagenous fishes, retain a complete and heavy chondrocranium. In most other vertebrates, the chondrocranium begins in larval or fetal stages as cartilage but is mostly replaced by bone in adult animals. Notable bony elements that derive from ossification of the chondrocranium include the sphenoid, ethmoid, prootic, and occipital.

The splanchnocranium (or visceral skeleton) is derived from the pharynx, the expanded part of the digestive system situated between the oral cavity and the stomach. Early in vertebrate evolution, the pharynx become perforated laterally by paired gill slits that functioned in both feeding and respiration. The gills were supported by a series of paired cartilagenous or bony bars, with associated muscles, and respiratory tissues called visceral arches, so called because they are derived from part of the digestive system. Several of these are retained to support the gills in fishes and are thus referred to as gill or branchial arches. However, the anterior-most visceral arch was modified to form the initial components of the jaws in gnathostomes. The palatoquadrate is the first key element of the upper jaw while the mandibular cartilage forms the lower jaw. The second, or hyoid, arch has been modified to form the hyoid apparatus that controls and supports the tongue and vocal structures in tetrapods. A key element of the hyoid arch in gnathostomes is the hyomandibular.

The remaining bones of the vertebrate skull are membrane bones, derived from the skin (or dermis) and are referred to as dermal elements or collectively as the dermatocranium. An amazing range of shapes and sizes of dermal bones occurred in early vertebrates, but a pattern of medium-sized elements stabilized by the time the bony fishes emerged. The basic pattern evident in the bony fishes has been modified in the different classes through various fusions and deletions.

An especially important region of cranial variation in fishes concerns the jaw elements involved in feeding. A wide range of adaptations for such structures is represented in taxa native to western North America. Lacking hinged jaws, the lampreys are the most primitive agnathan form. They have instead an oral disk, bearing pointed, toothlike structures derived from keratin. All other native fishes are within the class Actinopterygii and thus have hinged jaws with the mandibular arch supported primarily by the hyomandibular (= hyostyli), although this class has substantial variations in jaw structures.

Representing a less specialized jaw structure, trout and salmon (*Oncorhynchus*) have large mouths with the hinge located far back (posterior) on the skull (Figure 2.1). And although the upper jaw does not move forward in relation to the rest of the skull (i.e., it is not protrusible), the maxilla bears impressive teeth and pivots downward as the lower jaw opens, preventing prey from escaping out the corners of the mouth. Most other more derived fishes, such as those in the minnow family (Cyprinidae; Figure 2.1), have smaller mouths as the hinge of the mandible has migrated forward under the orbit. These taxa are also equipped with mobile premaxilla and max-

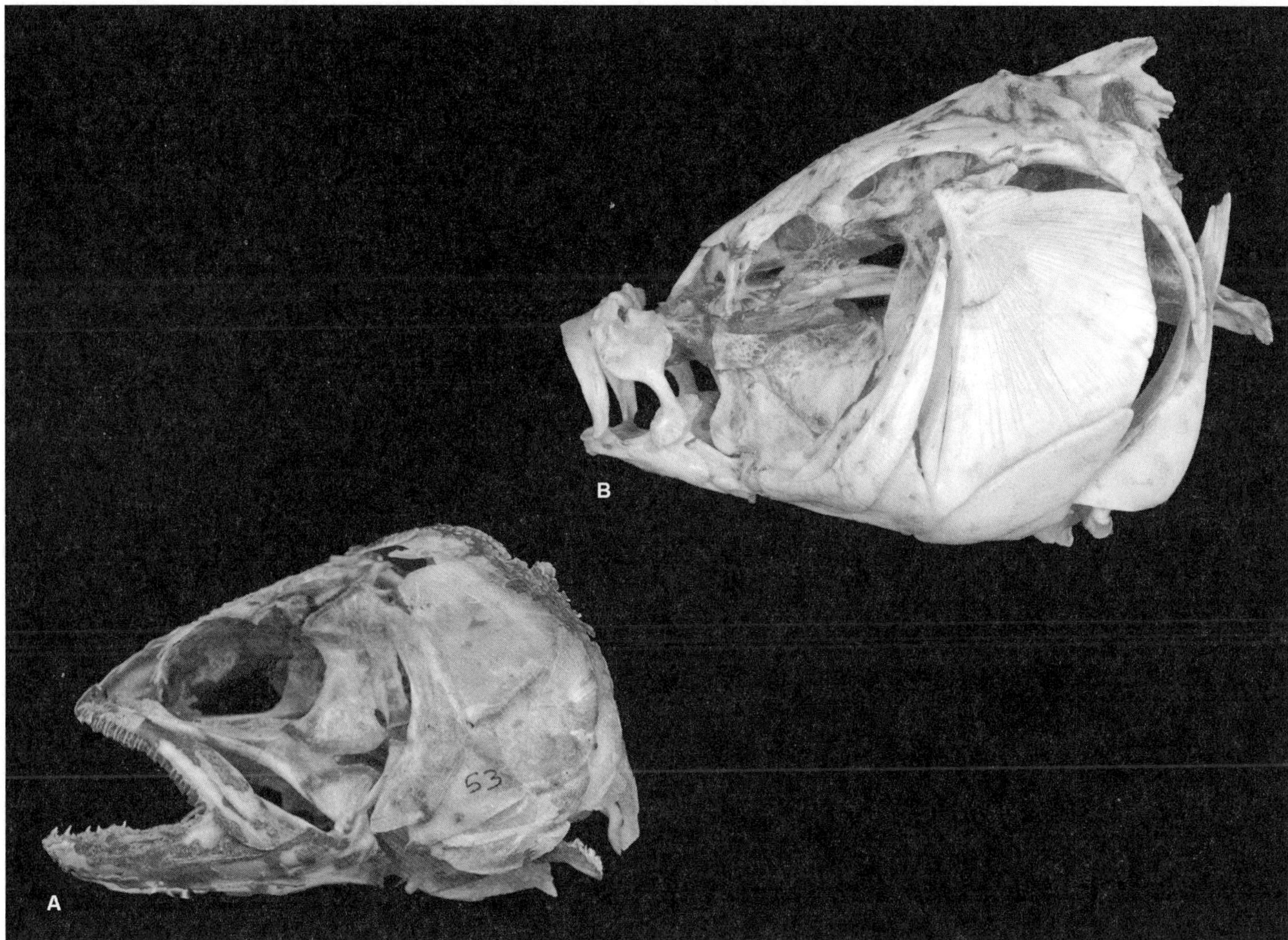

FIGURE 2.1. Different jaw forms in fishes. The jaw structure of Cutthroat Trout (*Oncorhynchus clarkii*) (A) is less specialized compared to suction feeders with protrusible jaws, as in Common Carp (*Cyprinus carpio*) (B).

illa and protrusible jaws that convert the mouth into a small tube that allows for suction feeding (Figure 2.1). Suction feeders have less need for grasping prey, and most have reduced teeth in their mouths or none at all. Several groups that lack teeth in their mouths, however, have evolved toothed structures from gill arches that function to process food. Most bony fishes have acrodont tooth attachments with homodont patterns of tooth variation.

Other noteworthy aspects of the fish skeleton are related to adaptations for life in a water medium. A series of dermal bones provide bony support for a large flattened structure, the operculum, which covers and protects the gills on the posterolateral portion of the skull (Figures 2.2–2.3). Appendicular elements are generally attenuated, and the pectoral girdle is strongly connected to the posterior cranium by the post-temporal and supracleithrum bones (Figure 2.3). Additional elements of the pectoral girdle—cleithrum, coracoid, scapula—provide support for pectoral fins (Figure 2.14). The pelvic girdle consists of a single bone, the basipterygium or pelvic bone (Figure 2.14). The fins associated with these two girdles, as well as dorsal, anal, and caudal fins, typically consist of bony fin rays (lepidotrichia) that are supported by small rods of bone, the radials (or epurals and hypurals in the case of the caudal fin). The amphicoelus (hollowed on each side) vertebrae do not bear body weight, connect only with opposing centra, and lack well-developed accessory articulations that characterize these elements in the tetrapods. Most fishes possess a series of ventrolaterally projecting ribs, known as ventral or subperitoneal ribs, but some also have an additional series of smaller, more delicate dorsal (or intramuscular) ribs. A few species have only a series of dorsal ribs.

Remarks

As aquatic organisms, the fossil record of fishes is unparalleled among vertebrates. And since many taxa have unique adaptations to specific habitats and are sensitive to differences in water flow level, turbidity, salinity, and temperature, fish assemblages can provide fine-grained barometers of paleoaquatic conditions. Fish resources were widely used by prehistoric peoples, but given the typically smaller size of fishes—relative to large mammals, for instance—and higher capture costs associated with extraction from a water medium, fishes tend to be added to regional dietary sequences relatively late in time. However, in such contexts, they were also subject to intensive utilization, including harvest rates that provided temporary surpluses that were processed heavily for storage. Such intensive fish use may be related, in part, to the emergence of complex hunter-gatherer societies in certain areas such as in the Pacific Northwest Coast.

As noted, fish are ectothermic and exhibit indeterminate growth. For temperate taxa, seasonal variation in growth rates is recorded in their bony structures, much like the rings of a tree. These visible rings typically form during a winter growth arrestation period, and are referred to as annuli. Analyses of patterning in annuli can provide a wealth of information on the lives of fishes, including age, growth rates, and season of death. These can in turn inform a wide range of topics, from change in past water temperatures to prehistoric human harvest intensity.

Studies of fish annuli to determine age and growth in both modern and prehistoric fishes are most commonly conducted on structures known as otoliths (Figure 2.8). Otoliths are small, acellular, calcified structures that function in both sound detection and maintenance of equilibrium and are located at the posterior base of the skull (within the dorsal basioccipital). While they occur widely in vertebrates they are largest and most distinctive in fishes. They vary in number taxonomically and are individually named. Most fish contain three per side: the sagitta, lapillus, and astericus. Because otoliths are heavily mineralized, they typically represent the highest density structures in the fish skeleton and are thus less prone to density-mediated attrition, or the differential loss of skeletal parts due to variation in their density. And since otoliths are acellular and are thus not subject to resorption as in the case of bone, annuli studies from these structures generally provide the most accurate information on age and growth. They also have distinctive shapes and sizes in different species, and thus allow species-level identification when present.

Of the six living classes of fishes currently recognized, only two occur in inland waters of western North America: Cephalaspidomorphi and Actinopterygii. The Cephalaspidomorphi include the lampreys and as noted represent the only agnathan vertebrates in our area. In addition to their jawless oral discs, lampreys have elongated eel-like bodies but, unlike other fishes, lack bony skeletons and paired fins. They are distant relatives of a long-extinct group of agnathans, the ostracoderms, which sucked organic material from the bottoms of ancient oceans and lakes.

The Actinopterygii, or ray-finned fishes, are far more diverse and represent one of the two living groups of bony fishes. The ray-finned fishes are so named for the presence of bony or horny spines (lepidotrichia) that support the fins; these are not present in the other class of bony fishes, the Sarcopterygii. The ray-finned fishes are by far the most dominant living class of vertebrates and comprise nearly 95% of the 32,000 named fish species worldwide. A total of nine orders occur in inland settings of western North America, although four of these have very restricted ranges or are represented by very few species. With these exceptions (Cyprinodontiformes, pupfishes and killifishes; Esociformes, pikes; Osmeriformes, smelts; and Gasterosteiformes, sticklebacks) we examine each of the main fish orders.

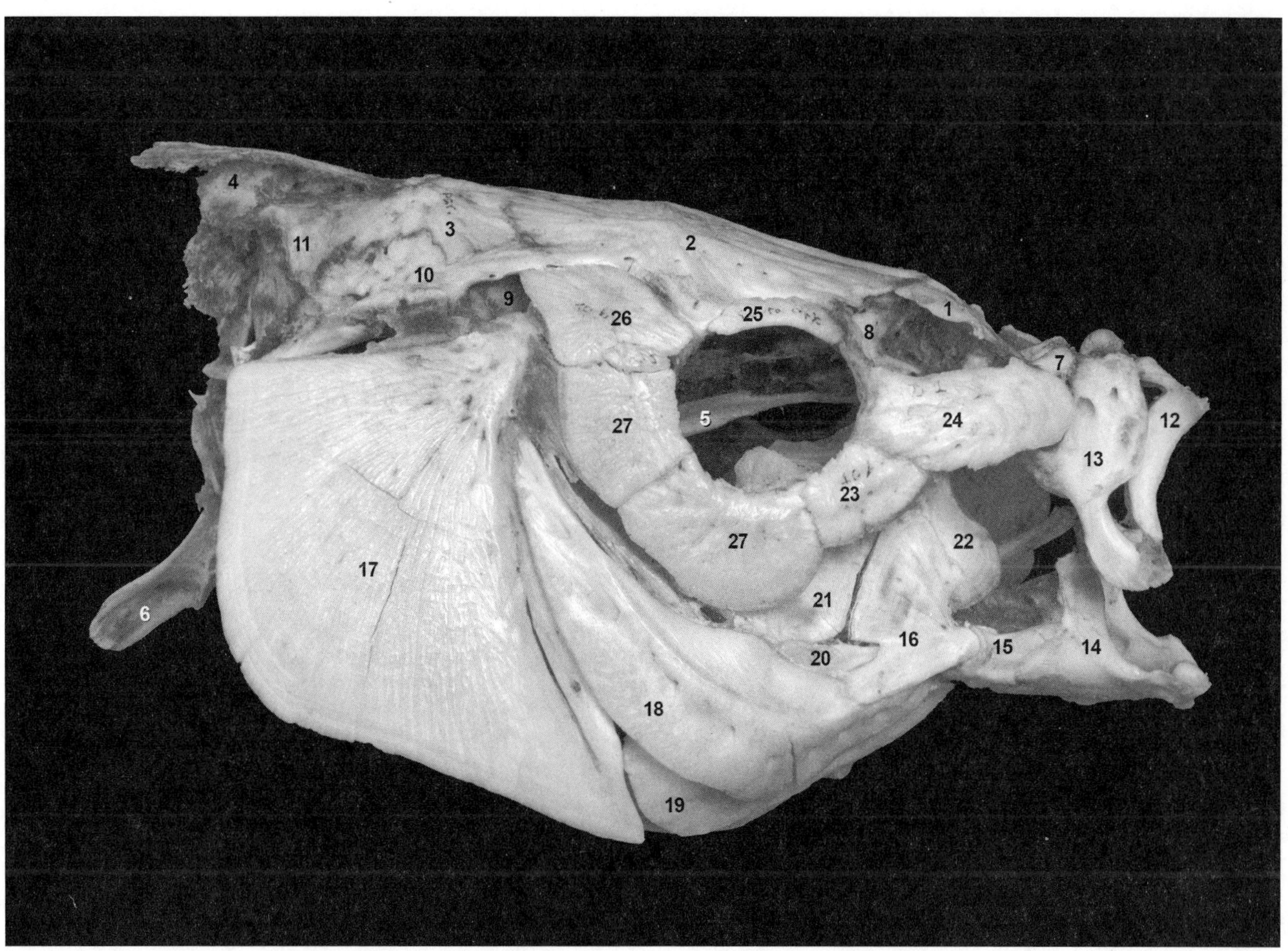

FIGURE 2.2. Cranium of Common Carp (*Cyprinus carpio*), lateral view.

1. Ethmoid
2. Frontal
3. Parietal
4. Supraoccipital
5. Parasphenoid
6. Basioccipital
7. Palatine
8. Lateral ethmoid
9. Sphenotic
10. Pterotic
11. Epiotic
12. Premaxilla
13. Maxilla
14. Dentary
15. Articular
16. Quadrate
17. Opercle
18. Preopercle
19. Interopercle
20. Symplectic
21. Metapterygoid
22. Pterygoid
23. Jugal
24. Lacrimal
25. Supraorbital
26. Dorsal sphenotic
27. Suborbital

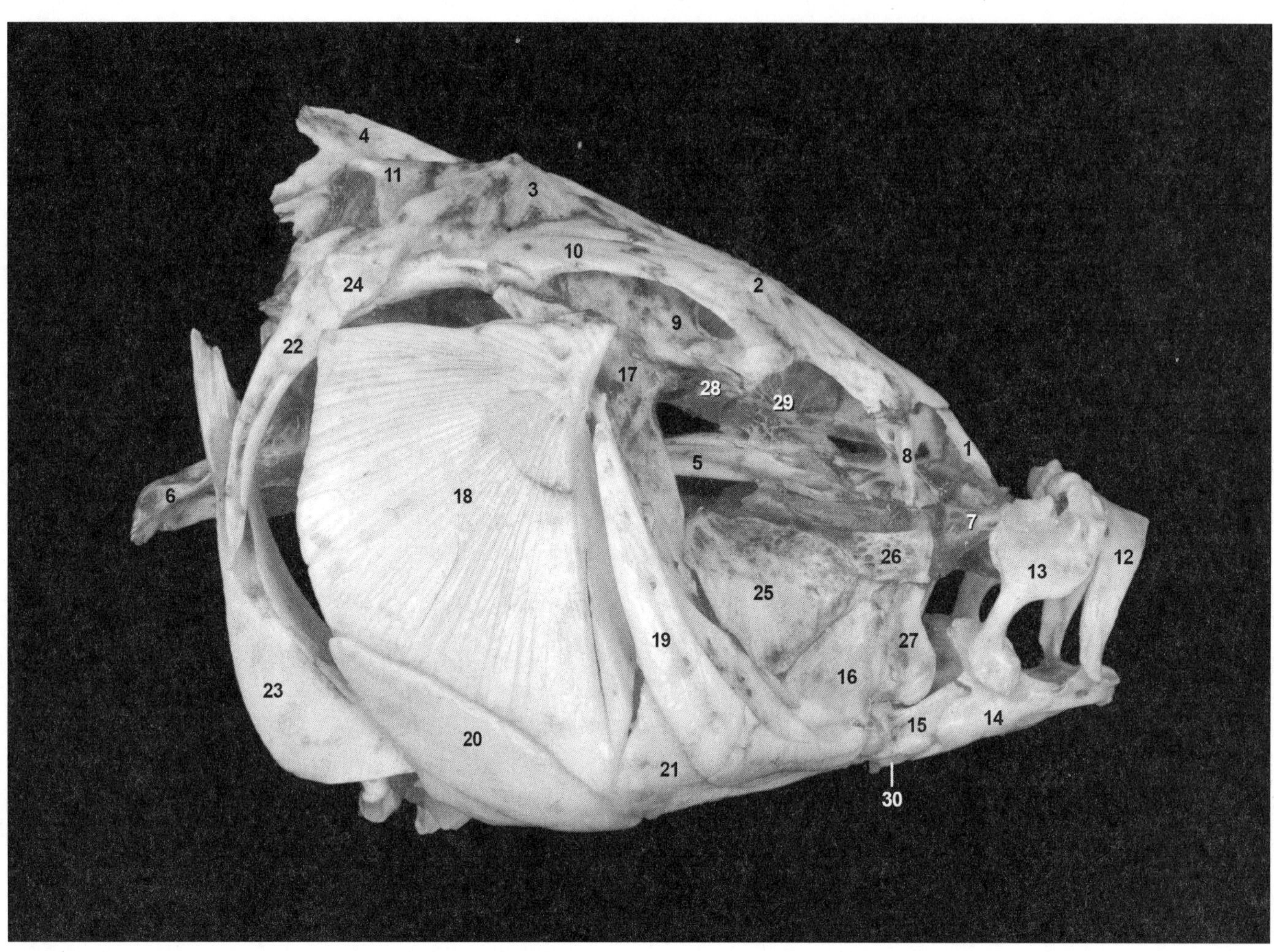

FIGURE 2.3. Cranium of Common Carp, lateral view.

1. Ethmoid
2. Frontal
3. Parietal
4. Supraoccipital
5. Parasphenoid
6. Basioccipital
7. Palatine
8. Lateral ethmoid
9. Sphenotic
10. Pterotic
11. Epiotic
12. Premaxilla
13. Maxilla
14. Dentary
15. Articular
16. Quadrate
17. Hyomandibular
18. Opercle
19. Preopercle
20. Subopercle
21. Interopercle
22. Supracleithrum
23. Cleithrum
24. Post-temporal
25. Metapterygoid
26. Entopterygoid
27. Pterygoid
28. Alisphenoid
29. Orbitosphenoid
30. Angular

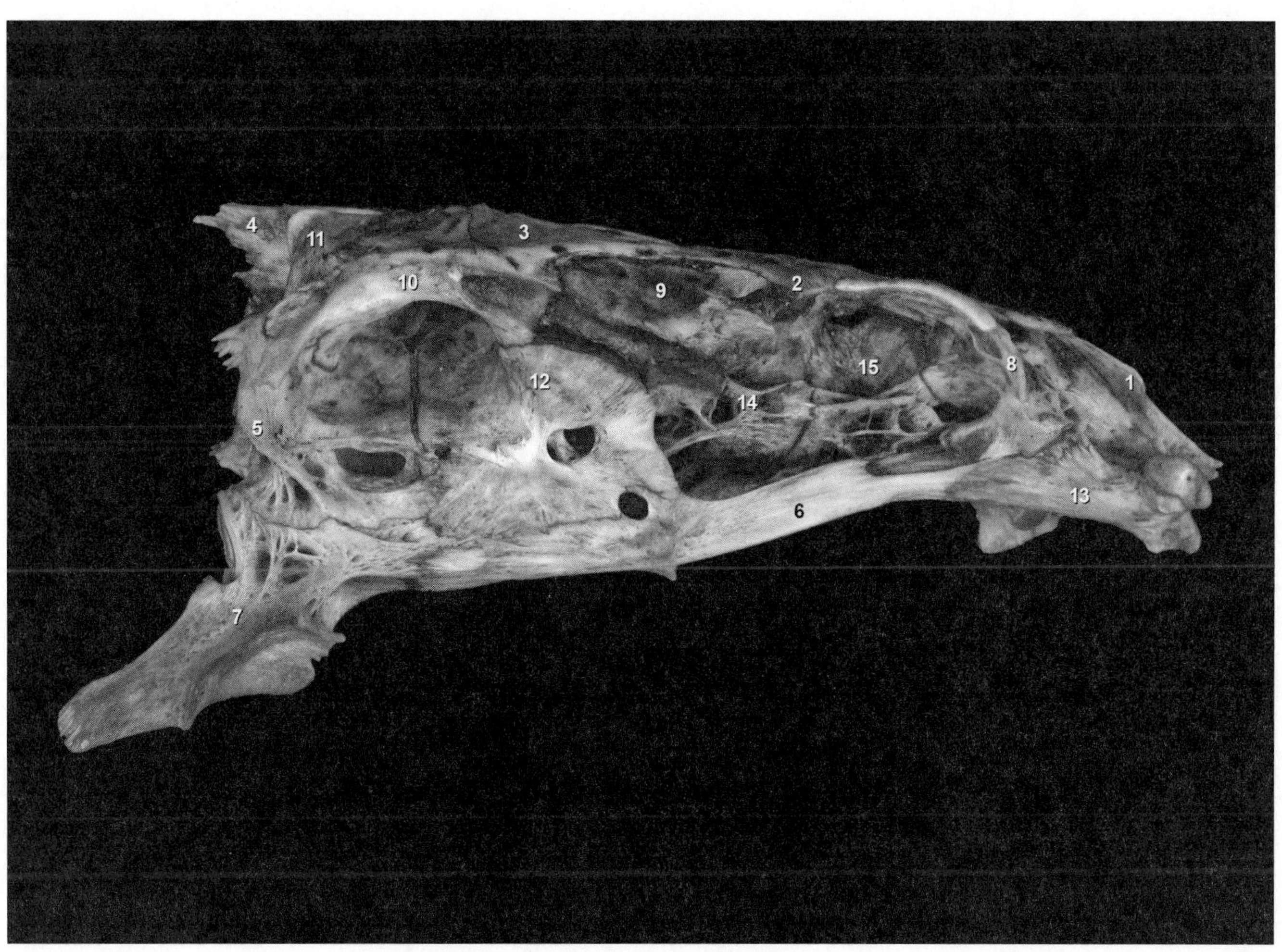

FIGURE 2.4. Cranium of Common Carp, lateral view.

1. Ethmoid
2. Frontal
3. Parietal
4. Supraoccipital
5. Exoccipital
6. Parasphenoid
7. Basioccipital
8. Lateral ethmoid
9. Sphenotic
10. Pterotic
11. Epiotic
12. Prootic
13. Vomer
14. Alisphenoid
15. Orbitosphenoid

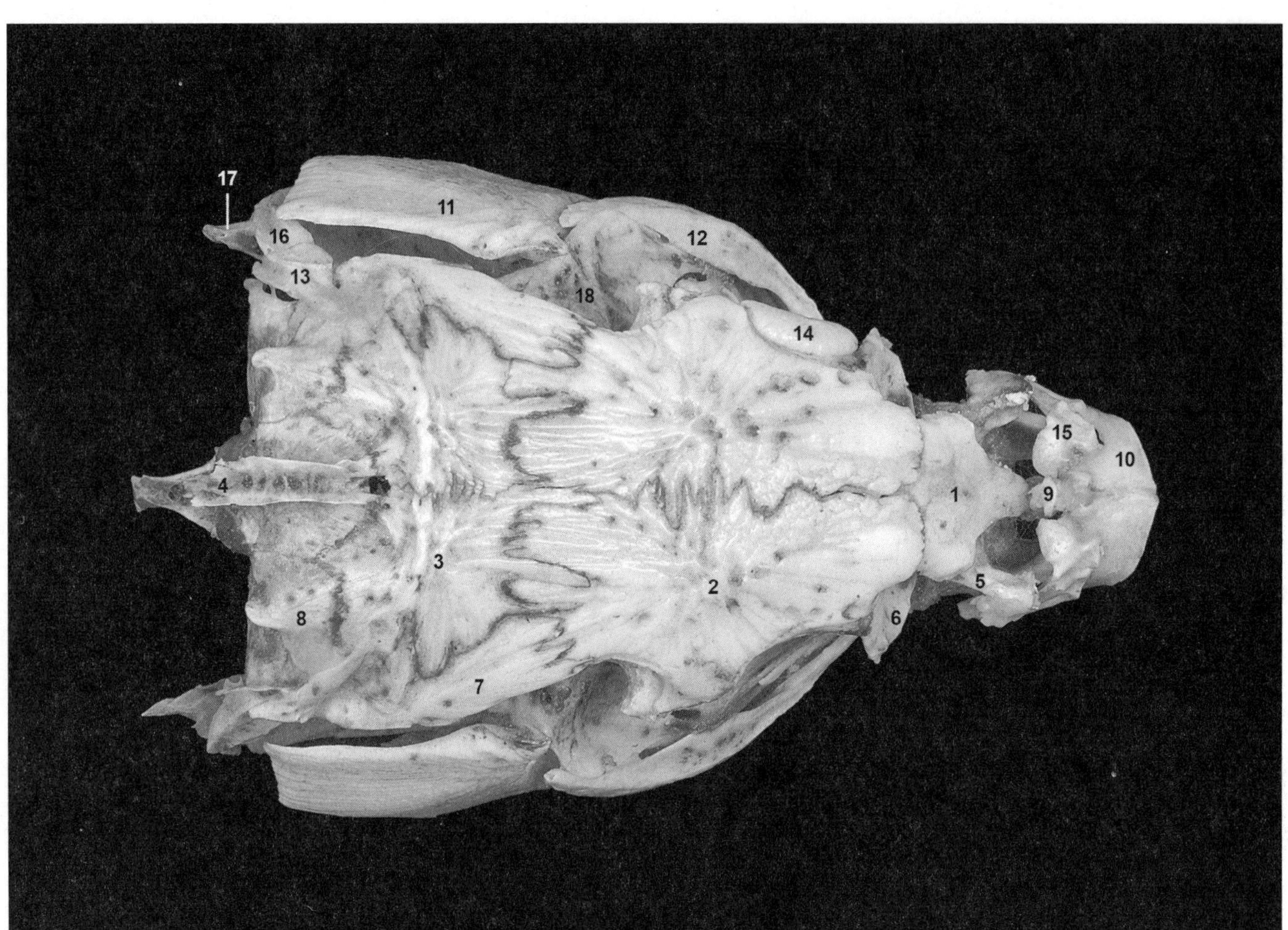

FIGURE 2.5. Cranium of Common Carp, dorsal view.

1. Ethmoid
2. Frontal
3. Parietal
4. Supraoccipital
5. Palatine
6. Lateral ethmoid
7. Pterotic
8. Epiotic
9. Prevomer
10. Premaxilla
11. Opercle
12. Preopercle
13. Post-temporal
14. Supraorbital
15. Maxilla
16. Supracleithrum
17. Cleithrum
18. Hyomandibular

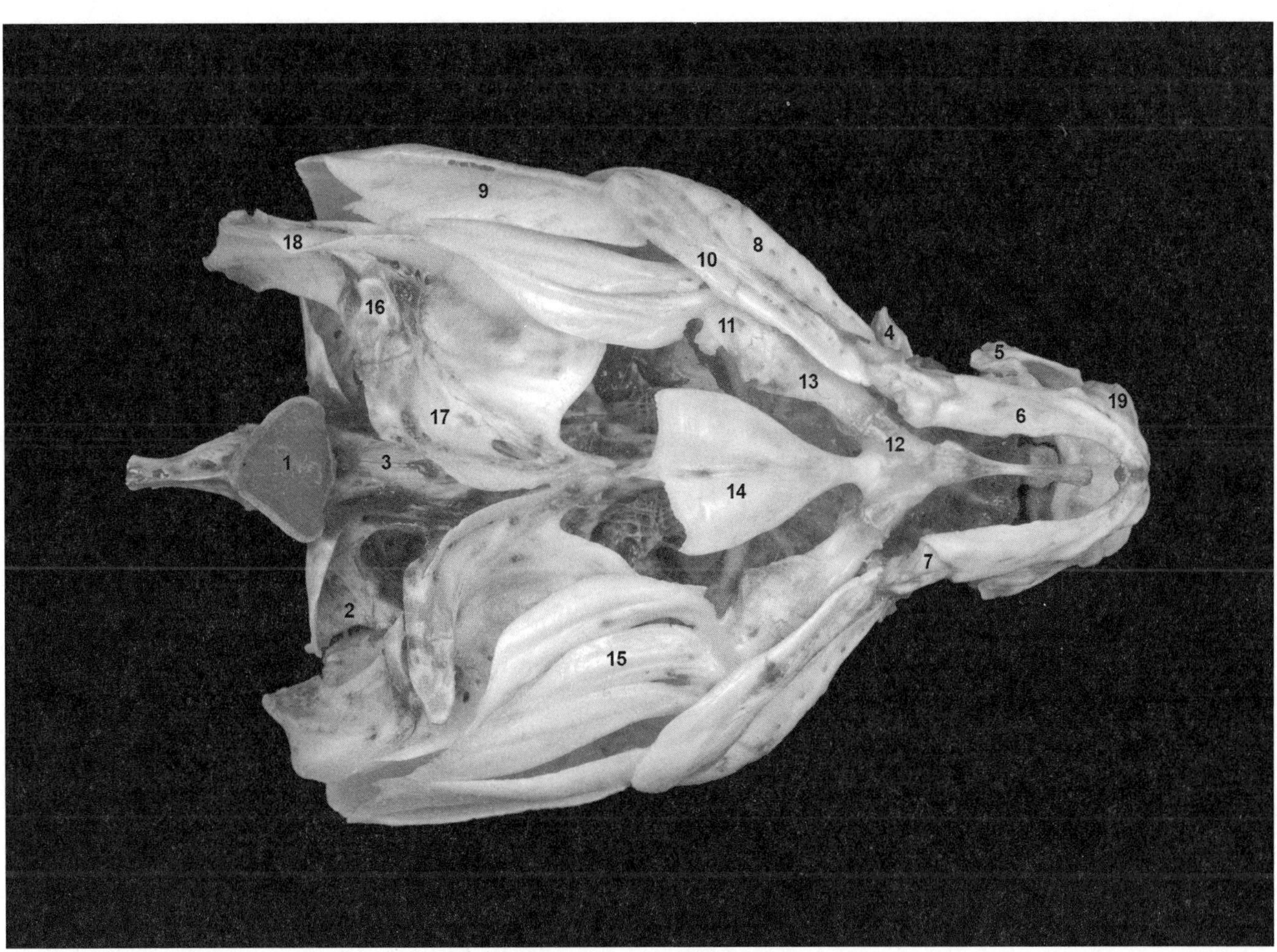

FIGURE 2.6. Cranium of Common Carp, ventral view.

1. Basioccipital
2. Exoccipital
3. Parasphenoid
4. Lateral ethmoid
5. Maxilla
6. Dentary
7. Angular
8. Preopercle
9. Subopercle
10. Interopercle
11. Epihyal
12. Hypohyal
13. Ceratohyal
14. Urohyal
15. Branchiostegal rays
16. Scapula
17. Coracoid
18. Cleithrum
19. Premaxilla

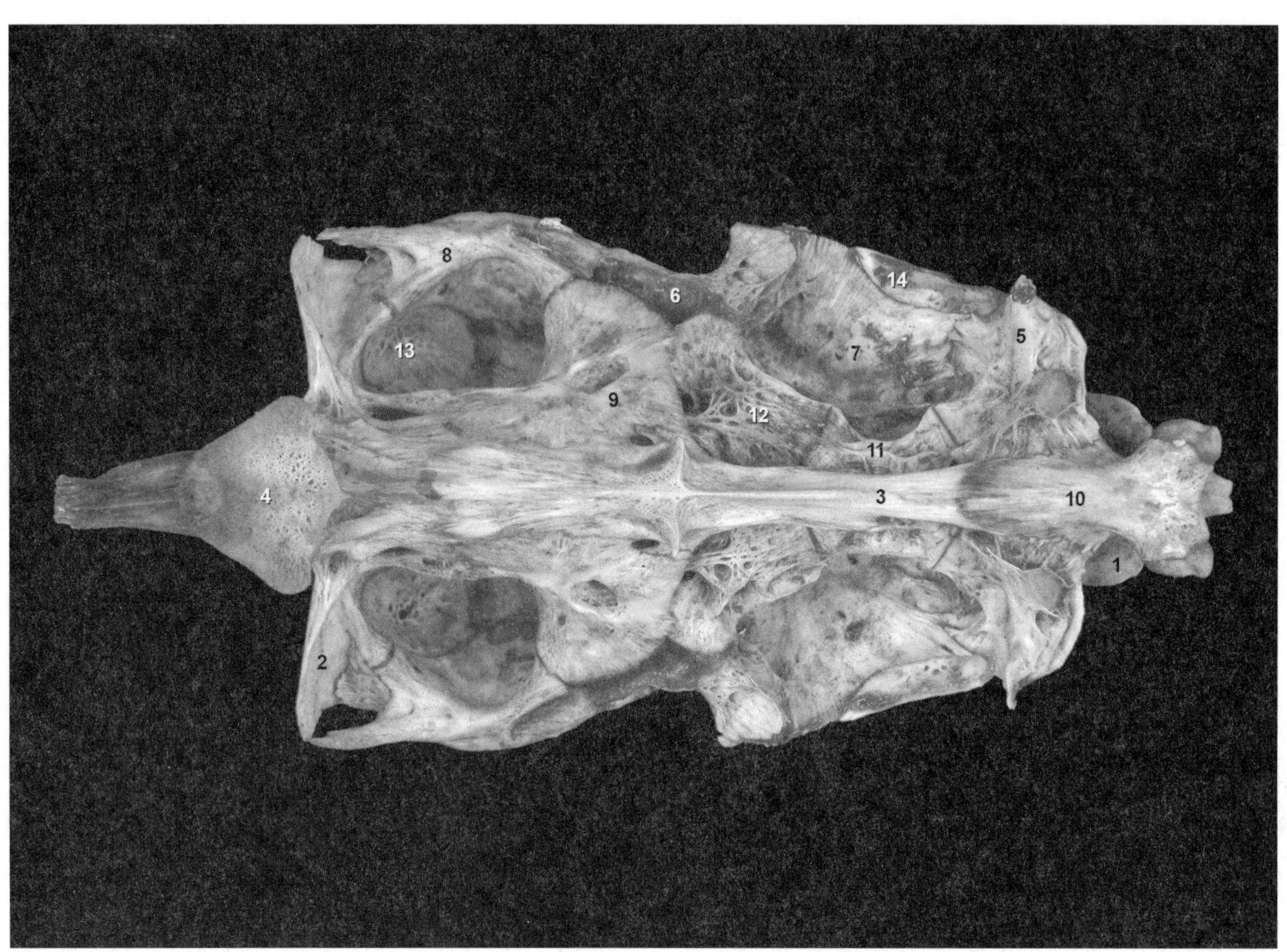

FIGURE 2.7. Cranium of Common Carp, ventral view.

1. Ethmoid
2. Exoccipital
3. Parasphenoid
4. Basioccipital
5. Lateral ethmoid
6. Sphenotic
7. Frontal
8. Pterotic
9. Prootic
10. Vomer
11. Orbitosphenoid
12. Alisphenoid
13. Epiotic
14. Supraorbital

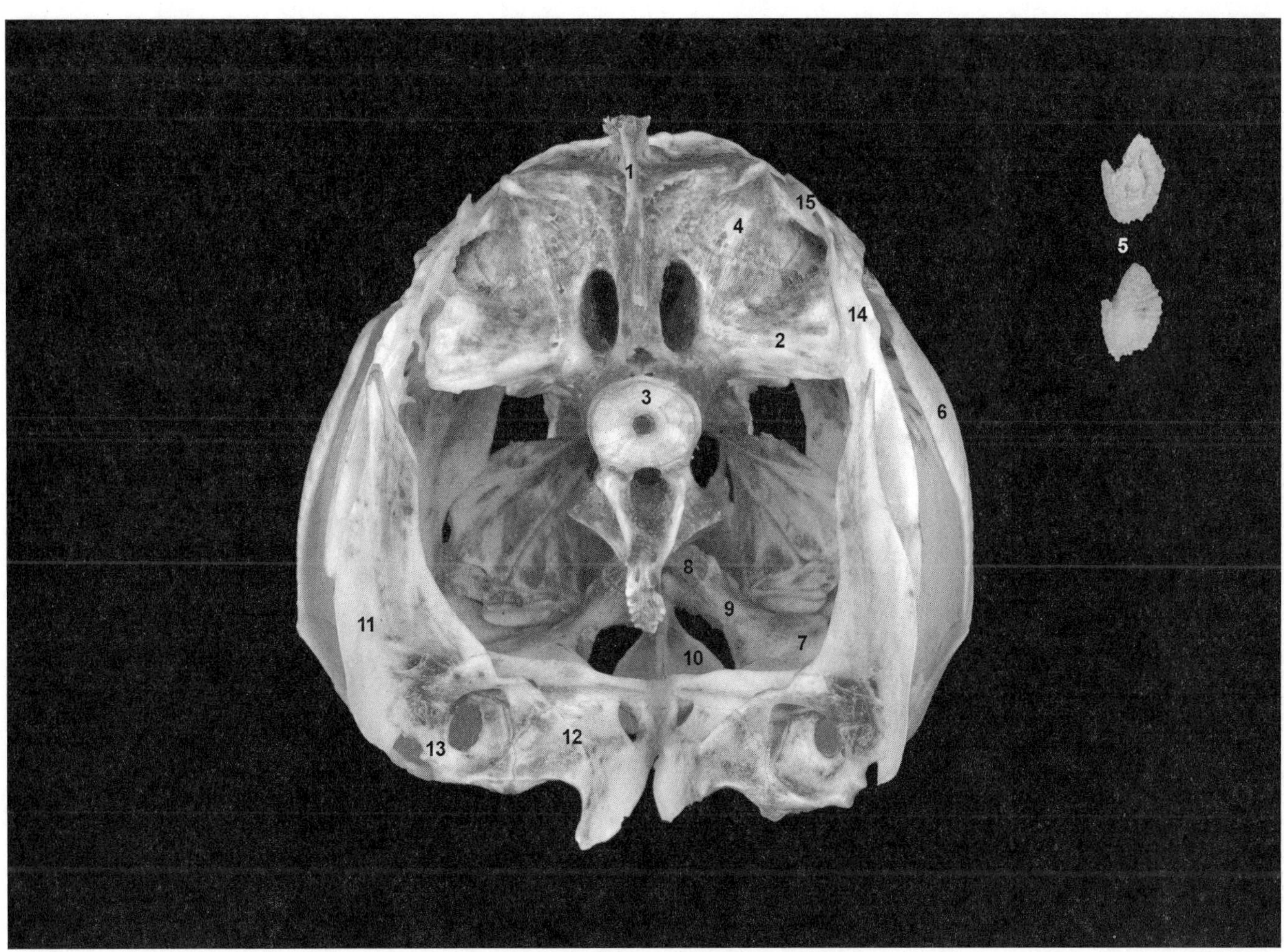

FIGURE 2.8. Cranium of Common Carp, posterior view.

1. Supraoccipital
2. Exoccipital
3. Basioccipital
4. Epiotic
5. Otoliths
6. Opercle
7. Epihyal
8. Hypohyal
9. Ceratohyal
10. Urohyal
11. Cleithrum
12. Coracoid
13. Scapula
14. Supracleithrum
15. Post-temporal

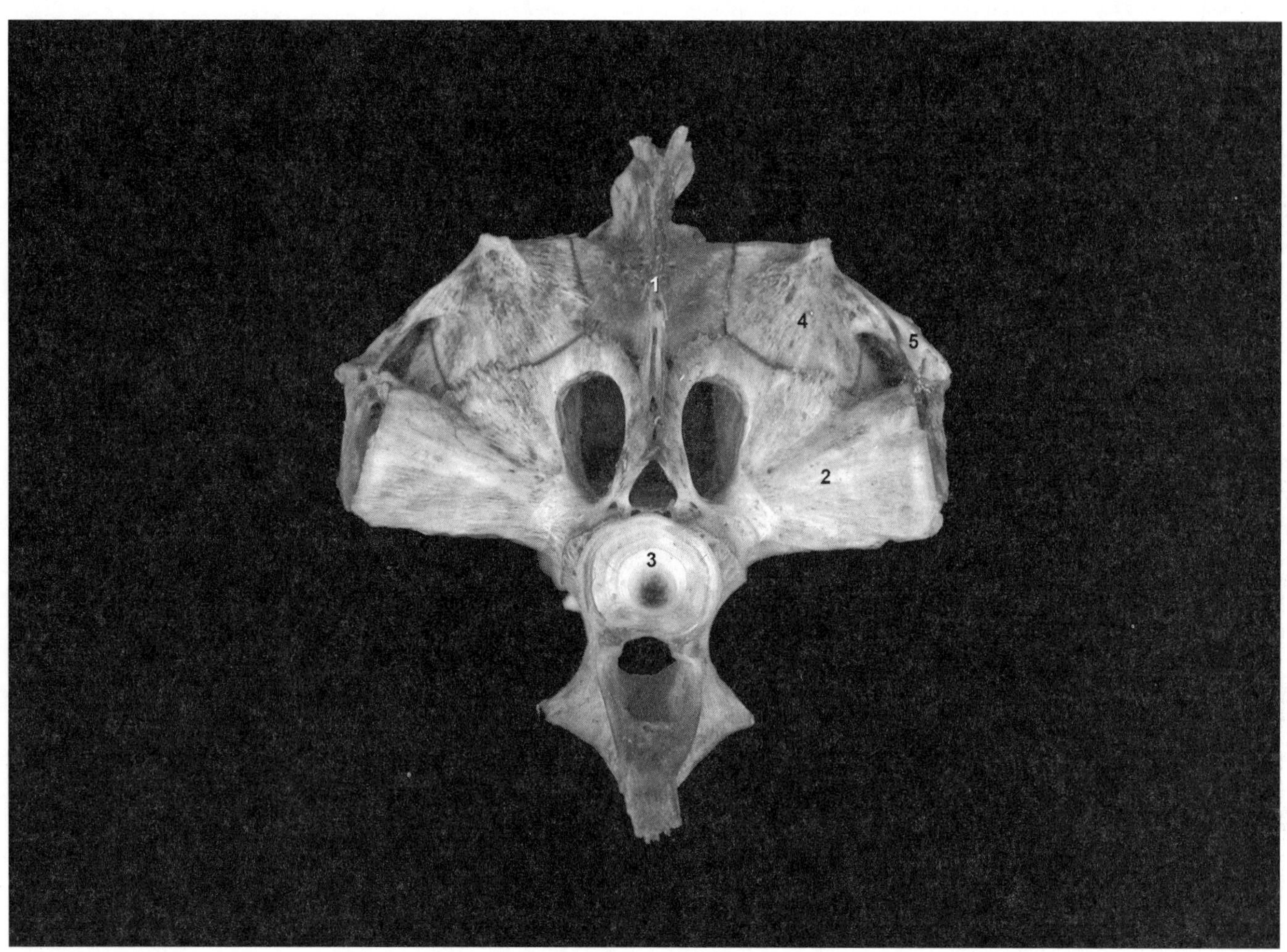

FIGURE 2.9. Cranium of Common Carp, posterior view.

1. Supraoccipital
2. Exoccipital
3. Basioccipital
4. Epiotic
5. Post-temporal

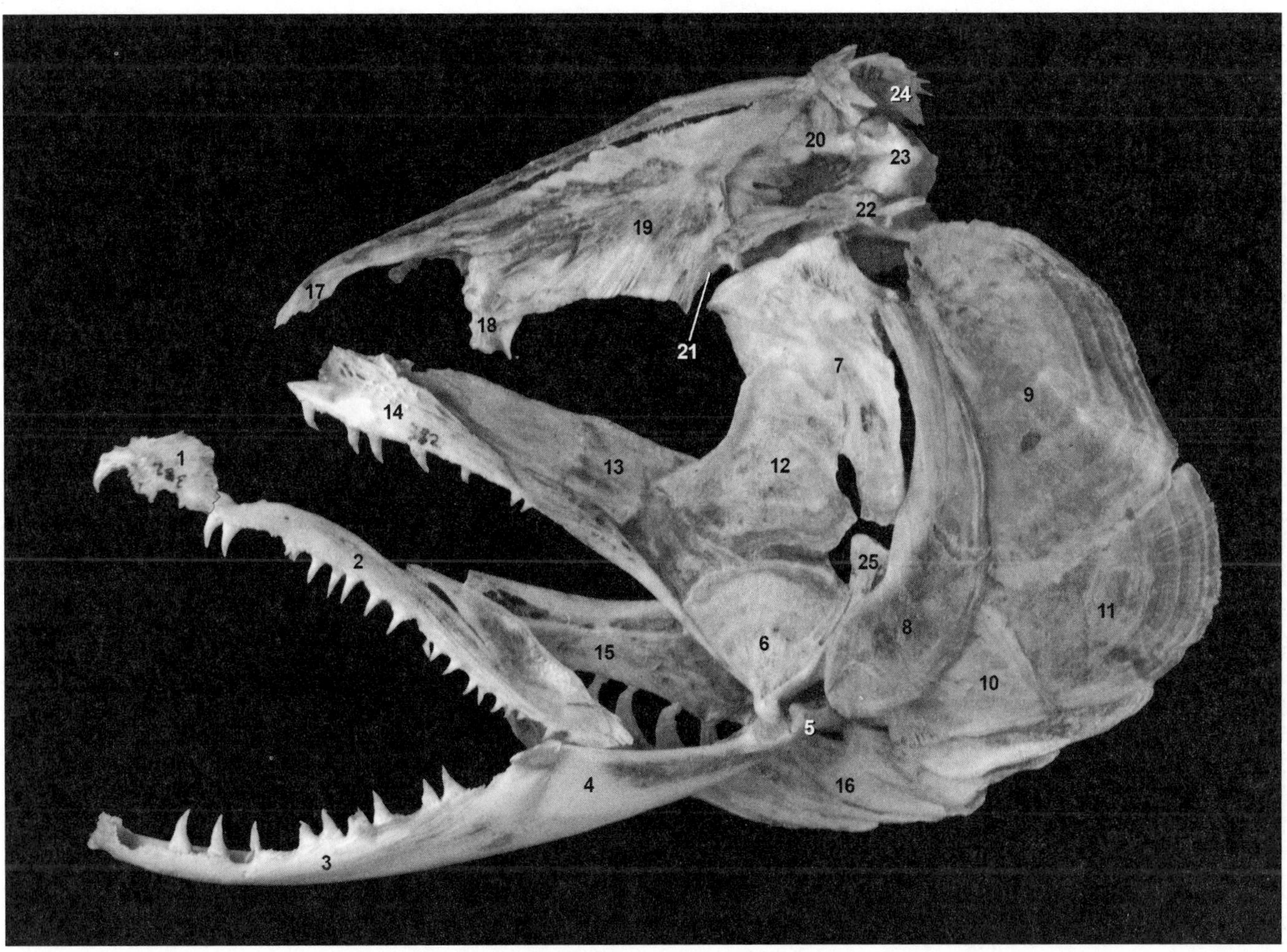

FIGURE 2.10. Cranium of Cutthroat Trout (*Oncorhynchus clarkii*), lateral view.

1. Premaxilla
2. Maxilla
3. Dentary
4. Articular
5. Angular
6. Quadrate
7. Hyomandibular
8. Preopercle
9. Opercle
10. Interopercle
11. Subopercle
12. Metapterygoid
13. Entopterygoid
14. Palatine
15. Ceratohyal
16. Branchiostegal ray
17. Ethmoid
18. Lateral ethmoid
19. Frontal
20. Parietal
21. Sphenotic
22. Pterotic
23. Epiotic
24. Supraoccipital
25. Symplectic

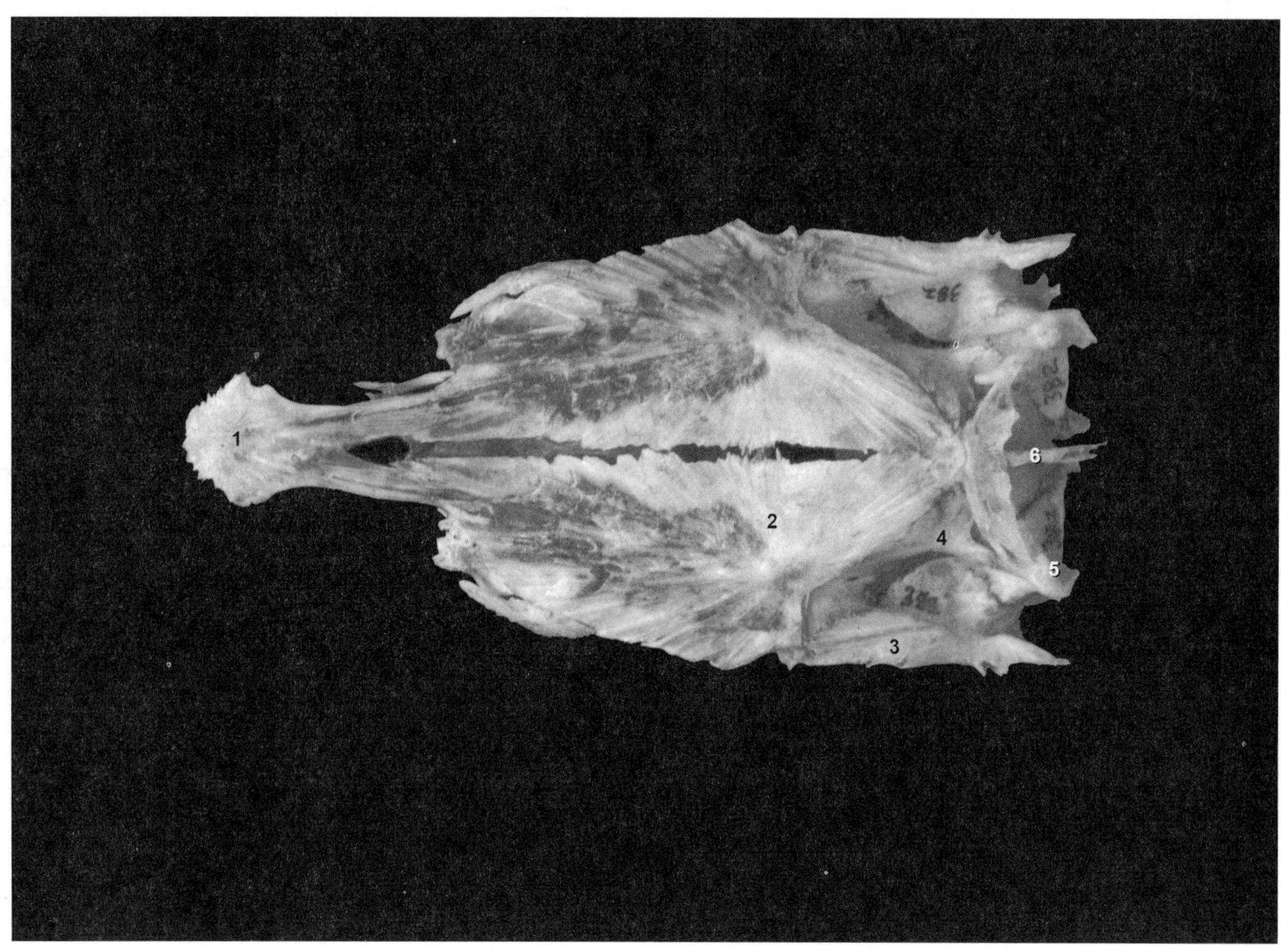

FIGURE 2.11. Cranium of Cutthroat Trout, dorsal view.

1. Ethmoid	4. Parietal
2. Frontal	5. Epiotic
3. Pterotic	6. Supraoccipital

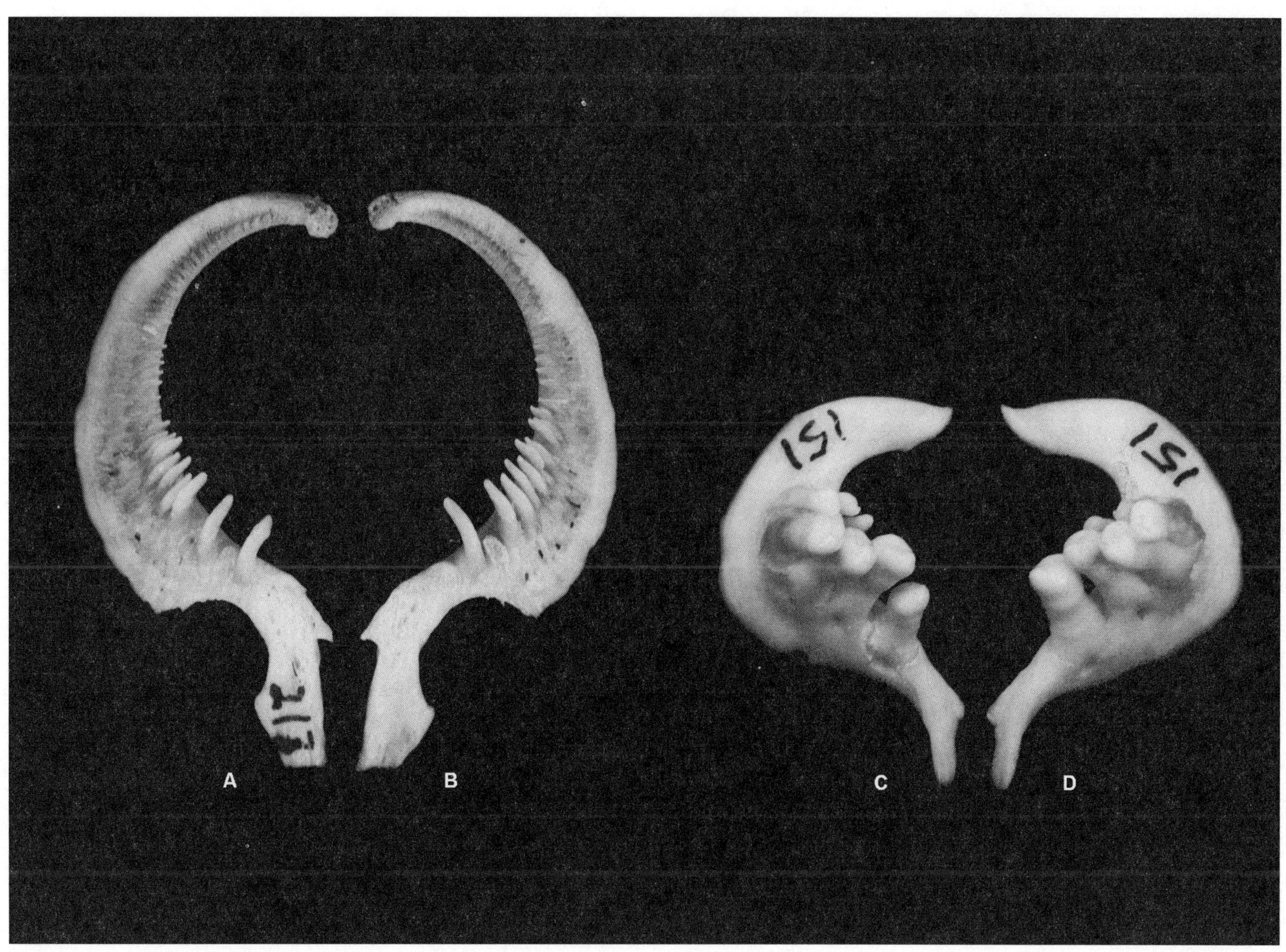

FIGURE 2.12. Utah Sucker (*Catostomus ardens*) pharyngeal teeth, ventral view, left (A) and right (B). Hardhead (*Mylopharodon conocephalus*) pharyngeal teeth, ventral view, left (C) and right (D).

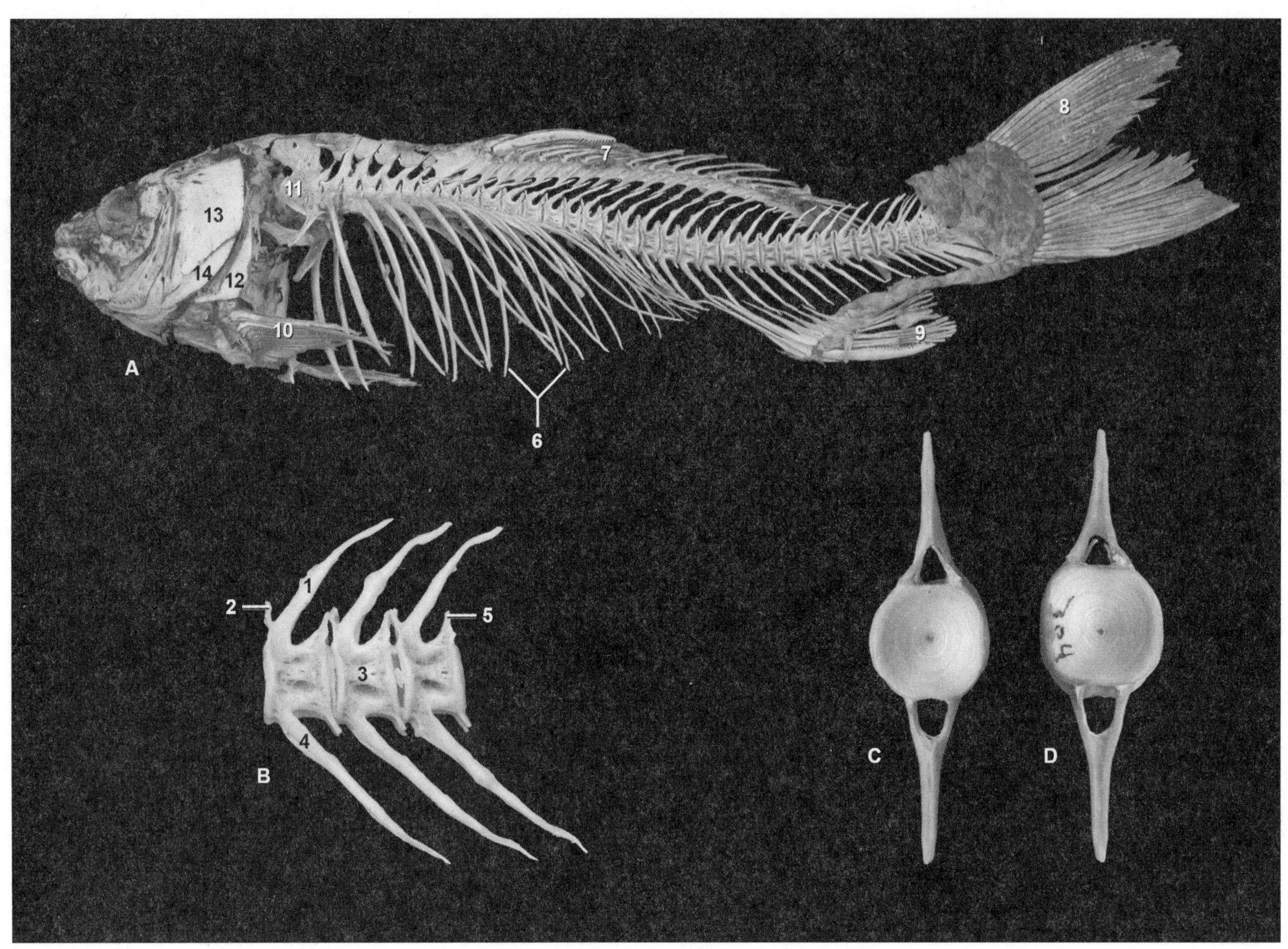

FIGURE 2.13. Common Carp skeleton (A) and lateral (B), anterior (C), and posterior (D) views of vertebrae.

1. Neural spine
2. Prezygapophysis
3. Centrum
4. Haemal spine
5. Postzygapophysis
6. Ventral rib
7. Dorsal fin rays
8. Caudal fin rays
9. Anal fin rays
10. Pectoral fin rays
11. Weberian apparatus
12. Cleithrum
13. Opercle
14. Subopercle

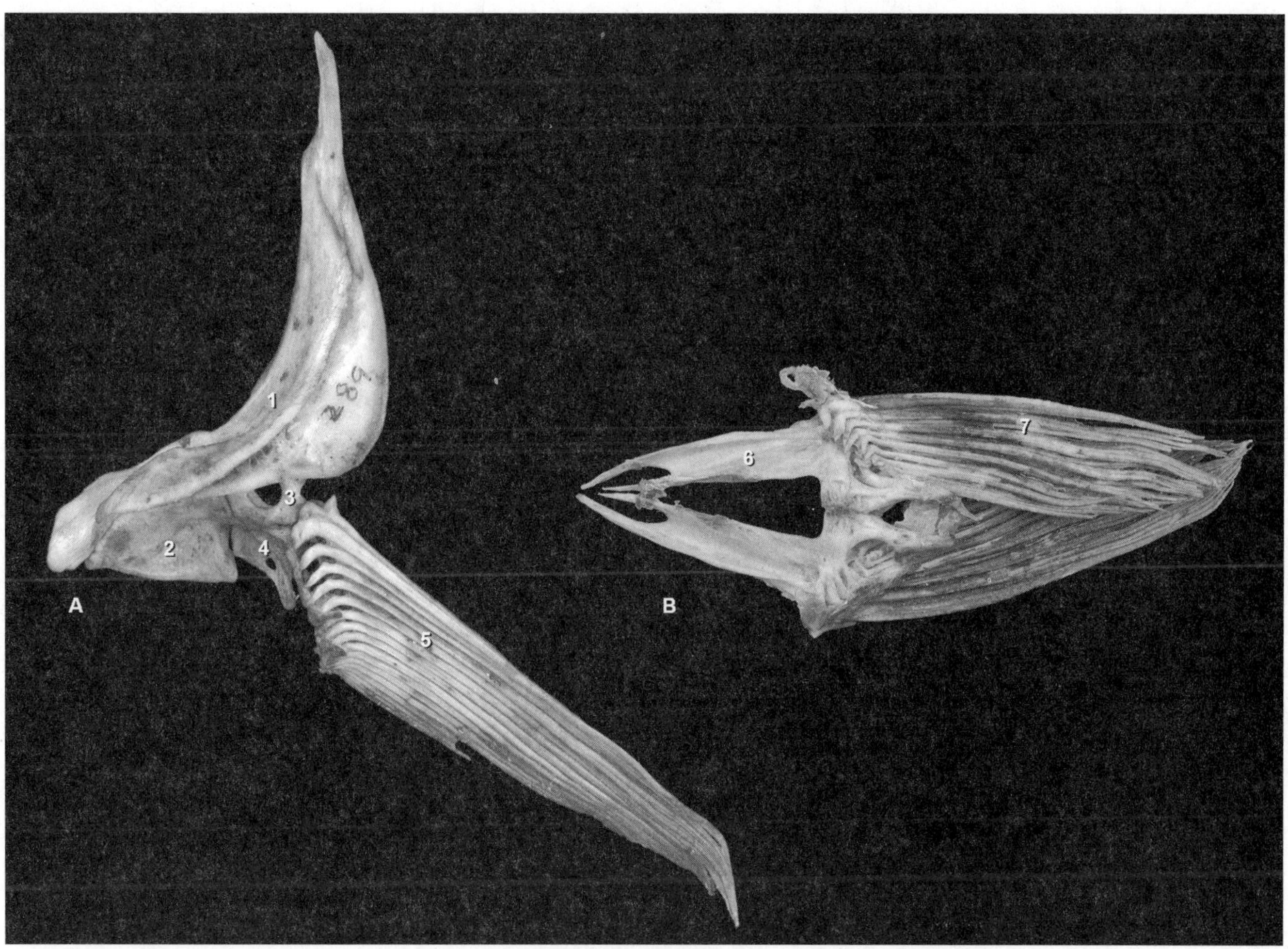

FIGURE 2.14. Left pectoral girdle (A), lateral view, and right and left pelvic girdles (B), dorsal view, of Common Carp.

1. Cleithrum
2. Coracoid
3. Scapula
4. Radials
5. Fin rays (pectoral fin)
6. Pelvic bone (basipterygium)
7. Fin rays (pelvic fin)

II. Taxonomy and Osteological Variation of Western Fishes

CLASS Cephalaspidomorphi (Lampreys)
ORDER Petromyzontiformes (Lampreys)
FAMILY Petromyzontidae (Northern Lampreys)
Genus *Lampetra*

The order Petromyzontiformes (L., petra = stone; Gr., myzon = suckle; L., forma = shape) is represented by six species, all within the genus *Lampetra*, in our region (Figure 2.15). Lampreys have two separate life-history stages. Larvae or ammocoetes hatch from eggs laid in freshwater streams and are carried downstream to settle along sand- or mud-bottomed areas under slower-moving water. Here, they burrow into sediments where they filter-feed on detritus and algae and remain from two to six years. Metamorphosis into the adult stage involves the formation of eyes and an oral disc with a powerful tongue bearing sharp, pointed, tooth-like structures derived from keratin. Adults adopt either parasitic or nonparasitic strategies. Parasitic forms are usually anadromous and spend most of their adult lives in marine waters off the coast where they latch onto the sides of other fish with their sucker-like mouths and rasp a hole with their toothed tongue and oral disk. They feed by extracting blood and bodily fluids and release when they are satiated. Parasitic forms typically return to rivers to spawn, after which most die. Nonparasitic forms spend much more time in the larval stage and often die within a month of their adult transformation. Larger species (i.e., Pacific Lamprey, *Lampetra tridentata*; Figures 2.15–2.16) can reach over 70 cm in length and weigh over .5 kg. They were widely used by ethnographically recorded native peoples living along rivers and streams connected to the Pacific Ocean and were primarily harvested during upstream spawning migrations. However, they have yet to be identified from prehistoric archaeological sites in the region due either to poor preservation, lack of knowledge of lamprey anatomy by faunal analysts, or limited prehistoric use.

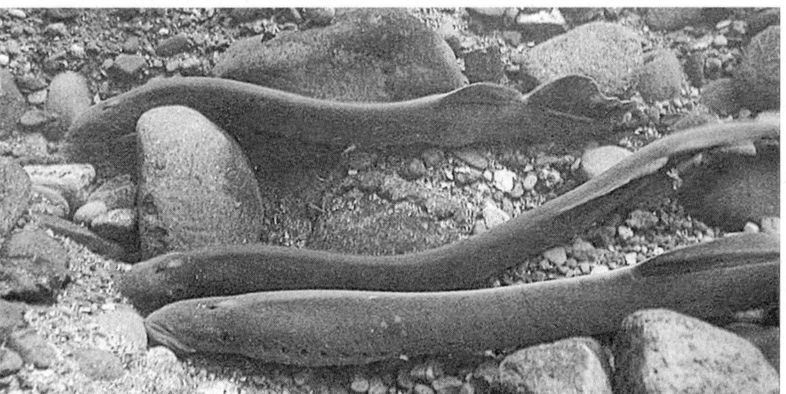

FIGURE. 2.15. Pacific Lamprey (*Lampetra tridentata*). Photo by United States Fish and Wildlife

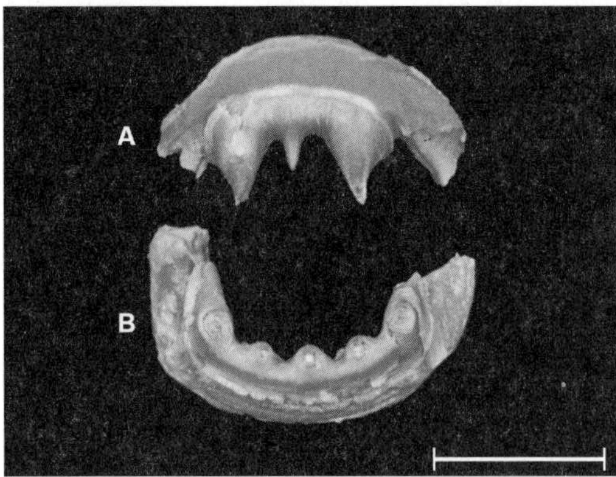

FIGURE 2.16. Pacific Lamprey (*Lampetra tridentata*) supraoral (A) and infraoral (B) laminae, anterior views. Osteological features (*Lampetra*): 1, contains no bony skeleton; 2, mouth with supraoral and infraoral laminae with keratinized "teeth." Scale bar equals 1 cm unless otherwise stated.

CLASS Actinopterygii (Ray-finned Fishes)
ORDER Acipenseriformes (Sturgeons)
FAMILY Acipenseridae (Sturgeons)
Genus *Acipenser*

The order Acipenseriformes (L., Acipenser = sturgeon) is represented by two native species: *Acipenser transmontanus* (White Sturgeon; Figure

FIGURE 2.17. White Sturgeon (*Acipenser transmontanus*). Photo by Thomas Taylor

2.17) and *Acipenser medirostris* (Green Sturgeon; Figure 2.18). Sturgeon are the largest and longest-lived fishes native to freshwater habitats of western North America. The record size for white sturgeon is nearly 6 m in length and 820 kg in weight. Monsters of this size probably lived over 100 years. Today, however, individuals over 2 m and older than 27 years are rare. Sturgeon are not ancestral to the other bony fishes but instead represent a highly specialized offshoot. Much of the bony skeleton has been secondarily replaced by cartilage, although the cranium retains substantial bony elements. Sturgeon are adapted to prey on bottom-dwelling animals, primarily invertebrates such as mollusks and crustaceans, and use a row of sensitive barbels near their huge, ventrally oriented and toothless mouths to detect them. Both native species are anadromous, spending most of their lives in the ocean or coastal estuaries and moving up major river tributaries to spawn in the spring. Although they range in salt water from northern Baja to the Bering Sea, they only penetrate inland through large rivers on the Pacific Coast from central California northward. Landlocked populations now exist above major dams in the Columbia-Snake River system. With an extended life history pattern, large size, fat-rich flesh, and sluggish nature, they have been depressed the world over by human foraging activities in both prehistoric and historic times.

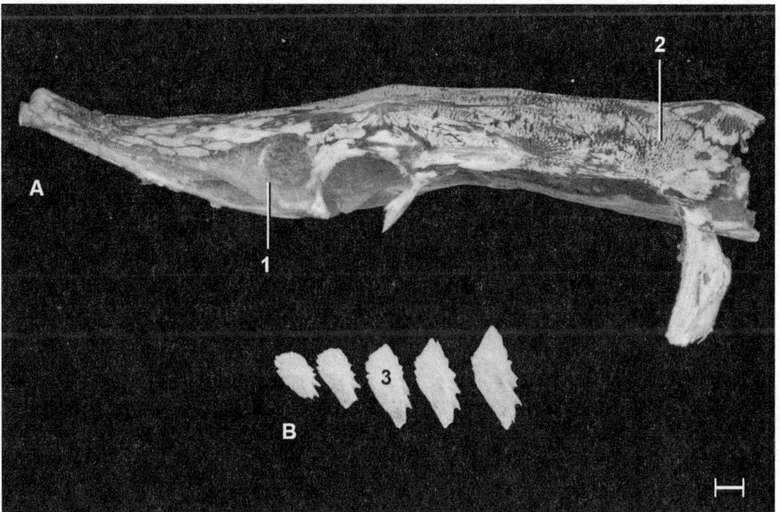

FIGURE 2.18. Green Sturgeon (*Acipenser medirostris*) cranium (A) and scutes (B), lateral views. Osteological features (*Acipenser*): 1, reduced ossification of postcranial endoskeleton, well-developed cranial dermal skeleton—vertebral elements cartilaginous; 2, unique texture of exterior skeleton—bone surface with pits, tiny projections, "goose bump-like" appearance; 3, bony scutes in five rows (two lateral, two ventral, and one dorsal) extend along trunk—scutes with circular outline, prominent medial pitch.

ORDER Cypriniformes (Minnows and Suckers)

Cypriniformes (Gr., kyprinos = goldfish) are represented by two families in the west, Cyprinidae (minnows) and Catostomidae (suckers). Considering the number of native species it is easily the most dominant order of fishes in western North America. All species have protrusible jaws and are suction feeders that lack teeth in the jaw elements but possess paired pharyngeal teeth—modifications of the fifth gill arch that appear as sickle-shaped bony platforms from which a variable number of teeth extend. The teeth are structurally similar to mammalian teeth, with an underlying core of dentin covered by a coat of enamel-like (enameloid) material. These paired structures are used to process food items against a hard plate on the roof of the buccal cavity and come in different shapes and sizes, reflecting the unique dietary adaptations of different species; such distinctive morphologies almost always allow for species-level identifications. Pharyngeal teeth have contributed to the great diversification of the Cypriniformes in much the same way that highly specialized teeth in the jaws of mammals allowed for their success in terrestrial environments. Cypriniformes also have a well-developed sense of hearing that is enabled in part by the Weberian apparatus, a complex of bones surrounding the first four vertebrae. These elements connect the auditory system and labyrinth structures of the inner ear to the swim bladder. Swim bladders are gas filled structures used in buoyancy regulation in ray-finned fishes. The bladders also readily intercept sound waves passing through the water. The vertebrae of Cypriniformes are distinctive to the order level with one or two large, robust, antero-posterior strut(s) that run along the centrum. However, distinguishing these elements between different genera, or even families, is extremely challenging.

FAMILY Cyprinidae (Minnows)

Minnows are elongated silvery fish with prominently forked caudal fins which lack true spines. They have scales evenly distributed over the body but not on the head (Color Plate 1; Figure 2.19). Minnows range in length from less than 2.5 cm to more than 90 cm—obviously, not all minnows are small fish. Cyprinidae is one of the most widespread and speciose families in the world and is represented by 15 genera in the west. The greatest diversity occurs in the Central Valley of California, which is home to nine genera and eleven native species, including the now-extinct Thick-tailed Chub (*Gila crassicauda*). In this setting, the different minnow species have adapted to a wide range of habitats and feeding niches and can provide excellent indicators of lentic (slow water) or lotic (fast water) habitats. Tui Chub (*Siphateles bicolor*) and Utah Chub (*Gila atraria*; Figure 2.20) are the largest and best-known taxa of the western and eastern Great Basin, respectively. These two

FIGURE 2.19. Lahontan Redside (*Richardsonius egregious*).

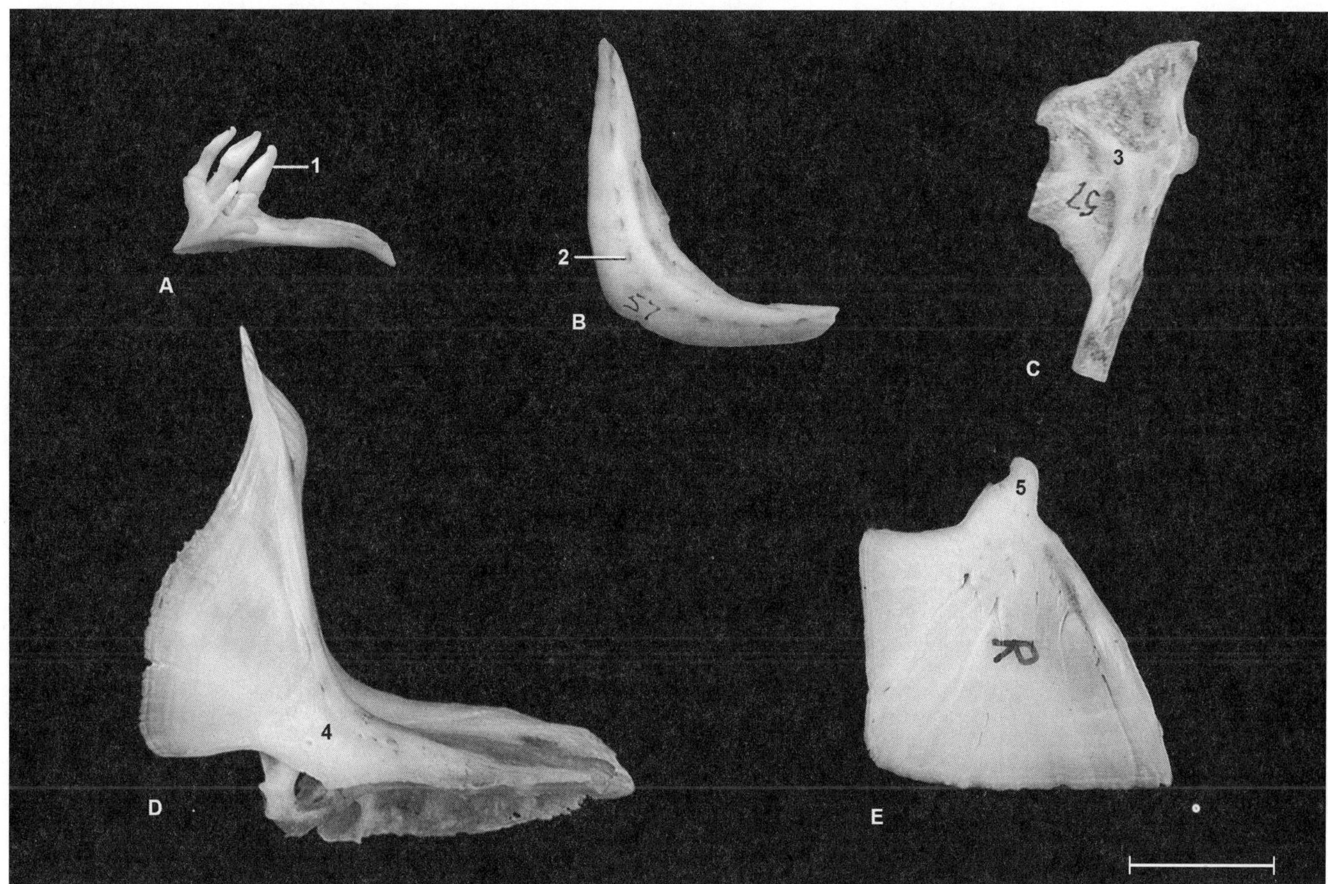

FIGURE 2.20. Utah Chub (*Gila atraria*) r. pharyngeal, medial view (A); lateral views of r. preopercle (B); r. hyomandibular (C); r. cleithrum (D); and r. opercle (E). Osteological features (Cyprinidae): 1, pharyngeal teeth, diagnostic to species, with two rows of heavily enameled teeth—not comb-like, as in catostomids; 2, small foramina in rows on many cranial elements (e.g., dentary, frontal, preopercle); 3, hyomandibular broad, deep, trapezoid-shaped; 4, narrow medial-lateral depth of cleithrum at mid-section; 5, opercle with blunt, rounded projection dorsal to hyomandibular fossa.

species are generalists and can adapt to a variety of water qualities. Utah Chub, for example, is found not only in cold, high-elevation lakes, but warm, moderately saline sinks (up to 2.5% NaCl) and springs on the desert floor. Several genera are endemic to the Columbia River (e.g., *Mylocheilus caurinus*, Peamouth; *Acrocheilus alutaceus*, Chiselmouth) and the Colorado River drainages (*Plagopterus argentissimus*, Woundfin; *Lepidomeda mollispinus*, Virgin Spinedace). Common Carp (*Cyprinus carpio*), introduced from Asia, thrive in many aquatic settings across the west.

FAMILY Catostomidae (Suckers)

Three genera (*Catostomus*, *Xyrauchen*, and *Chasmistes*) of suckers inhabit the inland waters of the west (Figures 2.21–2.22). These fishes have an overall body form similar to minnows, but unlike the latter, suckers have more rounded heads and prominent, ventrally oriented, protrusible and fleshy lips. The lips are adaptations to bottom feeding where they suck up a range of small invertebrates, algae, and detritus in settings ranging from large lakes to small, fast-moving creeks. For small species (e.g., *Catostomus platyrhynchus*, Mountain Sucker), adult sizes rarely exceed about 20 cm; for large species (e.g., *Xyrauchen texanus*, Razorback Sucker), record catches exceed 1 m in length and 7.3 kg. Although these benthic browsers lack the species diversity of the minnows, they are widely distributed across the west. Utah Sucker (*Catostomus ardens*; Figure 2.22) and June Sucker (*Chasmistes liorus*; currently threatened) are the largest and most well-known and utilized suckers in the Bonneville Basin. In the Lahontan Basin (western Great Basin) the

FIGURE 2.21. Longnose Sucker (*Catostomus catostomus*). Photo by National Park Service

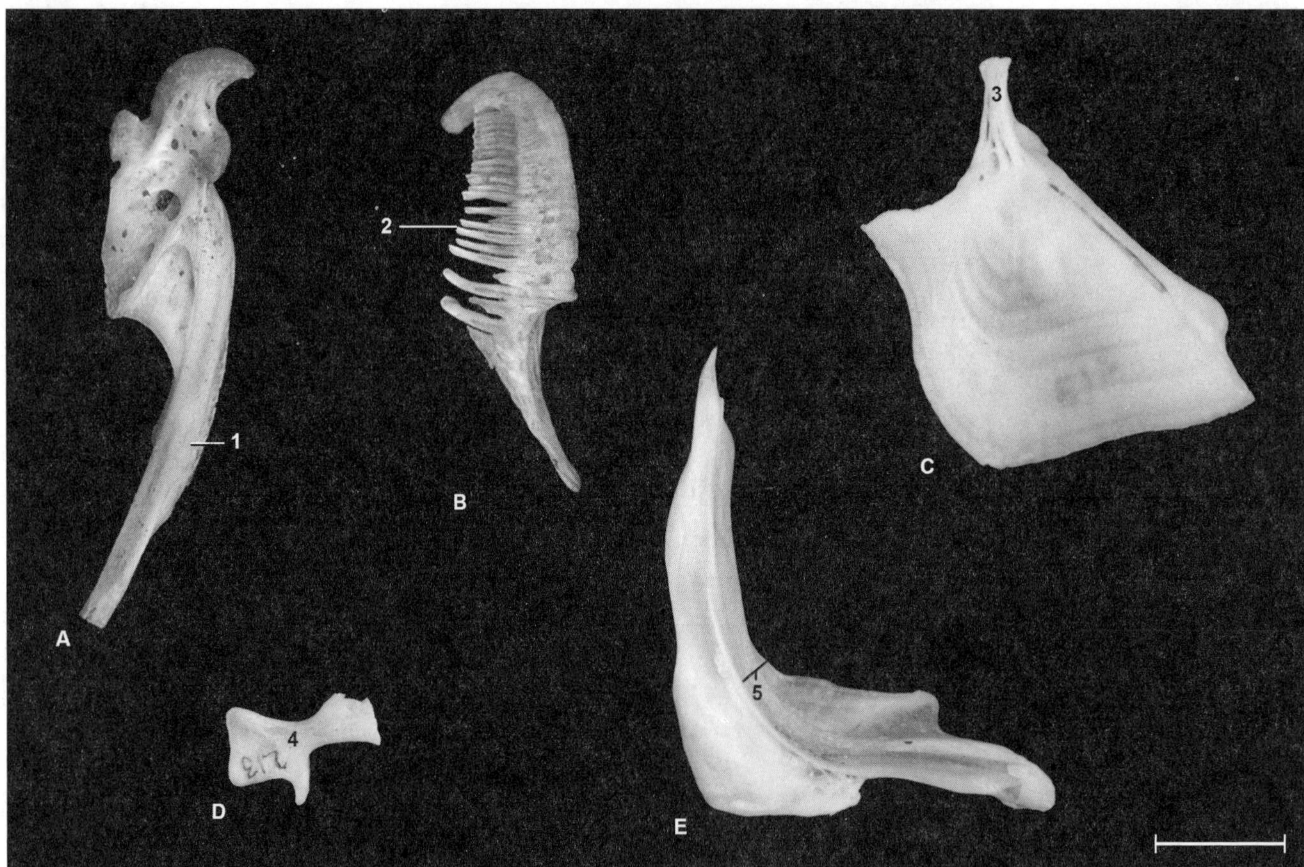

FIGURE 2.22. Utah Sucker (*Catostomus ardens*) r. hyomandibular, medial view (A); r. pharyngeal, lateral view, (B); r. opercle, lateral view, (C); r. palatine, lateral view, (D); r. cleithrum, lateral view (E). Osteological features (Catostomidae): 1, hyomandibular, thin and elongate; 2, comb-like pharyngeal with 20+ individual peg-like teeth in single, curved row; 3, opercle with finger-like projection dorsal to hyomandibular fossa; 4, short and squat palatine with medial projection terminating in crescent-shaped bowl (not visible here); 5, wide medial-lateral depth of cleithrum at mid-section.

Cui-ui (*Chasmistes cujus*) and Tahoe Sucker (*Catostomus tahoensis*) are similarly recognized. The Sacramento Sucker (*Catostomus occidentalis*) is the most abundant and has the largest range west of the Sierra Nevada. The Longnose Sucker (*Catostomus catostomus*; Figure 2.21) is the most widespread catostomid in North America.

ORDER Salmoniformes (Trout, Salmon, Whitefish, and Grayling)

FAMILY Salmonidae (Trout, Salmon, Whitefish, and Grayling)

Three subfamilies of Salmoniformes (L., salmo = salmon), Salmoninae (trout and salmon), Thymallinae (grayling), and Coregoninae (whitefish), five genera (*Thymallus*, *Oncorhynchus*, *Salvelinus*, *Prosopium*, *Coregonus*), and 27 species occur in our region. Members of the family are widely recognized as among the most desired native gamefish in the west (Figures 2.23–2.27). Originating during the Eocene, the salmonids are restricted to the Northern Hemisphere and are identifiable by a fusiform body, forked tail, adipose fin, and axillary scales at the base of the pelvic fins. They are normally associated with cool, well-oxygenated waters. We treat the two subfamilies that have more widespread distributions in the west below.

SUBFAMILY Salmoninae (Trout and Salmon)

Most native trout and salmon species are in the genus *Oncorhynchus* (Figures 2.23–2.25; Color Plates 2–3). The five species of pacific salmon are anadromous, with a few exceptions where populations have become landlocked. The most widespread native species of trout—Rainbow Trout (*Oncorhynchus mykiss*; Figure 2.23) and Cutthroat Trout (*Oncorhynchus clarkii*; Color Plate 2)—have both anadromous races and resident freshwater forms.

The life-history cycle for the anadromous forms involves an extended growth phase in the ocean with some individuals in some species spending six or more years in the marine environment. Upon reaching sexual maturity, adult fish move into natal freshwater rivers and streams to spawn. This involves a major physiological transformation to a spawning morphology with dramatic color changes and, in males, the development of a dorsal hump and hooked upper and lower jaws; the latter is the source for the generic

FIGURE 2.23. Rainbow Trout (*Oncorhynchus mykiss*).

FIGURE 2.24. Chinook Salmon (*Oncorhynchus tshawytscha*).

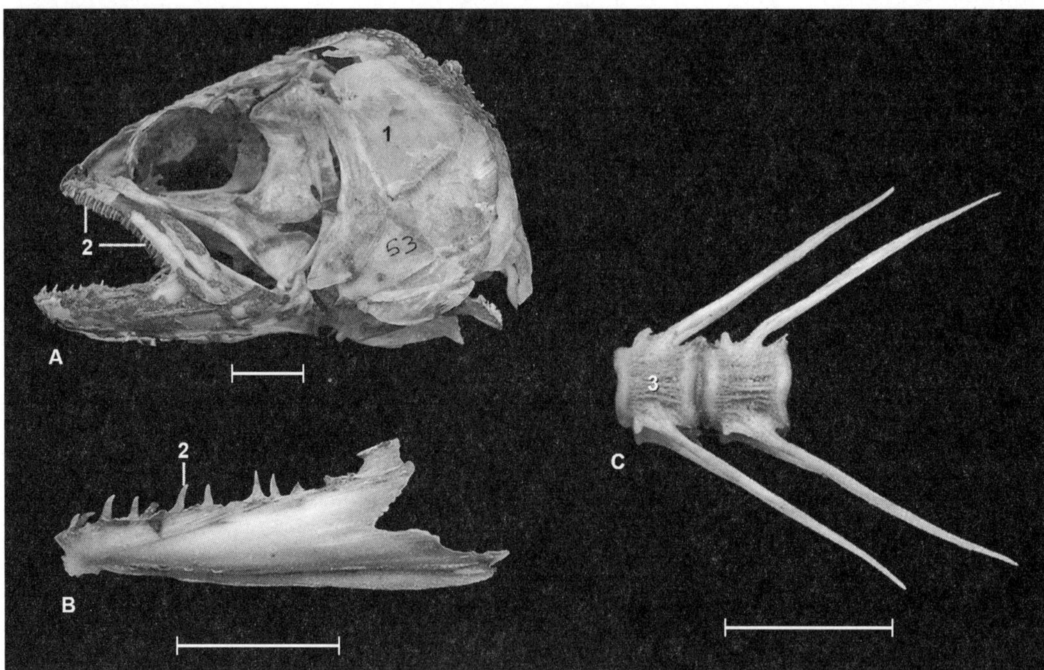

FIGURE 2.25. Cutthroat Trout (*Oncorhynchus clarkii*) cranium (A); l. dentary (B); and vertebrae (C), lateral views. Osteological features (Salmoninae): 1, cranium cartilaginous, porous, spongy; 2, well-developed teeth on dentary, premaxilla, maxilla, palatine, vomer, and base of tongue (lingual plate); 3, vertebral centra, reticulate, mesh-like.

name *Oncorhynchus* (Gr., onkos = hook, rhynchos = nose). Spawning occurs within specially prepared gravel beds or redds, often located hundreds of miles upstream.

All trout and salmon are carnivorous, their prey varying by local availability. Younger individuals feed more on invertebrates, with piscivory increasing with age. Small crustaceans, especially copepods and krill, are often important in the diet of salmonines, and they produce the orange flesh found in many species.

East of the Sierra Nevada, the primary native trout of the interior west is the Cutthroat Trout (Figure 2.25), which is divided into many regional subspecies. In most locations within its range, Cuthroat Trout occur in cold, freshwater lakes, rivers, and streams. Certain populations have adapted to higher-salinity conditions; some coastal populations are even anadromous, spending summer months in the ocean. The Lahontan Cutthroat Trout has adapted to the moderately saline waters of Pyramid Lake. Cutthroat Trout was a key member of the Pleistocene Lake Bonneville fish fauna. They can reach substantial sizes; records for Nevada and Utah are 41 (18.6 kg) and 26 lbs (12.1 kg), respectively. Trout and, especially salmon, were the source of major prehistoric and historic fisheries in the west, but with the exception of the widely stocked Rainbow Trout, most species have declined dramatically over the last century, and many populations are threatened.

SUBFAMILY Coregoninae (Whitefish)

Genus *Prosopium*

Although a number of whitefish species in the genus *Coregonus* occur in the high arctic region of the west, in lower latitudes only four native whitefish species occur. All are in the genus *Prosopium*. The name *Prosopium* means "masked" and refers to the prominent first suborbital (or lacrimal) bone located anterior to the eye. This element is evident in both living fish and in the skeleton. Its shape is useful for species identification that is notoriously challenging for this group—even with whole fish. In *Prosopium*, most individuals are relatively small, measuring less than 30 cm and weighing less than 1 kg; record sizes however extend over 2.3 kg. Three species—Bonneville Cisco (*Prosopium gemmifer*; Figure 2.26; Color Plate 4), Bear Lake Whitefish (*Prosopium abyssicola*), and Bonneville Whitefish (*Prosopium spilonotus*; Figure 2.27)—are endemic to Bear Lake, a high-elevation, cold, deep body of water on the Utah-Idaho border. All three species have been identified from the deposits of Pleistocene Lake Bonneville; Bonneville Cisco appears to have been the most abundant fish in the lake. The Mountain Whitefish (*Prosopium williamsoni*) is more widespread, occurring in cool rivers across northwest North America. The diet of whitefish includes a wide range of zooplankton and invertebrates; larger individuals become progressively more piscivorous.

FIGURE 2.26. Bonneville Cisco (*Prosopium gemmifer*).

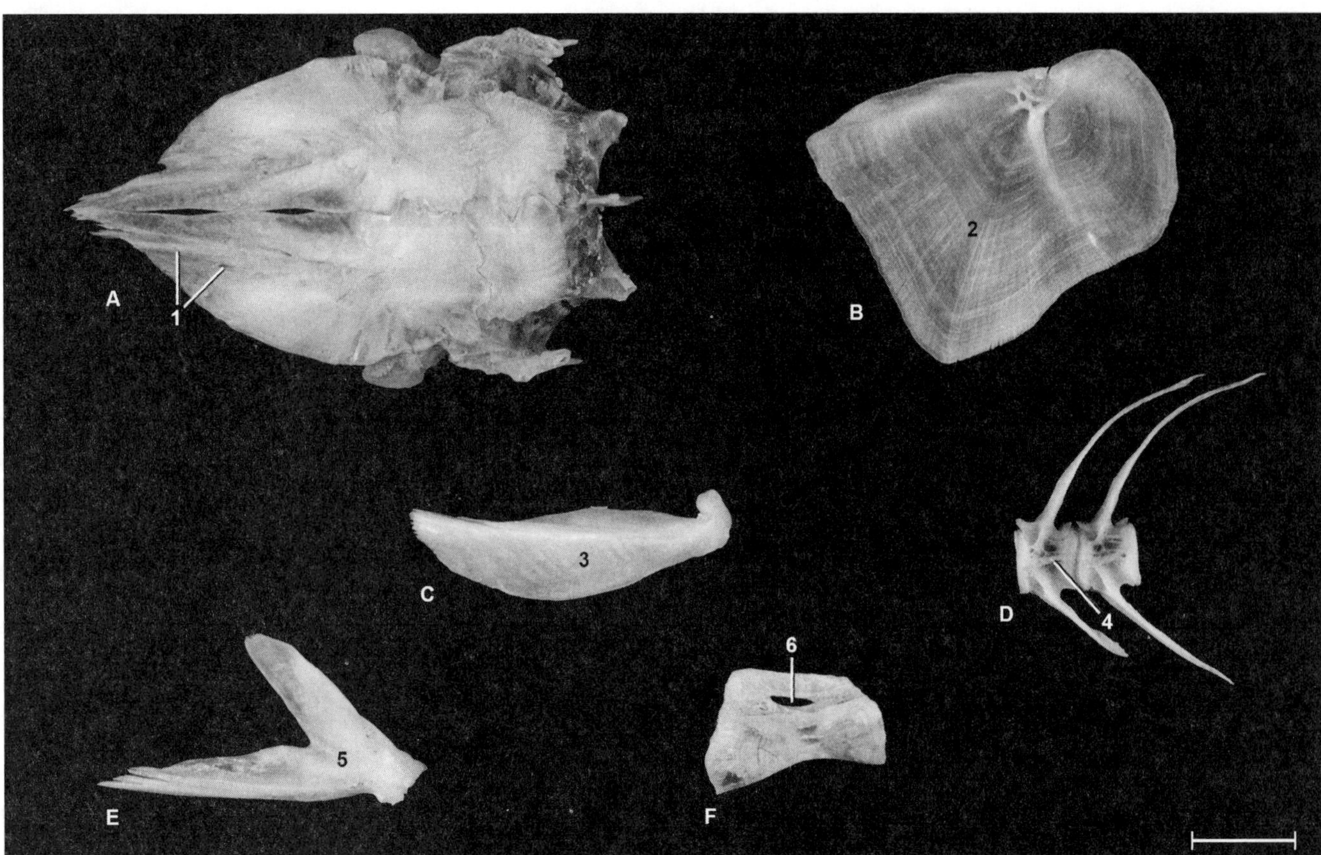

FIGURE 2.27. Bonneville Whitefish (*Prosopium spilonotus*) cranium, dorsal view (A); medial view of r. opercle (B); lateral view of r. maxilla (C); vertebrae (D); lateral view of r. dentary (E); lateral view of r. ceratohyal (F). Osteological features (*Prosopium*): cranium lacks teeth; 1, leaf-shaped frontal with pores opening from tube-like structures extending across antero-dorsal surface; 2, triangular opercle with hyomandibular fossa positioned farther anterior (left in this photo), relative to salmonines; 3, sickle-shaped maxilla; 4, delicate struts along vertebral centra; 5, V-shaped dentary; 6, ceratohyal with large dorsal foramen.

ORDER Scorpaeniformes (Scorpionheads, Rockfishes, Sculpins and Allies)

Scorpaeniformes (Gr., skorpaina = scorpion) is an enormous order that contains 36 families of mostly marine fishes. As a group, this worldwide order is referred to as the "mail-cheeked fishes" because the third suborbital bone extends posteriorly across the cheek to the preopercle. Only a single family (Cottidae) and genus (*Cottus*) occurs in freshwater settings in western North America.

FAMILY Cottidae (Sculpins)

Most members of this large and diverse family are bottom-dwelling marine fishes. All cottids lack swim bladders; have broad heads; prominent, fan-shaped pectoral fins; scaleless bodies; and preopercles with prominent posteriorly projecting spines.

Genus *Cottus*

Sculpins in the genus *Cottus* have invaded freshwater habitats, with a concentration of taxa in the northwestern United States. Freshwater sculpin are small (< 15 cm) and have dark, mottled coloration that is cryptic against the rocky bottoms of the cold, well-oxygenated streams or lakes that they typically inhabit (Figure 2.28). Sculpins in small streams can be readily collected by seining (Color Plates 5–6). Their marine ancestry is reflected in the ability of several freshwater species (e.g., *Cottus asper*, Prickly Sculpin; *Cottus aleuticus*, Coastrange Sculpin) to utilize brackish and saltwater habitats. They are commonly referred to as "bullheads" by anglers. The Utah Lake Sculpin (*Cottus echinatus*) was once endemic to Utah Lake but is now extinct. The Bear Lake Sculpin (*Cottus extensus*; Figure 2.29) that once occurred in ancient Lake Bonneville is now endemic to

FIGURE 2.28. Paiute Sculpin (*Cottus beldingii*).

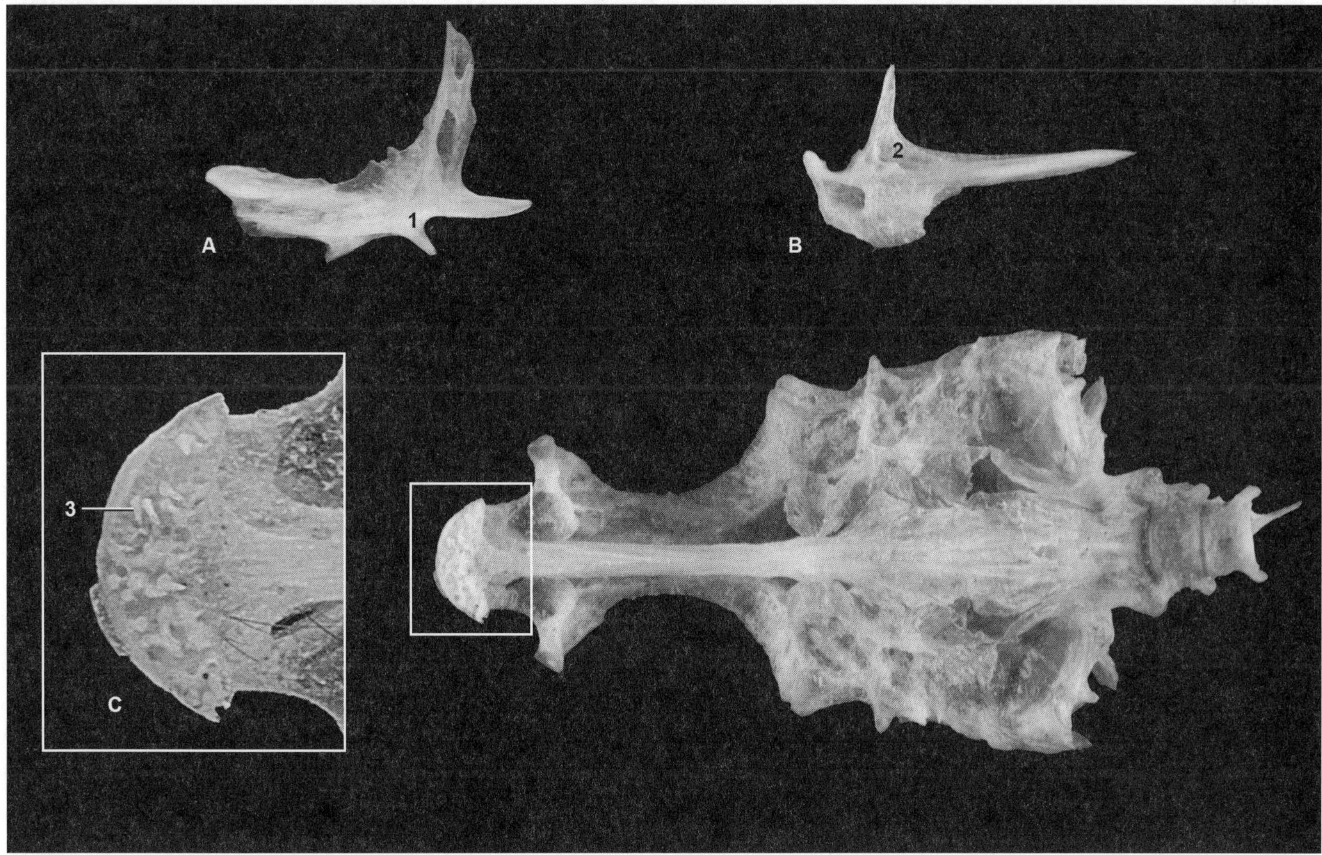

FIGURE 2.29. Bear Lake Sculpin (*Cottus extensus*) lateral views of r. preopercle (A), and r. articular (B), and ventral view of vomer (C). Osteological features (*Cottus*): 1, prominent spines on preopercle—arrangement diagnostic to species; 2, articular with single antero-dorsal spine; 3, vomer with crescent-shaped pad bearing multiple tiny teeth.

FIGURE 2.30. Tule Perch (*Hysterocarpus traski*). Photo by Thomas Taylor

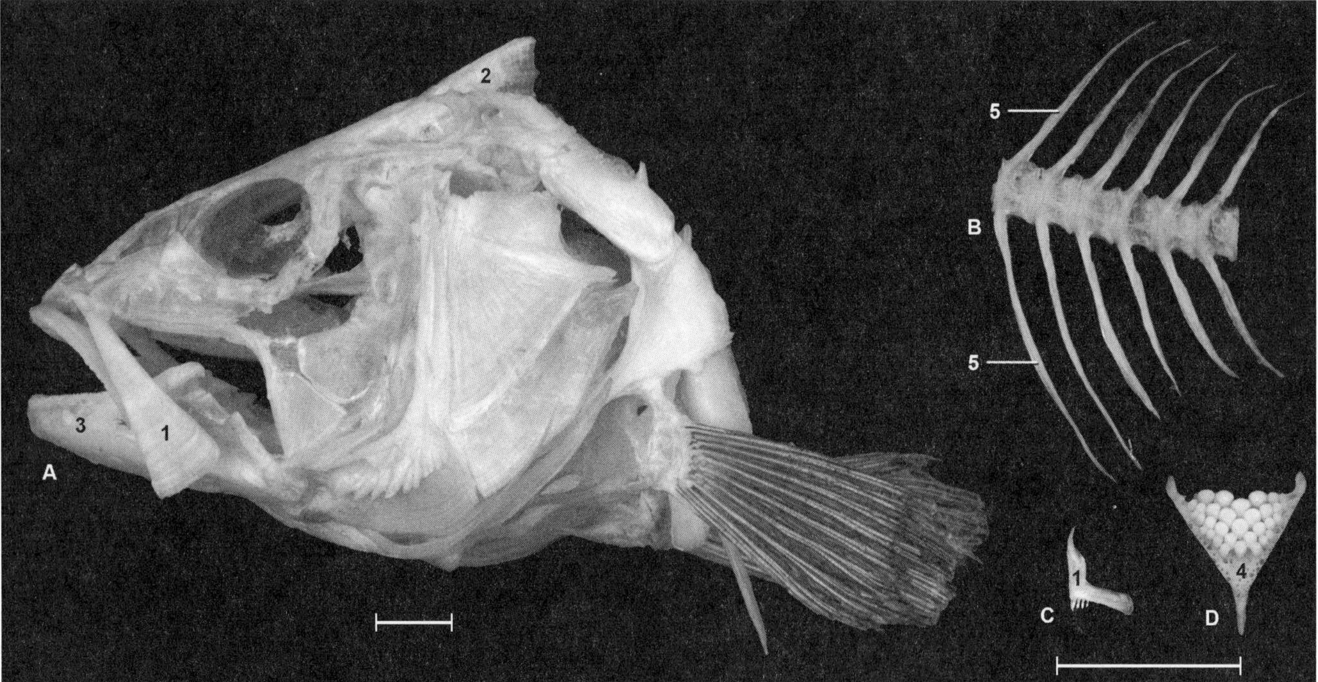

FIGURE 2.31. Yellow Perch (*Perca flavescens*) lateral views of cranium (A) and vertebral column (B); Tule Perch (*Hysterocarpus traski*), lateral view of l. premaxilla (C), and dorsal view of pharyngeal plate (D). Osteological features (Perciformes): 1, maxilla paddle-shaped and toothless—premaxilla toothed; 2, prominent mid-sagittal keel along supraoccipital and exoccipital; 3, dentary V-shaped with very small teeth; 4, pharyngeal teeth in plates, single in Embiotocidae, as pictured in (D), but paired in Centrarchidae, Percidae (not shown); 5, vertebrae with long neural and haemal spines, supporting deep body form.

Bear Lake, on the Utah-Idaho border. Due to their small size, these fish rarely occur in archaeological deposits.

ORDER Perciformes (Perch and Allies)

Perciformes (Gr., perke = perch) is the single largest order of vertebrates, represented by 135 families and 7,000 species worldwide. This giant order contains about 40% of all bony fish, yet despite this global dominance, the native west was represented by only two species, both occurring in the Sacramento-San Joaquin River system in central California. That situation has changed dramatically, however, with countless introductions over the last 150 years, so that members of this order—Bluegill (*Lepomis macrochirus*), Walleye (*Sander vitreus*), Largemouth Bass (*Micropterus salmoides*), and Striped Bass (*Morone saxatilis*)—are among the most common and desirable gamefish in the region. Perciformes are easy to recognize because their dorsal and anal fins are divided into an anterior portion with prominent bony spines and a posterior portion with only soft ray supports. In addition, the pelvic fins are positioned unusually far forward (anteriorly), just ventral and posterior to the pectoral fins. The two native species, Sacramento Perch (*Archoplites interruptus*) and Tule Perch (*Hysterocarpus traski*; Figures 2.30–2.31) are morphologically similar small-sized fish, with deep (dorsal-ventral depth goes 2½ times into standard length), medial-laterally compressed bodies and terminal mouths. These fishes also occupied similar native ranges in lowland freshwater habitats of central California, including lakes, estuarine sloughs, and clear streams and rivers. Both also are primarily carnivorous, feeding on a range of aquatic organisms from fish to zooplankton. These perch do however belong to separate families and are both standouts within them. The Sacramento Perch represents the only native sunfish (Centrarchidae) west of the Rocky Mountains while the Tule Perch represents the only strictly freshwater member of the surfperch family (Embiotocidae). Also noteworthy, the Tule Perch is viviparous (as are all surfperch) and gives birth to live young rather than laying eggs as most other fishes.

Notes

We have relied on Liem et al. (2001), Hildebrand (1995), and King and Custance (1982) for our discussion of fish characteristics, origins, and osteology. We follow Wheeler and Jones (1989) and Froese and Pauly (2015) for fish osteological and taxonomic nomenclature, respectively. Our discussion of the natural history and ecology of western fishes is derived primarily from Moyle (2002) but also Sigler and Sigler (1987) and Wydoski and Whitney (2003). See also Smith and Butler (2008) for a detailed discussion of the ethnographic use and hard-tissue anatomy of lampreys.

CHAPTER 3

Amphibians

I. General Osteology of Amphibians

Characteristics

Modern amphibians are characterized by glandular skin (capable of gas exchange) without external scales, hair, or feathers. They have gills in the aquatic, larval stage but lungs in the mostly terrestrial adults. Most amphibians are oviparous and deposit eggs externally within a gelatinous matrix that lacks an amniotic membrane, although some are ovoviviparous and retain fertilized eggs and give birth to developed larvae. There are 5,763 extant species.

Origins

The amphibians are the earliest of the tetrapods and evolved in the Devonian Period, nearly 400 million years ago, from rhipidistian fishes similar to modern lungfishes. These fish had developed multisegmented, leg-like fins that allowed them to crawl along the sea floor. Two extinct subclasses are recognized, and all recent species have been placed in the monophyletic subclass Lissamphibia.

Osteology

During the transition from a fully aquatic existence as fishes to one more terrestrially based, amphibian skeletal features underwent a number of changes (Figures 3.1–3.7). Most notably, dramatic modifications of the pectoral and pelvic girdles occurred—structures that in fishes acted as stable bases for the fins are now used in amphibians to transfer the body weight onto long, jointed limbs (Figures 3.4 and 3.6). Since the limbs are used as struts to lever the body off the ground in locomotion, much stronger attachments between them and the body developed. However, in comparison to more derived vertebrates such as mammals, amphibians have limbs that project sideways from the body, hence, their movement is comparatively awkward.

The cranium is also now unconnected to the pectoral girdle and is thus capable of greater mobility, enabling different methods of prey acquisition. Skull structure (Figures 3.1–3.2) tends to stabilize with the formation of the early tetrapod skull—there is in fact more variation in skull structure within the bony fishes alone than in all of the tetrapods combined. The complex of bones associated with fish gills (e.g., opercles, ceratohyals, epihyals, pharyngeals) was lost or adapted to different functions, and new features associated with sensory structures appropriate for terrestrial life also evolved.

Changes in jaw function and structure are also notable. For example, the hyomandibular in fish helps to connect the lower jaw to the braincase, but in amphibians that connection is formed through an articular-quadrate-squamosal articulation and the hyomandibular is modified to provide a sound-conducting structure, the columella (Figures 3.1–3.2). In addition, the vertebral column (Figure 3.3) now bears body weight and, as a result, the vertebrae are more firmly attached to each other with interlocking centra and additional strong articulations (zygapophyses). Anterior and posterior sections of adjacent vertebrae interlock with a bulge in one fitting into a concavity in the other. In amphibians, the concavity is anterior and the vertebrae are described as procoelus (pro = front, coel = hollow; Figure 3.3).

Finally, it should be noted that modern amphibians such as frogs and toads, although commonly depicted as representing the general amphibian morphology, do have much more specialized skeletons compared to early amphibians such as labyrinthodonts that fall more along the main line of tetrapod evolution. Labyrinthodonts are thus more similar to modern reptiles.

Remarks

Different amphibian taxa have distinct requirements relating to the amount, source, and quality of water that they inhabit, especially during the early stages of their life cycle. As a result, the presence of specific taxa can provide detailed indicators of local aquatic habitats. Indeed, because amphibians are especially sensitive to changes in their environment, they are often used in modern conservation studies as sentinel species or indicators of general ecosystem health. Unfortunately, recent estimates suggest that nearly one third of the world's amphibian species are threatened, due to a series of factors involving habitat destruction, climate change, competition from introduced species, and chemical contaminants.

Amphibian remains are not commonly encountered in western archaeological faunas for a variety of reasons: relatively small package size and low economic value, low densities, and aquatic habitats that decrease encounter rates and increase the cost of capture for human consumers. In addition, because toads utilize burrows, their presence in archaeological deposits need not have resulted from past human activities. Toad remains should thus be subjected to detailed taphonomic analyses to ascertain their source of deposition.

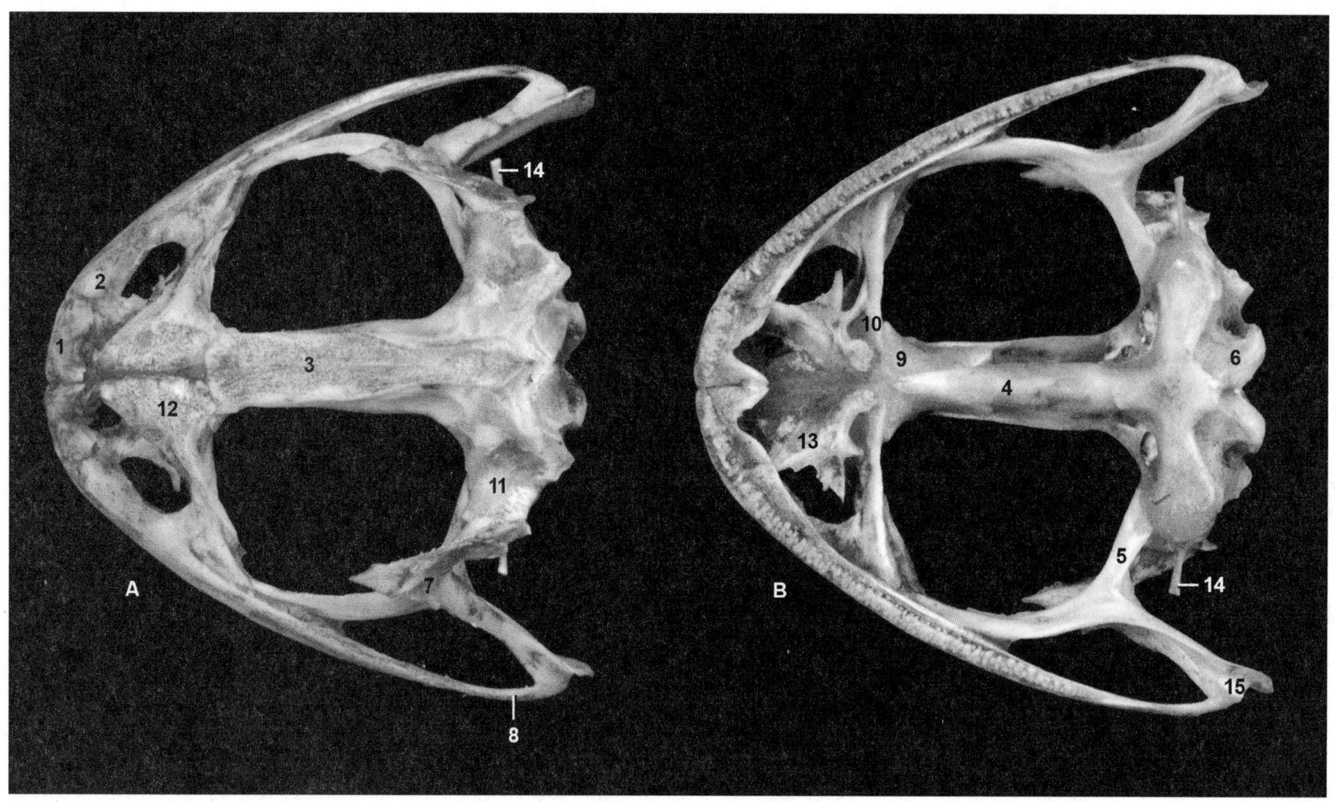

FIGURE 3.1. Cranium of Northern Leopard Frog (*Lithobates pipiens*), dorsal (A) and ventral (B) views.

1. Premaxilla
2. Maxilla
3. Fronto-parietal
4. Parasphenoid
5. Pterygoid
6. Exoccipital
7. Squamosal
8. Quadratojugal
9. Ethmoid
10. Palatine
11. Prootic
12. Nasal
13. Vomer
14. Columella
15. Quadrate

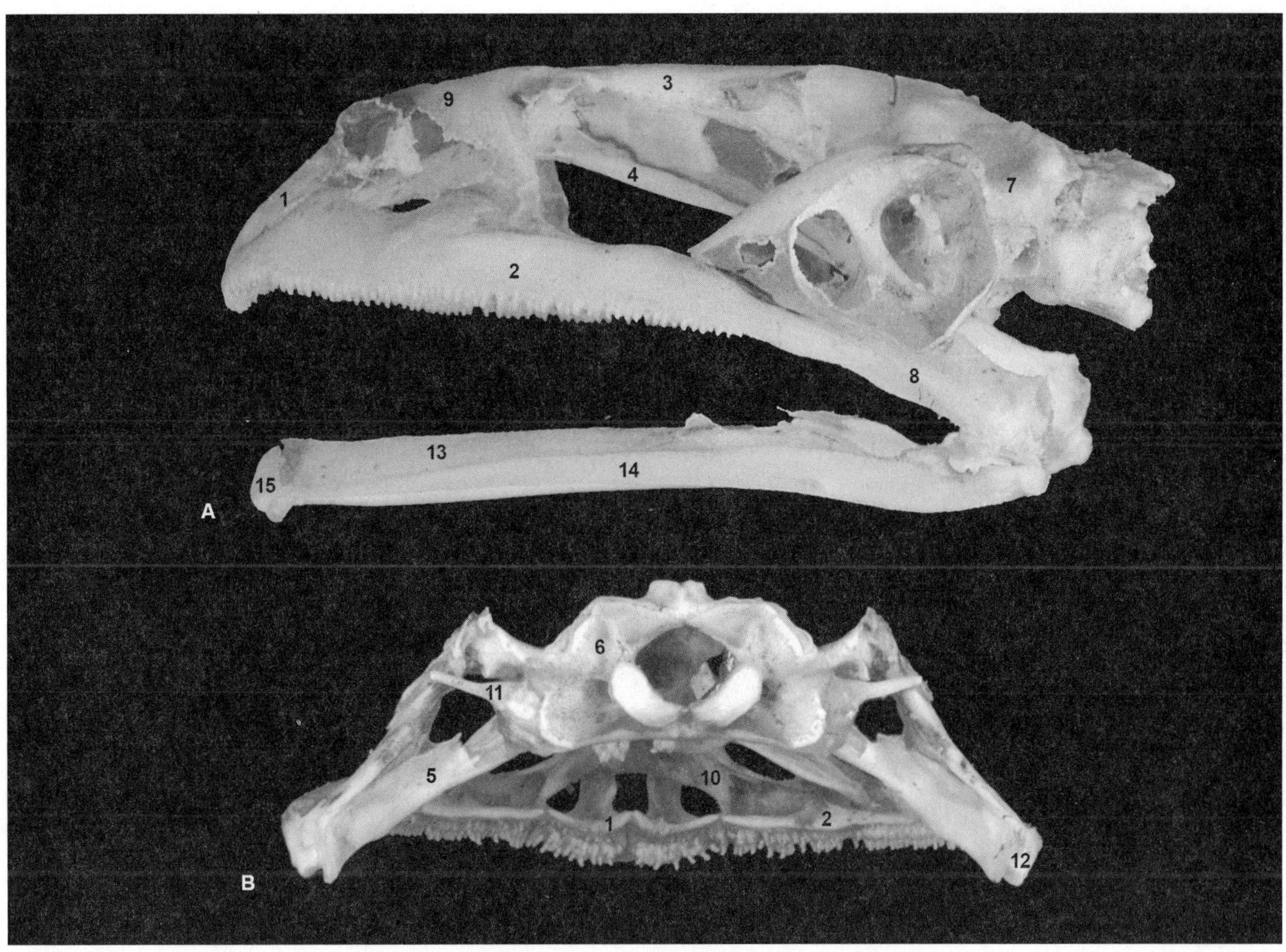

FIGURE 3.2. Cranium of Northern Leopard Frog, lateral (A) and posterior (B) views.

1. Premaxilla
2. Maxilla
3. Fronto-parietal
4. Parasphenoid
5. Pterygoid
6. Exoccipital
7. Squamosal
8. Quadratojugal
9. Nasal
10. Vomer
11. Columella
12. Quadrate
13. Dentary
14. Angulo-splenial
15. Mento-mekelian

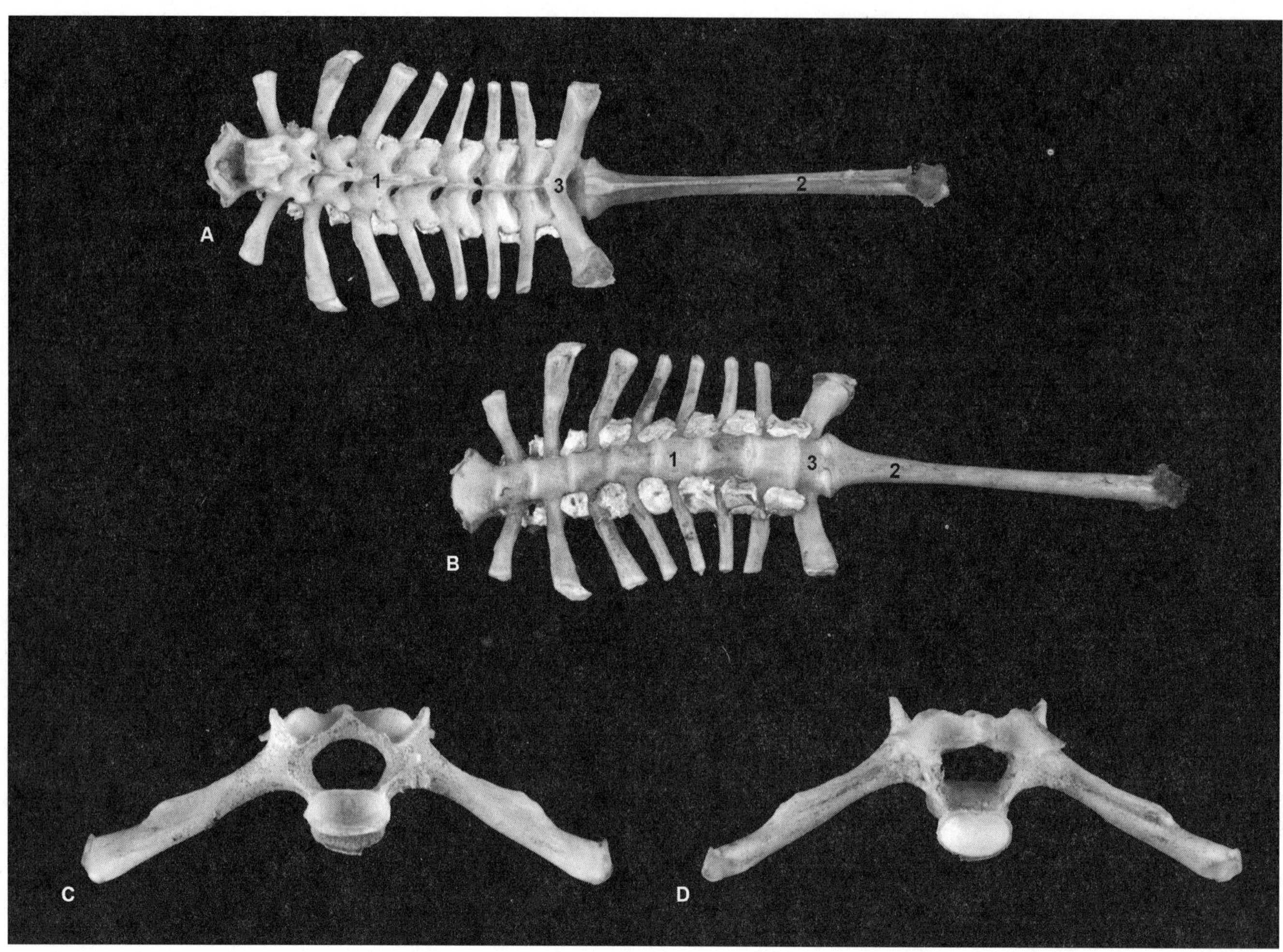

FIGURE 3.3. Northern Leopard Frog vertebral column and urostyle dorsal view (A) and ventral view (B) (anterior is left, posterior is right); isolated vertebrae, anterior (C) and posterior (D) views.

1. Vertebrae
2. Urostyle
3. Sacrum

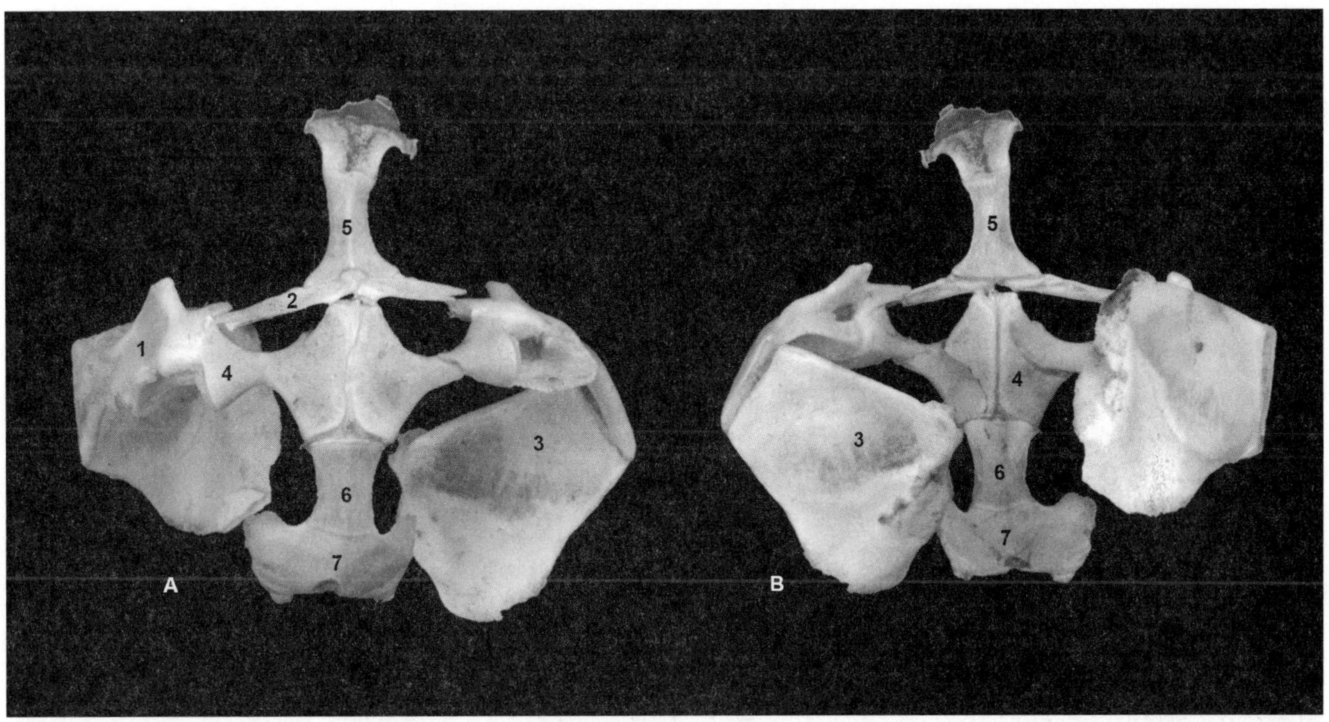

FIGURE 3.4. Pectoral girdle of Northern Leopard Frog, ventral (A) and dorsal (B) views, anterior at top, posterior at bottom.

1. Scapula
2. Clavicle
3. Suprascapula
4. Coracoid
5. Omosternum
6. Sternum
7. Xiphisternum

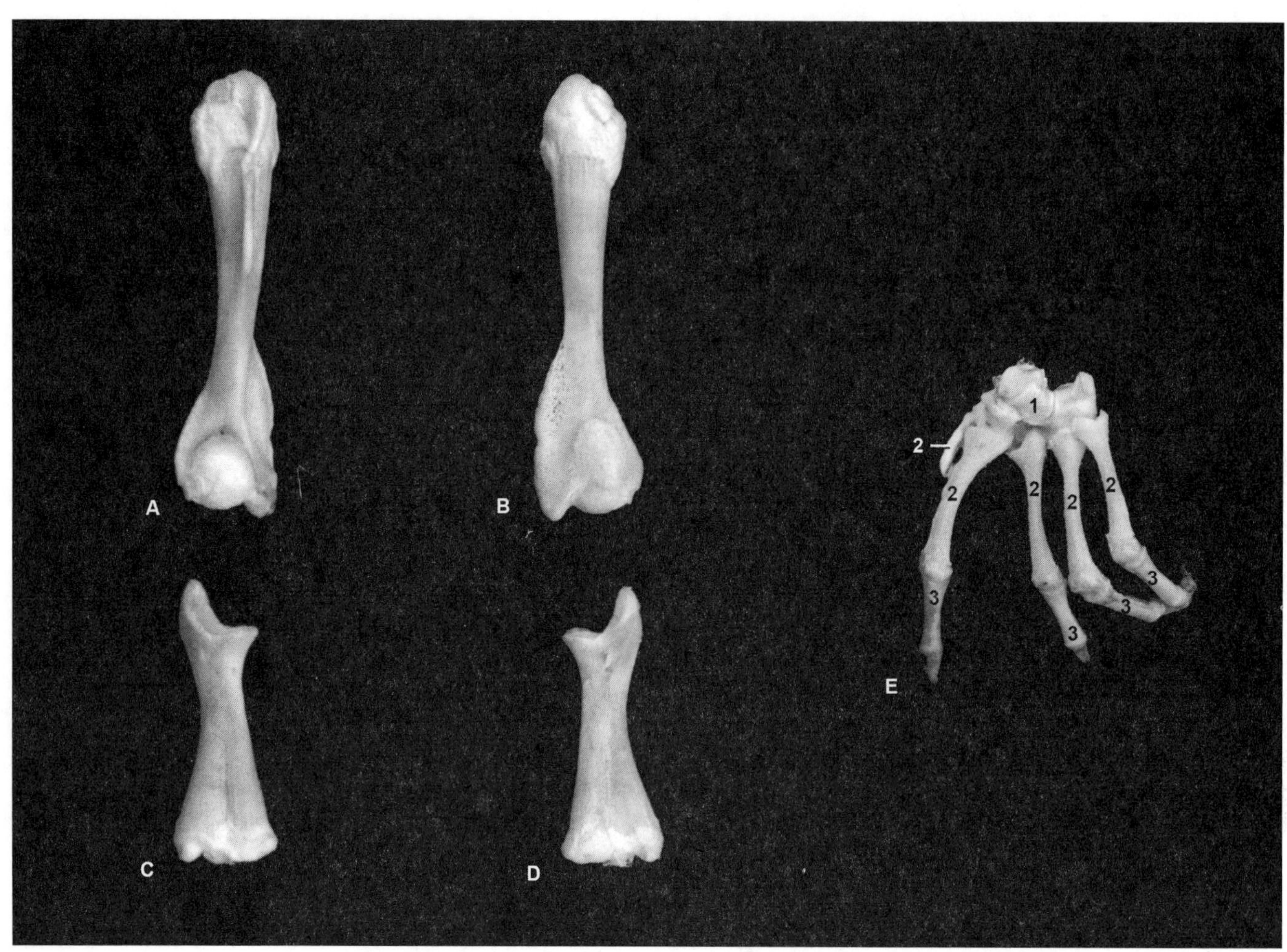

FIGURE 3.5. Northern Leopard Frog forelimb elements: r. humerus, anterior (A) and posterior (B) views; r. radio-ulna, anterior (C) and posterior (D) views; r. manus, dorsal view (E), left is medial. Top is proximal.

1. Carpals
2. Metacarpals
3. Phalanges

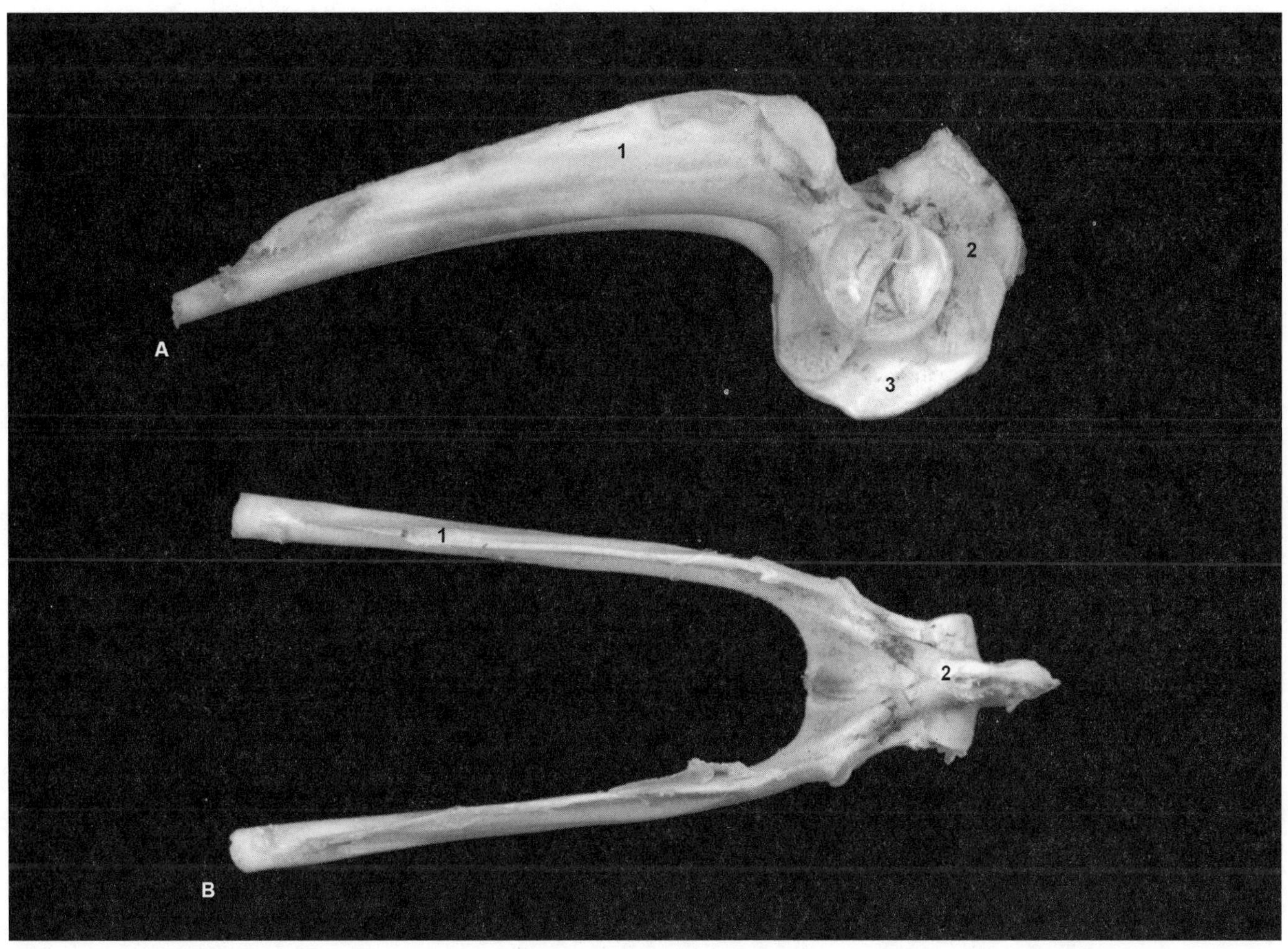

FIGURE 3.6. Pelvic girdle of Northern Leopard Frog, lateral (A) and dorsal (B) views. Anterior on left, posterior on right.

1. Ilium
2. Ischium
3. Pubis

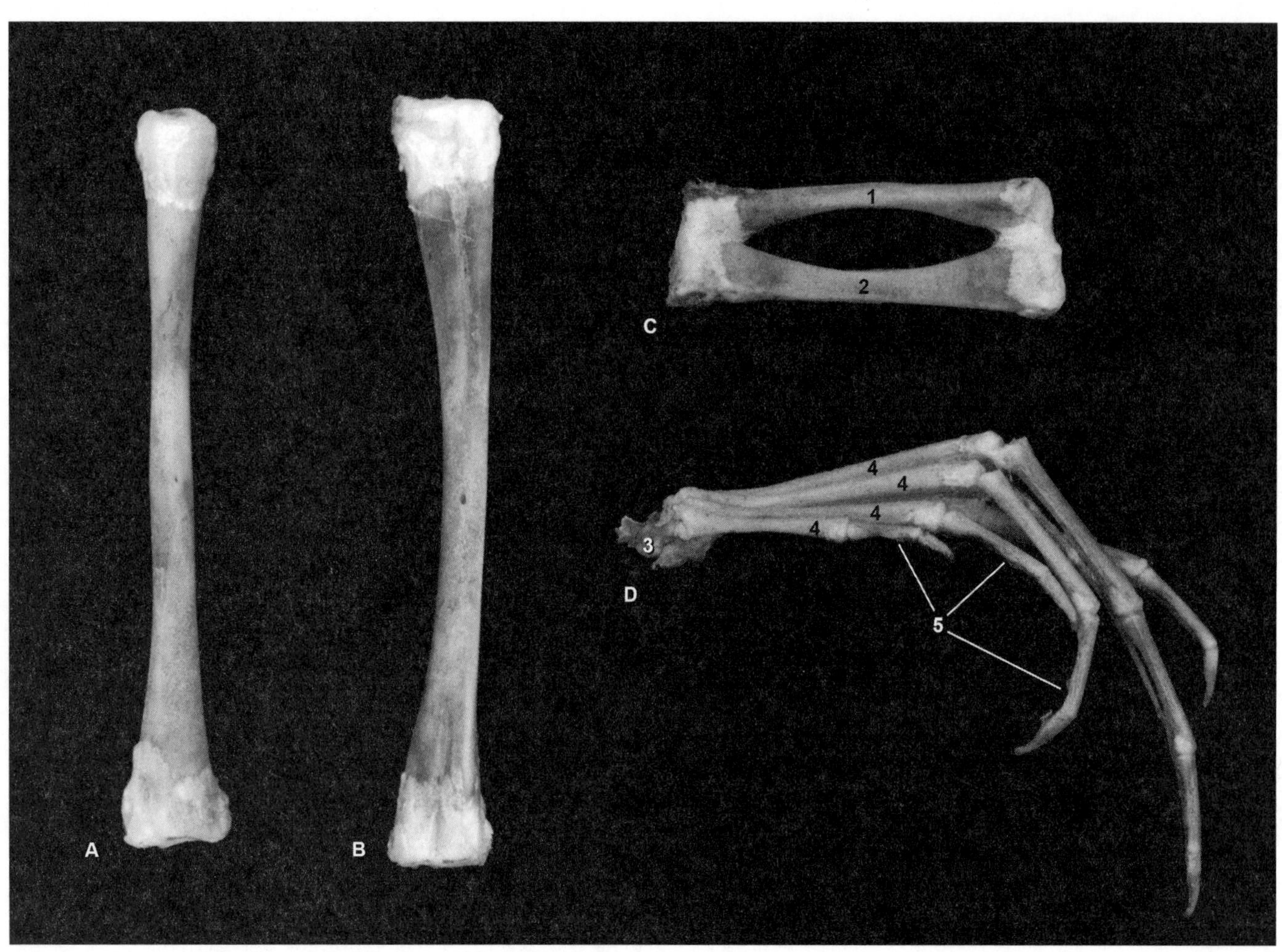

FIGURE 3.7. Hindlimb elements of Northern Leopard Frog: l. femur, anterior view (A); l. tibio-fibula, anterior view (B); proximal is top; l. calcaneus and astragalus, ventral view (C); and foot, lateral view (D), proximal is left.

1. Astragalus
2. Calcaneus
3. Tarsals
4. Metatarsals
5. Phalanges

II. Taxonomy and Osteological Variation of Western Amphibians

CLASS Amphibia
ORDER Anura (Frogs)

Anurans (L., an = without, Gr., oura = tail) are characterized by long hindlimbs, short bodies, webbed digits, bulging eyes, and the lack of tails (in adults). Enabled by long, powerful legs, they are widely known for their exceptional jumping ability. With permeable skin, frogs are often closely associated with aquatic habitats and require water at early stages of their life cycle—gelatinous egg masses are externally fertilized and are typically deposited in water. Lengthy excursions well away from water do, however, occur in many species. Mating takes place by amplexus in which the male dorsally mounts and clasps the female with his front legs and sheds sperm onto eggs as they are released from her cloaca.

Frogs are characterized by pronounced metamorphosis starting with limbless aquatic larvae, called tadpoles, which possess gills and long, medial-laterally compressed tails used for locomotion. As the tadpole develops, legs and lungs form, and the tail gradually disappears. Most tadpoles are herbivorous, feeding on algae and plant material, but some forms are omnivorous and will even feed on other tadpoles. Adult frogs are typically carnivorous, feeding on arthropods, worms, and snails. Frogs give species-specific calls, typically produced by males in mating season, and amplified by membranous vocal sacks located beneath the lower jaw.

The distinction between frogs and toads is not a taxonomic one but instead is based on the convergence of terrestrial adaptations that have independently evolved in different families. In general, anurans that have dry, rougher, warty skin and relatively short hind limbs are referred to as toads. Although six families of anurans are represented in western North America, with two exceptions (Western Barking Frog, [*Craugastor augusti*, Craugastoridae] and Coastal Tailed Frog [*Ascaphus truei*, Leiopelmatidae]), all native species fall within four families: Bufonidae, Hylidae, Ranidae, and Scaphiopodidae.

Distinct from the Caudata (salamanders), the anuran skeleton is characterized by anteropostero (lengthwise) compression with reduction in the number of vertebrae; a lack of ribs; elongate and anterior-oriented ilia; hindlimbs longer than forelimbs; fusion between frontals and parietal forming the fronto-parietal; toothless dentaries; caudal vertebrae fused to form a single elongate structure, the urostyle; radius and ulna fused to form a single element, the radio-ulna; elongated tarsals (calcaneus and astragalus) that form an additional limb segment; and tibia and fibula fused to form the tibio-fibula.

FAMILY Bufonidae (True Toads)

All bufonids are toads; the family is thus referred to as the "true toads." Two genera, *Anaxyrus* and *Incilius*, are represented by 15 species in our area (Figures 3.8–3.9). These anurans are stocky, warty,

FIGURE 3.8. Western Toad (*Anaxyrus boreas*).

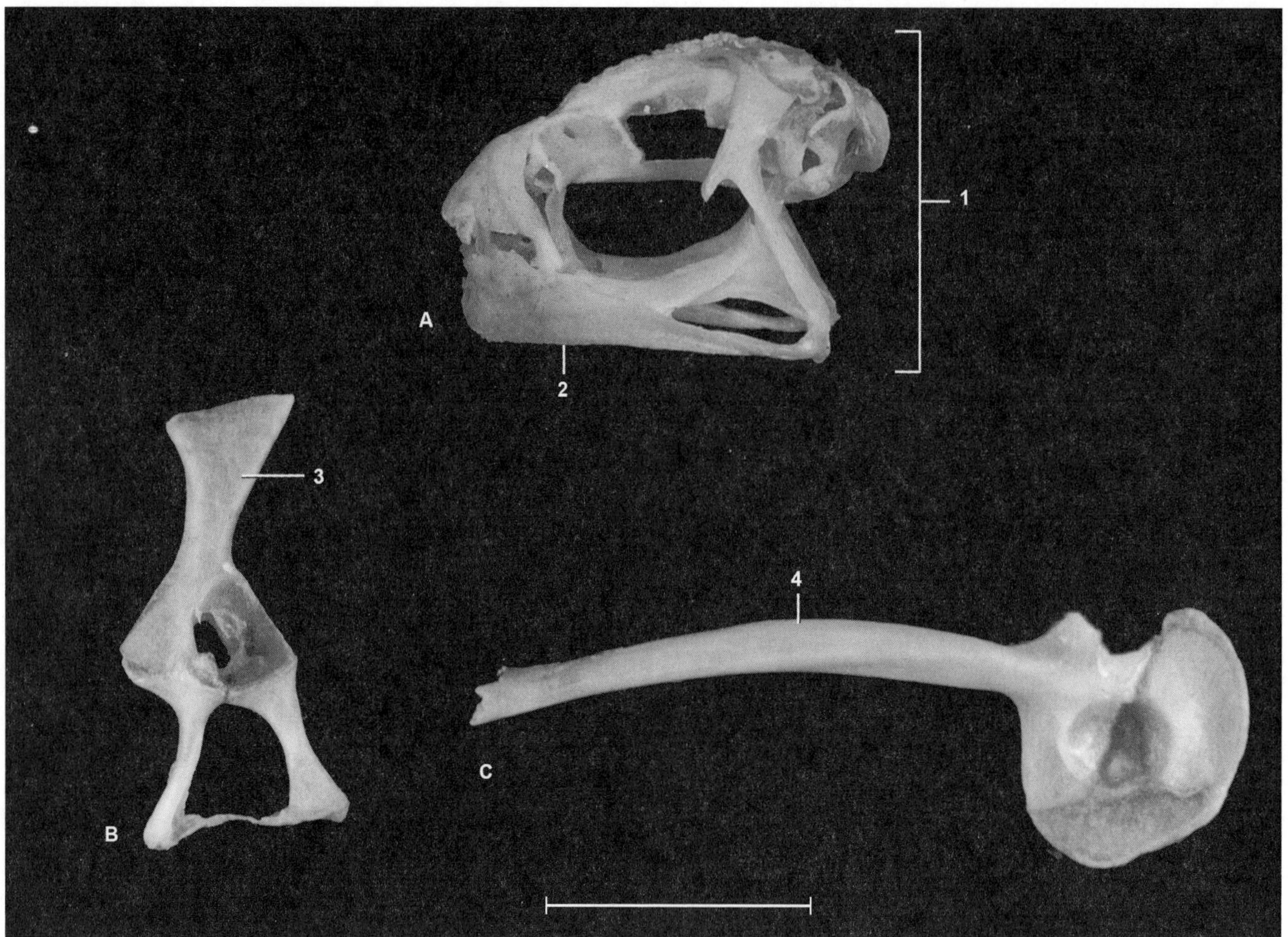

FIGURE 3.9. Woodhouse's Toad (*Anaxyrus woodhousii*), lateral views of cranium (A), pectoral girdle (B), and os coxa (C), anterior is left. Osteological features (Bufonidae): 1, cranium higher-domed, more enclosed than Hylidae, Ranidae; 2, cranium lacks teeth; 3, scapula proportionately long and narrow without antero-ventral flange; 4,ilium lacks blade-like crest present in ranids (true frogs).

short legged and possess parotoid glands, external skin glands on the back and sides just behind the eye. Most species also burrow and are nocturnal. Both the warts and parotoids secrete milky, alkaloid-based neurotoxins (bufotenin and bufotoxin) to deter predators. In some species (e.g., *Incilius alvarius*, Sonoran Desert Toad) these substances have hallucinogenic properties and have been utilized (by licking the skin or smoking dried skins) both by indigenous and modern peoples alike. Toxicity varies among species. In some bufonid species, these secretions have been known to kill dogs and other predators, but many animals eat toads with no adverse affects. We have been catching and handling Western Toads (*Anaxyrus boreas*; Figure 3.8; Color Plate 7) with bare hands for years with no problem. They also unload urine when handled, but they do not cause warts!

FAMILY Hylidae (Treefrogs)

Hylids are small-sized, slim-waisted, and long-legged frogs; many are arboreal and have well-developed, suction-cup-like toe pads that enhance clinging ability in trees and shrubs. Four genera represented by seven native species occur in the region. The genus *Pseudacris* (chorus frogs; Figures 3.10–3.11; Color Plate 8) is most widespread and familiar; all are small-sized, almost never reaching lengths more than 6.3 cm. Most can change colors to conform to their background.

FIGURE 3.10. Sierran Treefrog (*Pseudacris sierra*).

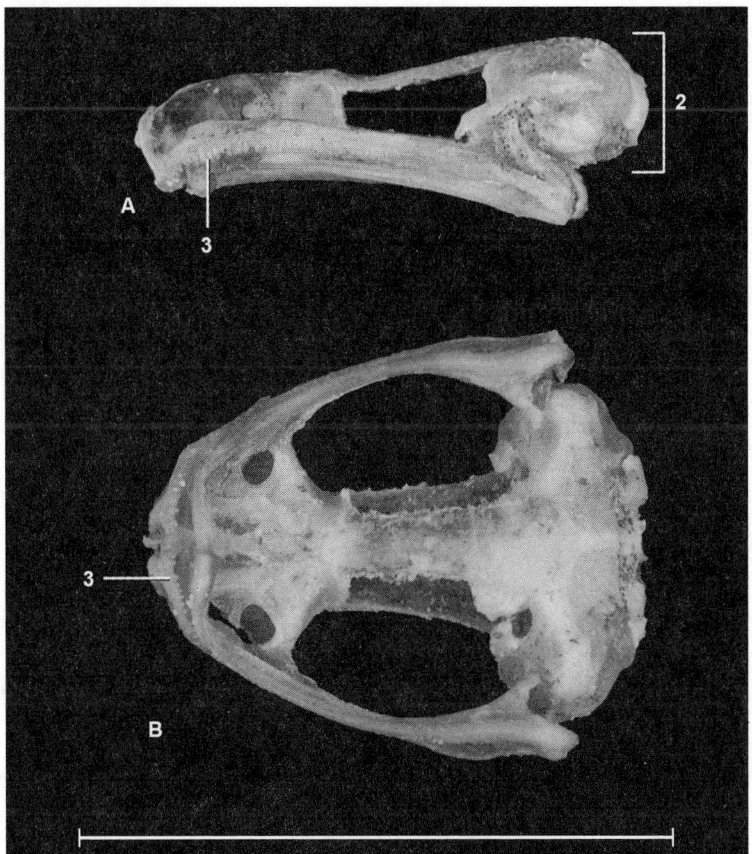

FIGURE 3.11. Chorus frog (*Pseudacris* sp.), lateral (A) and ventral (B) views of cranium. Osteological features (Hylidae): 1, all elements small; 2, cranium lower-domed, more open compared to Bufonidae and Scaphiopodidae; 3, maxilla, premaxilla with continuous rows of tiny teeth—most also with vomerine teeth.

FIGURE 3.12. American Bullfrog (*Lithobates catesbeianus*)

FAMILY Ranidae (True Frogs)

Two genera (*Rana*, *Lithobates*) and 11 native species occur in the region. Typically larger than hylids, the true frogs also have slim waists and long legs. Additionally, they have webbed hind feet, and most species have a pair of dorsolateral skin folds that extend from behind the eyes posteriorly to the lower back. Ranids also exhibit prominent paired dorsal humps on the back that are formed by anterodorsal extensions of the ilia. The American Bullfrog (*Lithobates catesbeianus*; Figure 3.12: Color Plate 9) is the largest, most widespread and familiar species in our area but was introduced from eastern North America. It can be found near larger, permanent water bodies such as ponds, lakes, and streams, where it is often encountered along the water's edge. It has had negative impacts on native frogs (e.g., Northern Leopard Frog, *Lithobates pipiens*; Figure 3.13).

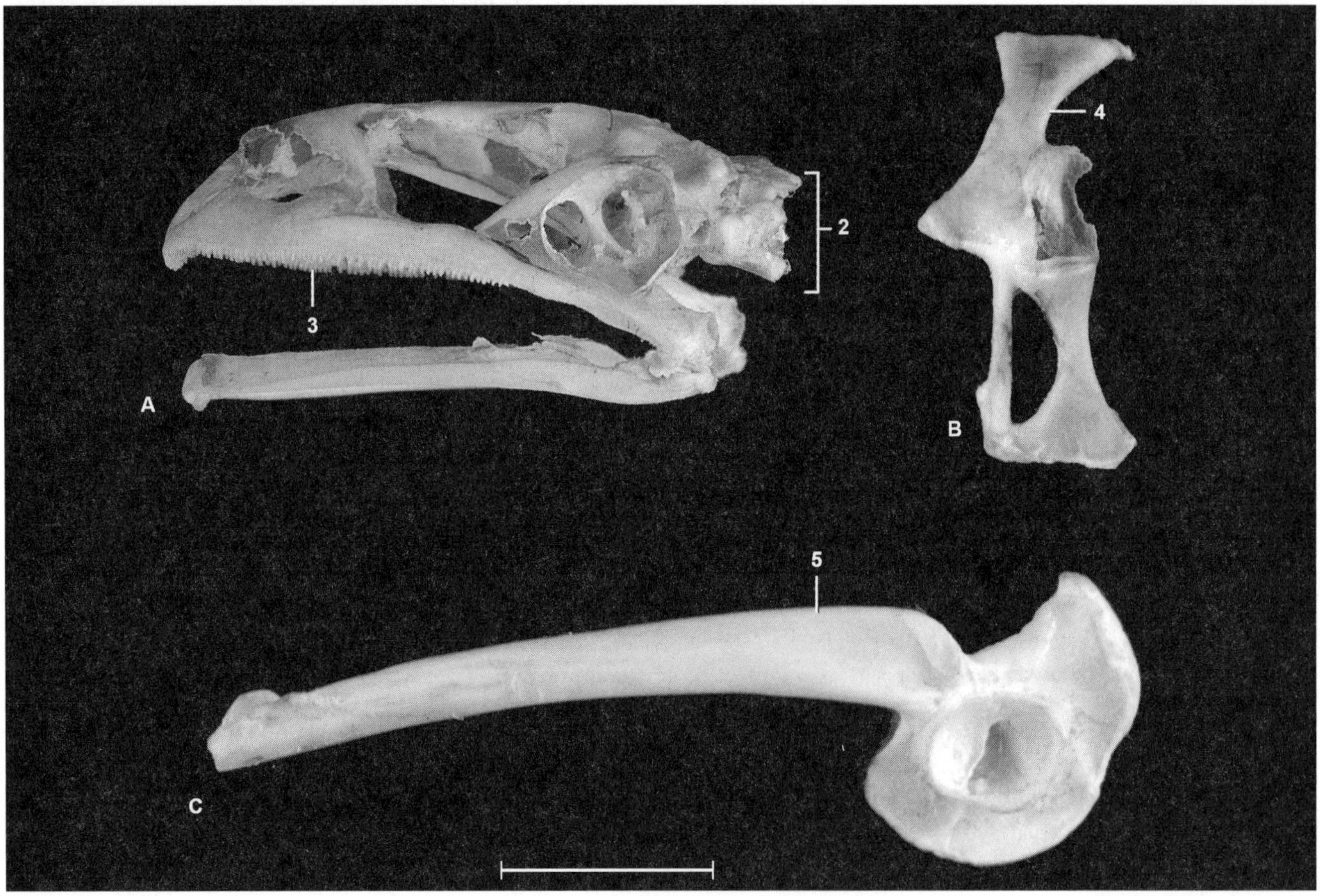

FIGURE 3.13. Northern Leopard Frog (*Lithobates pipiens*), lateral views of cranium (A), l. pectoral girdle (B), and l. os coxa (C), anterior is left. Osteological features (Ranidae): 1, all elements larger than hylids; 2, cranium lower-domed, more open compared to Bufonidae and Scaphiopodidae; 3, maxilla, premaxilla with continuous rows of small teeth—vomerine teeth present in some species; 4, scapula proportionately short, wide with deep anterior concavity—lacks antero-ventral flange present in scaphiopodids; 5, blade-like crest on dorsal ilium.

FAMILY Scaphiopodidae
(North American Spadefoots)

Two genera (*Scaphiopus* and *Spea*) and four spadefoot species occur in the west (Figures 3.14–3.15; Color Plate 10). Spadefoots are distinguished from the bufonids by their cat-like eyes (vertical pupils in bright light), the hard keratinous protrusions or "spades" on the inside of each back foot for digging, and reduced or absent parotoid glands. Spadefoot toads are most common in arid habitats and spend most of their lives underground in self-excavated burrows, typically beneath perennial ponds, creek beds, or other moisture-retaining areas. They are explosive breeders reacting immediately to standing water, which often results from summer thunderstorms.

FIGURE 3.14. Great Basin Spadefoot (*Spea intermontana*).

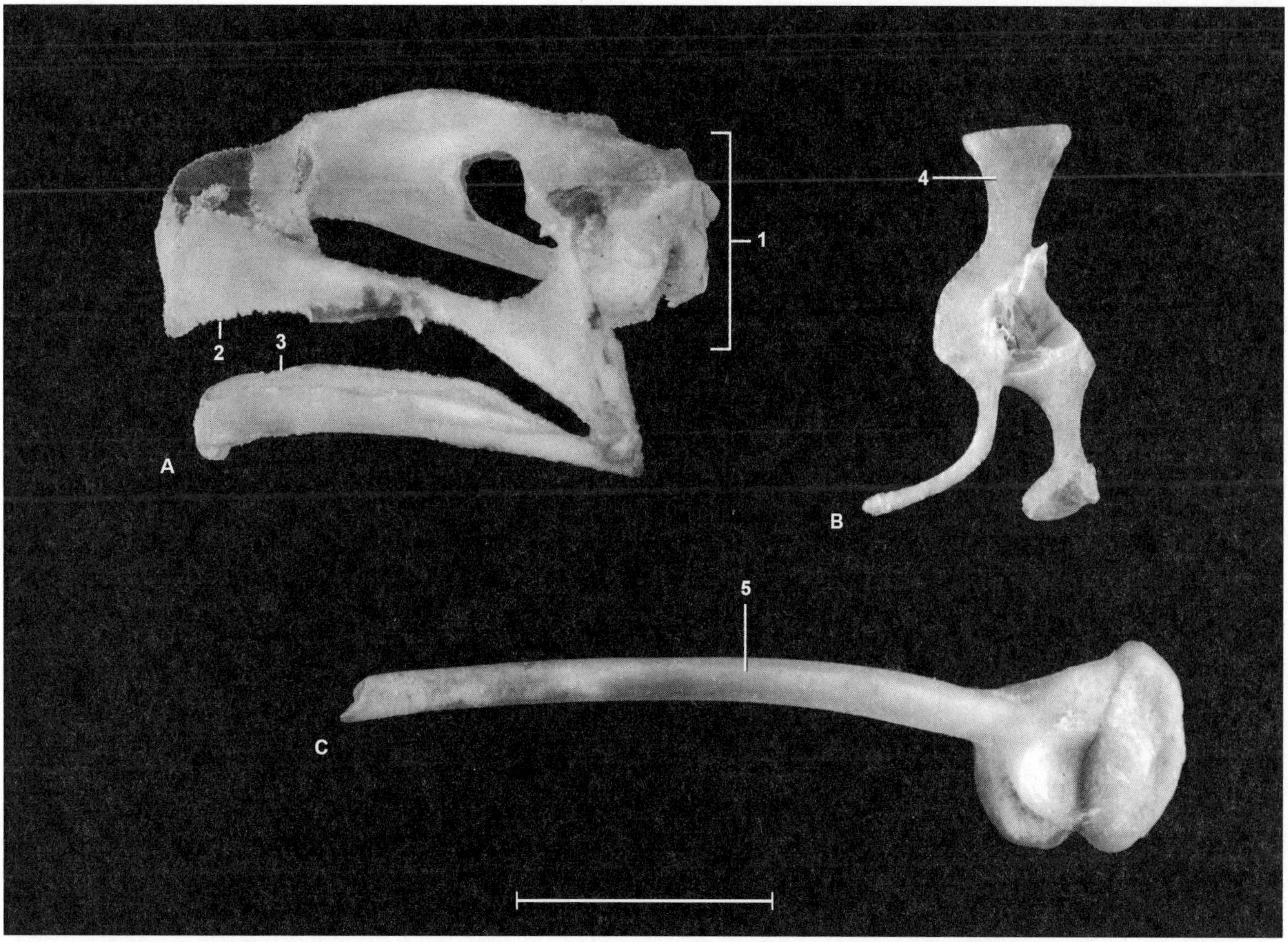

FIGURE 3.15. Cranium of Western Spadefoot (*Spea hammondii*), lateral views of cranium (A), l. pectoral girdle (B), and l. os coxa (C), anterior is left. Osteological features (Scaphiopodidae): 1, cranium higher domed, more enclosed than Hylidae and Ranidae; 2, maxilla and premaxilla with continuous rows of small teeth; 3, dentary toothless; 4, scapula proportionately long and narrow with antero-ventral flange; 5, ilium lacks blade-like crest present in ranids.

FIGURE 3.16. Western Tiger Salamander (*Ambystoma mavortium*).

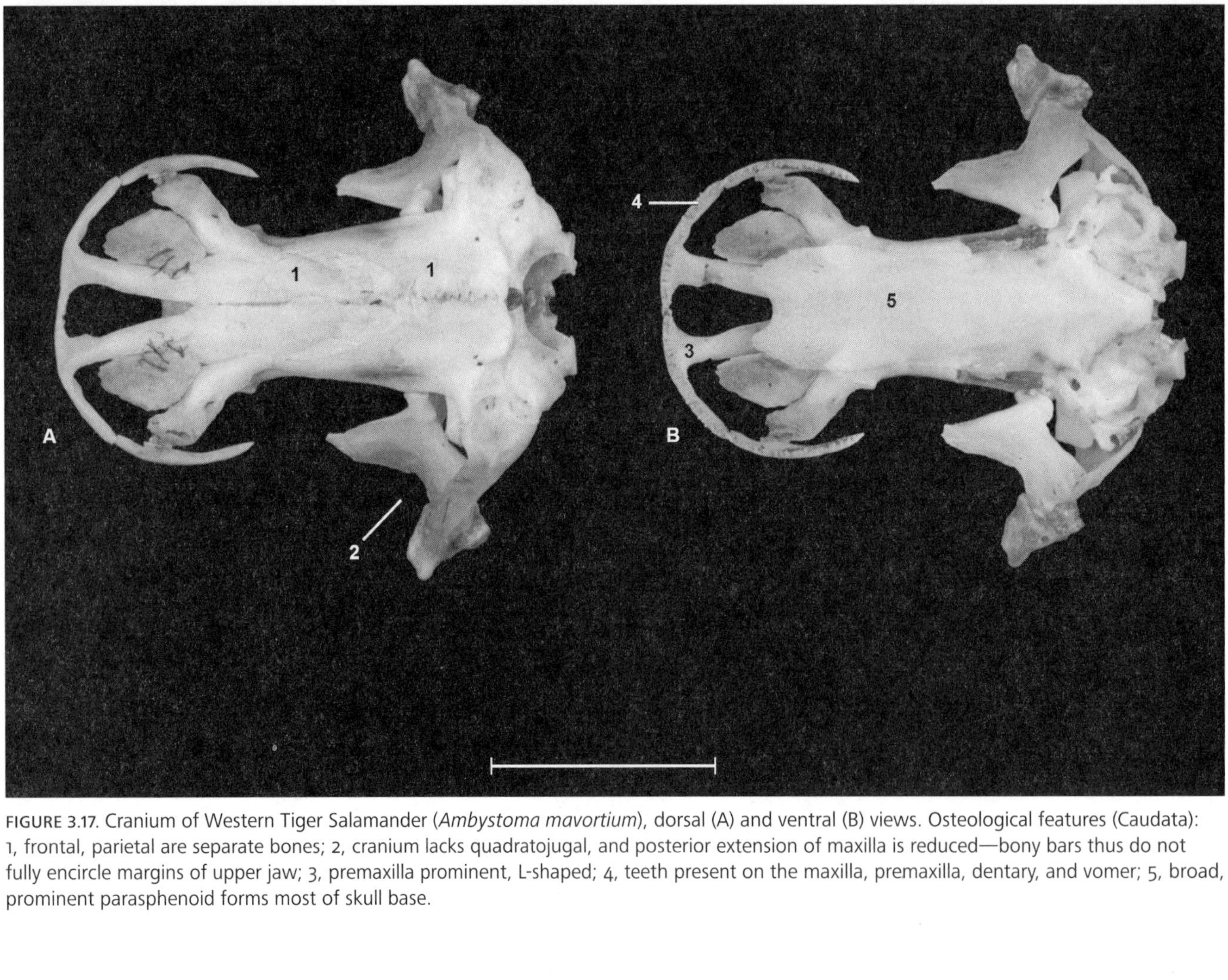

FIGURE 3.17. Cranium of Western Tiger Salamander (*Ambystoma mavortium*), dorsal (A) and ventral (B) views. Osteological features (Caudata): 1, frontal, parietal are separate bones; 2, cranium lacks quadratojugal, and posterior extension of maxilla is reduced—bony bars thus do not fully encircle margins of upper jaw; 3, premaxilla prominent, L-shaped; 4, teeth present on the maxilla, premaxilla, dentary, and vomer; 5, broad, prominent parasphenoid forms most of skull base.

ORDER Caudata (Salamanders)

The order Caudata (L., cauda = tail) is represented by four families and nine genera in our region. Unlike anurans, the appendicular skeleton follows the reptilian model with separate radius and ulna and tibia and fibula. With slender bodies, long tails, and four short limbs, salamanders resemble lizards but lack scales and claws and have moist, mucus-secreting skin. Many salamanders are brightly colored, and all native species are small in size. The largest reach only about 16.5 cm, while most species are less than 7.5 cm. Respiration can take many forms in salamanders: most taxa use external or internal gills or simple lungs. Some terrestrial species, however, lack lungs and gills altogether, with gas exchange taking place through the skin.

Life history is similar to anurans: most species exhibit an aquatic larval stage with variable metamorphosis patterns but in many cases involving the loss of gills, development of lungs, growth or increase in the size of legs, and change in the shape of the tail. Some salamander species are fully aquatic throughout life; some utilize water intermittently, such as for breeding; and some are entirely terrestrial, breed on land, and do not exhibit a larval stage. In addition, some taxa (e.g., Western Tiger Salamander, *Ambystoma mavortium*; Figures 3.16–3.17; Color Plate 11) are able to facultatively adopt different life histories in response to varying environmental conditions. Thus, *Ambystoma* may first adopt an aquatic larval stage and subsequently metamorphose into a terrestrial adult. Alternatively, they may follow a paedomorphic life history pattern—retaining the larval morphological features after reaching sexual maturity with continued residence in aquatic environments. Paedomorphosis is more prevalent in cooler, more mesic settings with higher pond permanence. Paedomorphic and metamorphosing individuals can be identified osteologically by differences in the maxilla, premaxilla, and vertebrae.

Notes

Our discussion of amphibian characteristics, origins, and osteology draws from Liem et al. (2001), Hildebrand (1995), and King and Custance (1982). For amphibian osteological nomenclature we follow Olsen (1968), and Liem et al. (2001) for more detail on the cranium. We follow Collins and Taggart (2015) for the taxonomic nomenclature of amphibians. Olsen (1968) discusses identification criteria for amphibian remains and taphonomic issues that may bias their representation in archaeological faunas. Our treatment of the natural history and ecology of amphibians relies heavily on Stebbins (2003). Our discussion of tiger salamander response to climatic variation is from Bruzgul et al. (2005).

CHAPTER 4

Reptiles

I. General Osteology of Reptiles

Characteristics

All modern reptiles are characterized by ectothermic metabolism, breathing that involves the ribs and associated musculature to draw air into lungs, and skin covered by epidermal scales or bony scutes. Relative to amphibians, the vertebral columns of reptiles are more differentiated into regions and more firmly attached to the pelvic girdle. In addition, and most notably, a major evolutionary breakthrough that occurred with the early reptiles is the membrane-encased amniotic egg. The amniotic embryos, whether deposited as eggs or retained and carried by the mother, are protected by a series of membranes. These prevent desiccation and allow the egg to breathe and cope with waste. The reptiles and all subsequent vertebrates that inherited the amniotic embryos (birds and mammals) are referred to collectively as amniotes. The membrane and shelled egg allowed the reptiles to develop without an aquatic larval stage and to colonize drier terrestrial habitats. Reptiles have either oviparous or ovoviviparous reproduction. There are 5,956 extant species.

Origins

Reptiles evolved from advanced amphibian ancestors (labyrinthodonts) in the steamy swamps of the late Carboniferous period just over 300 million years ago. The first reptiles are placed within the subclass Anapsida whose forms are characterized by solid skulls that lack temporal holes or fenestrae. Subsequently, several additional evolutionary lineages of reptiles emerged. One lineage, the Synapsida, was characterized by a pair of prominent fenestrae, one on each side of the skull in the temporal region just posterior to each orbit. This group contains both mammal-like reptiles and the earliest mammals themselves (Therapsida). A second branch is characterized by pairs of temporal fenestrae posterior to each orbit and is known as the diapsids. Early in the Permian period, diapsids also split into three other main lineages. The first is the turtles. Although most fossil and all modern turtles exhibit the anapsid condition (no temporal fenestrae), recent evidence suggests the earliest forms were diapsids, and fenestral closures evolved more recently. The second, the archosaurs, would lead ultimately to crocodiles, dinosaurs, and birds. The third, the lepidosaurs, represent the ancestors of modern snakes, lizards, and tuataras. These temporal fenestrae apparently evolved as early reptiles adapted to different terrestrial feeding niches, which created new selective pressures on jaw structures and the associated musculature and ultimately different forces placed on the temporal region. Temporal fenestrae have undergone substantial modification in the skulls of most modern diapsids: only crocodilians and tuataras retain the original two.

Osteology

Reptiles experienced dramatic diversification during the Mesozoic ("The Age of Reptiles") including the evolution of the dinosaurs, the largest land animals that have ever lived. Osteologically, a dazzling diversity of forms evolved to specialized niches—from winged pterosaurs with forelimb architecture adapted to flying in many ways convergent with modern bats, to ichthyosaurs that reverted to aquatic habitats and evolved body forms and fin-like structures analogous to fishes and dolphins. Although much of this diversity was lost with the mass extinctions near the end of Mesozoic (i.e., the Cretaceous-Tertiary extinction event at 65 mya), persisting reptiles still represent the greatest post-cranial osteological diversity of any class of vertebrates (Figures 4.1–4.4). Turtles and tortoises, for instance, evolved bony armored shells derived from the fusion of ribs and dermal scutes (Figure 4.4). As an adaptation to burrowing, snakes, with few exceptions, have lost their entire appendicular skeletons. Lizards represent the conservative tetrapod body plan (Figures 4.1–4.3) and are most similar to the labyrinthodont amphibian ancestors of the class—most reptiles also retain procoelus vertebrae.

Yet reptiles are also distinct in a number of ways. These distinctions include a deeper (dorsoventrally) and narrower skull, single occipital condyles, prominent (often toothed) pterygoids, and enlarged mobile quadrates (Figure 4.1). With the quadrates free from surrounding bones and able to readily move back and forth, another segment is effectively added to the lower jaw and the maximum gape of the mouth is increased. Additionally, some lizards have a pineal foramen on the top of the skull that can be helpful in making taxonomic identifications. The limb girdles are similar to the more generalized amphibians (not frogs) but the cleithrum is lost from the pectoral girdle and the ilia flare posteriorly. Finally, the hindlimbs are longer and more powerful than the forelimbs and can raise the body off the ground during running, with long tails acting as a counterbalance.

Remarks

Traditionally, the class Reptilia has included the turtles, tortoises, snakes, lizards, tuataras, and crocodilians but not birds. By not including birds, which represent a living clade of theropod dinosaurs, the class reptilia is a paraphyletic group: it includes a most recent common ancestor but not all of the descendants. Following modern systematic principles, reptilia should include all amniote tetrapods having epidermis-based scales which would include birds. We follow the traditional polyphyletic taxonomy here and our use of the taxon "reptilia" focuses on the nonavian reptiles. Along with amphibians, reptiles are generally underrepresented in archaeological faunas of western North America, owing to their typically small size and low population densities compared to fishes, birds, and mammals. As a result, far fewer researchers have specialized expertise in the analysis of archaeological and paleontological herpetofaunas in the region. Certain reptilian taxa do occur in higher densities in local contexts and were of greater economic importance accordingly, for example the intensive use of the Mohave Desert Tortoise (*Gopherus agassizii*) among prehistoric peoples of the southwestern Great Basin and adjacent areas of the Southwest. Western Pond Turtles (*Actinemys*) were also heavily utilized in aquatic settings within their range. Also, similar to amphibians, reptiles may occur as intrusive burials as they typically occupy subterranean burrows.

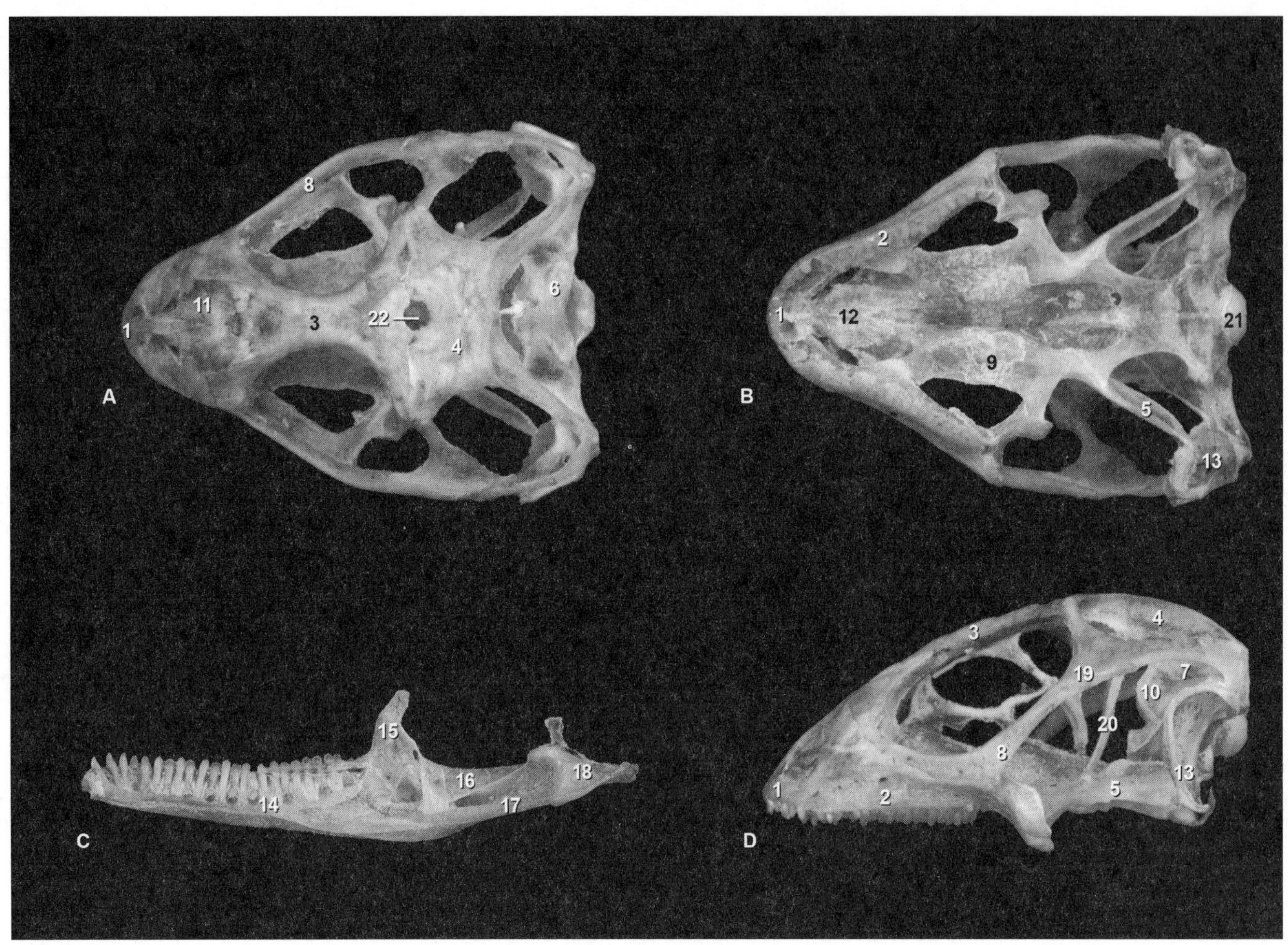

FIGURE 4.1. Cranium of Chuckwalla (*Sauromalus ater*), dorsal (A), ventral (B), and lateral (C, D) views.

1. Premaxilla
2. Maxilla
3. Frontal
4. Parietal
5. Pterygoid
6. Exoccipital
7. Squamosal
8. Jugal
9. Palatine
10. Prootic
11. Nasal
12. Vomer
13. Quadrate
14. Dentary
15. Coronoid
16. Surangular
17. Angular
18. Articular
19. Post-orbital
20. Epipterygoid
21. Occipital condyle
22. Pineal foramen

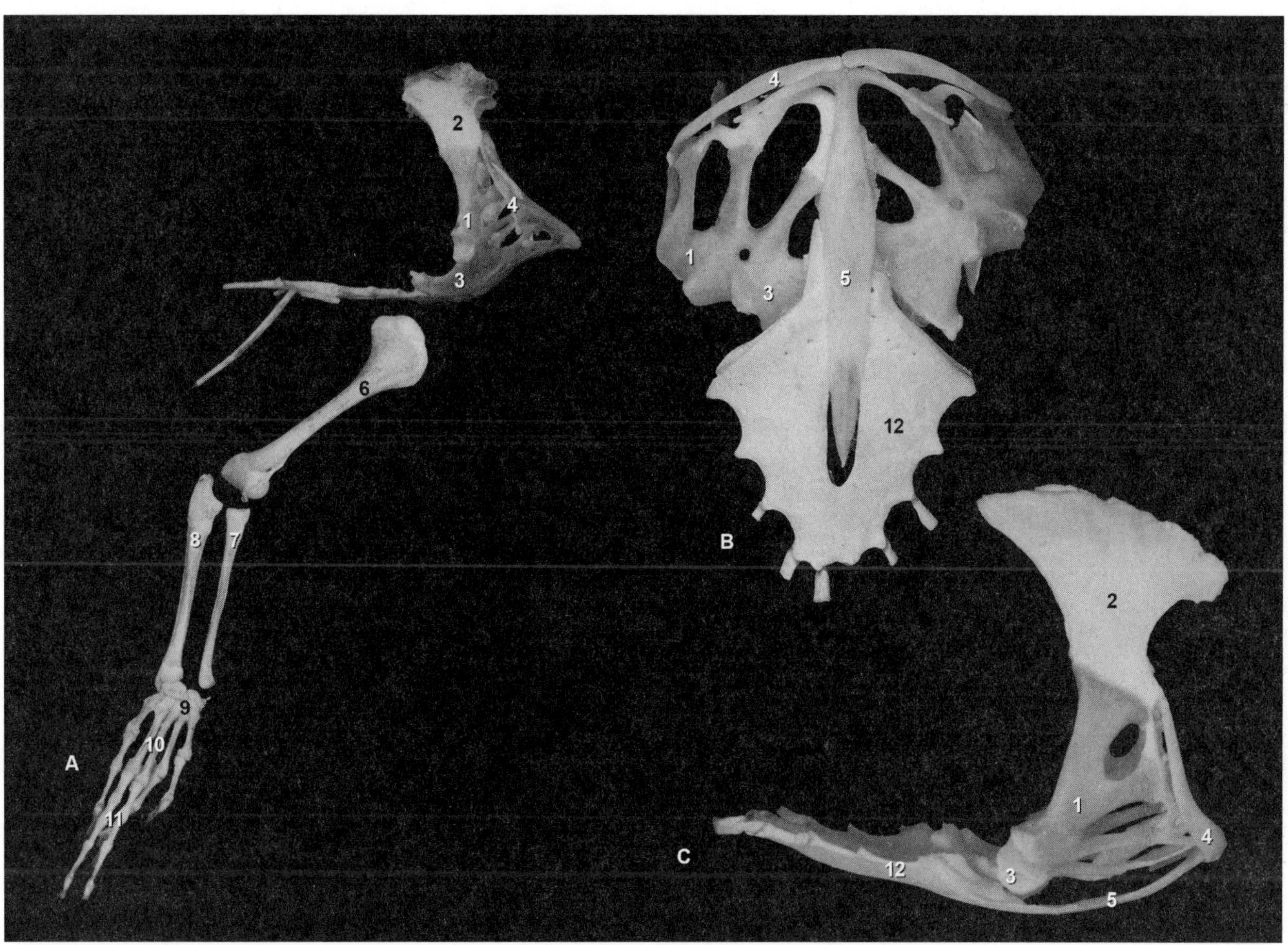

FIGURE 4.2. Left pectoral girdle and forelimb elements (A) of Tiger Whiptail (*Aspidoscelis tigris*); dorsal is top, anterior is right. Pectoral girdles of Green Iguana (*Iguana iguana*), ventral (B) and lateral (C) views; for ventral view, anterior is top; for lateral view, anterior is right.

1. Scapula
2. Suprascapula
3. Coracoid
4. Clavicle
5. Interclavicle
6. Humerus
7. Radius
8. Ulna
9. Carpals
10. Metacarpals
11. Phalanges
12. Sternum

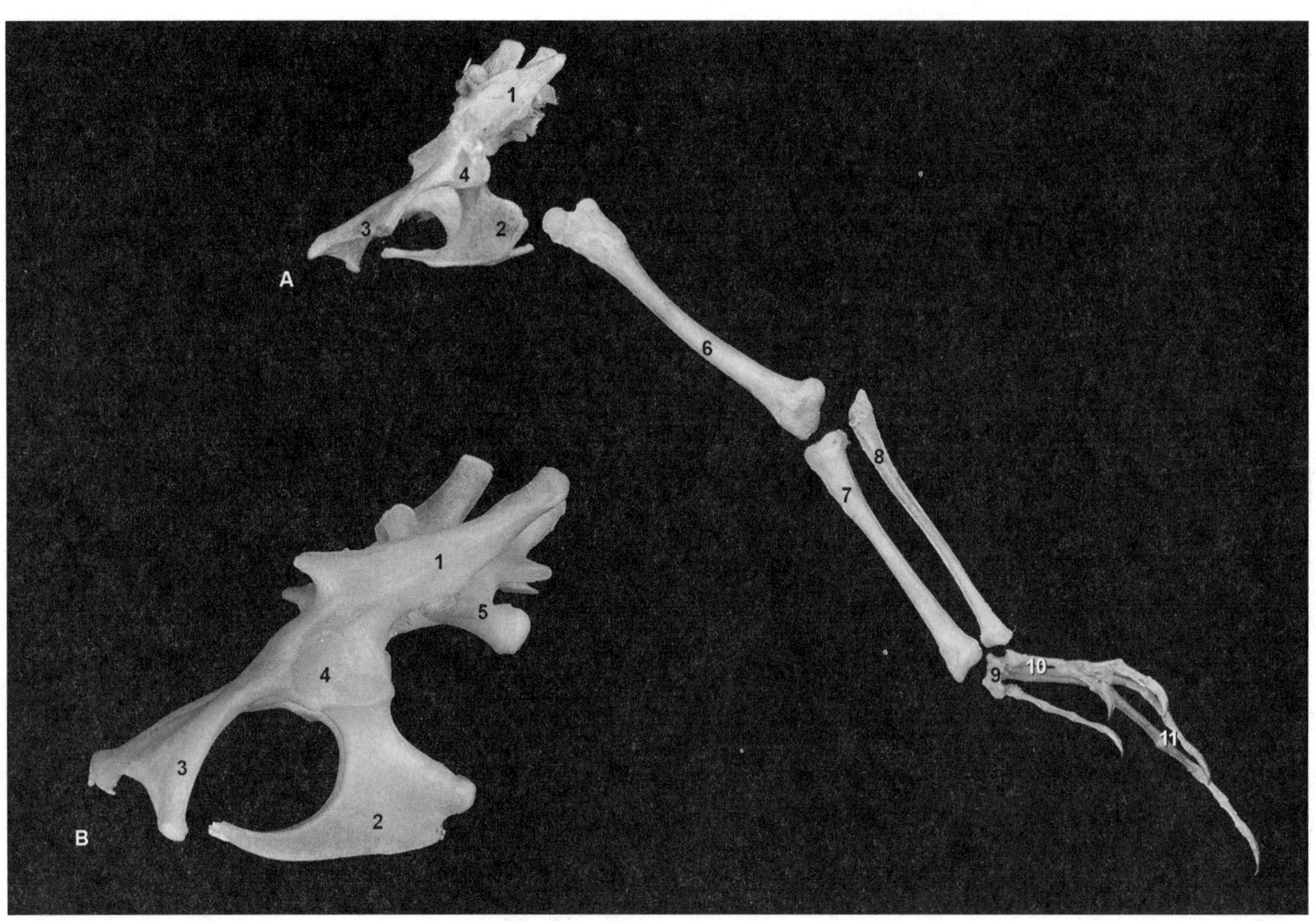

FIGURE 4.3. Left pelvic girdle and hindlimb elements (A) of Tiger Whiptail and l. pelvic girdle of Green Iguana (B); anterior is left, dorsal is top.

1. Ilium
2. Ischium
3. Pubis
4. Acetabulum
5. Sacrum
6. Femur
7. Tibia
8. Fibula
9. Tarsals
10. Metatarsals
11. Phalanges

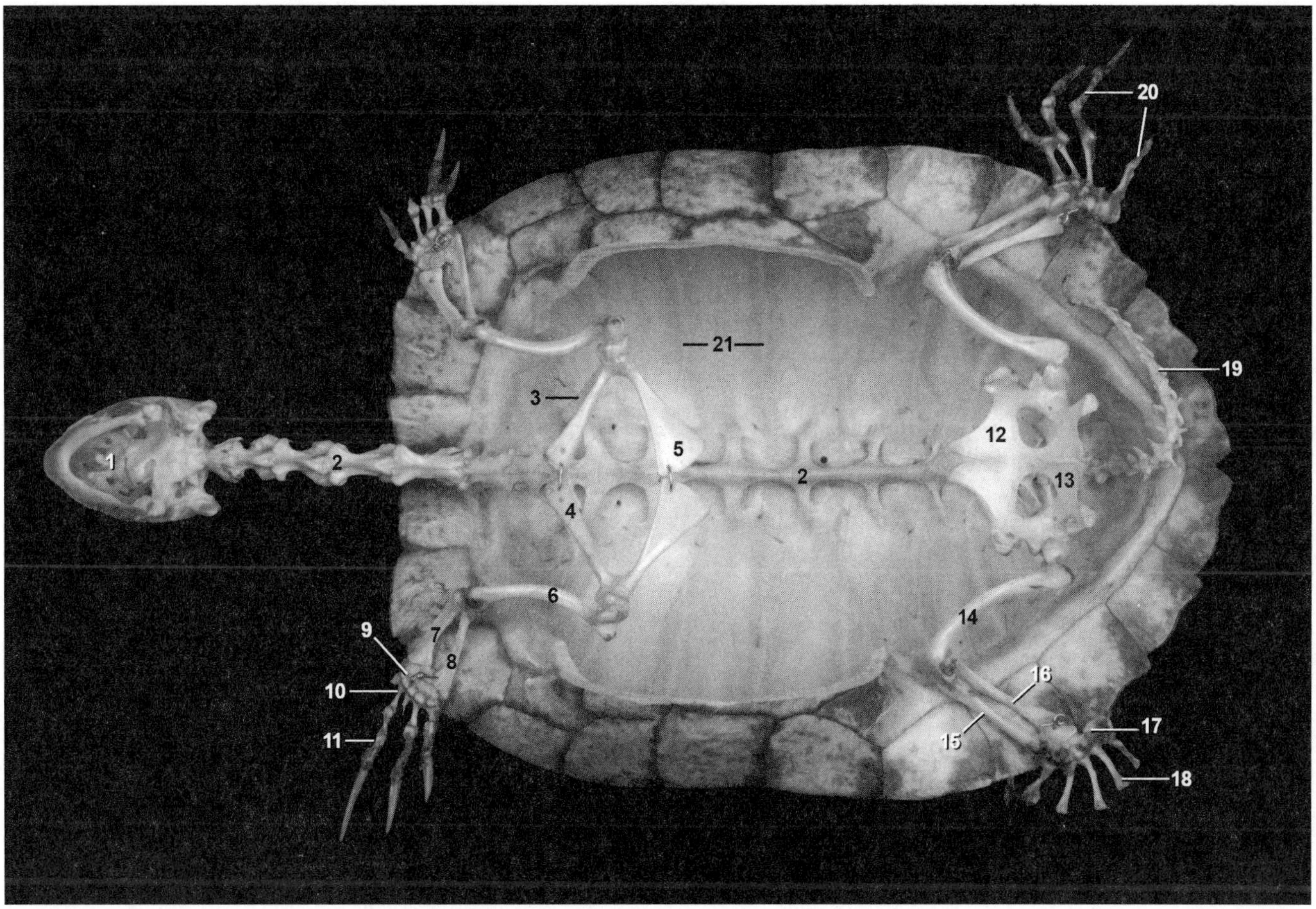

FIGURE 4.4. Skeleton of painted turtle (*Chrysemys* sp.), ventral view.

1. Cranium
2. Cervical vertebrae
3. Scapula
4. Precoracoid
5. Coracoid
6. Humerus
7. Ulna
8. Radius
9. Carpals
10. Metacarpals
11. Phalanges
12. Pubis
13. Ischium
14. Femur
15. Tibia
16. Fibula
17. Tarsals
18. Metatarsals
19. Caudal vertebrae
20. Phalanges
21. Ribs

II. Taxonomy and Osteological Variation of Western Reptiles

CLASS Reptilia
ORDER Testudines (Turtles)

Order Testudines (L., testudo = tortoise) is represented by four families and five genera in Western North America. With bodies encased in a bony carapace (dorsally) and plastron (ventrally), turtles have an unmistakable appearance (Figures 4.5–4.7). The protective shell formed by these structures allows the head and legs to be withdrawn when the turtle is threatened. The multi-segmented shell is formed primarily from ossifications in the skin but the dorsal shell—the carapace—also incorporates the ribs; many of the posterior vertebrae also fuse to the ventral surface of the carapace. This puts the ribs dorsal to the appendicular girdles, unlike the medial orientation found in all other vertebrates.

FIGURE 4.5. Western Pond Turtle (*Actinemys* sp.).

FIGURE 4.6. Mohave Desert Tortoise (*Gopherus agassizii*).

Although many turtles spend considerable time in or around water, others such as the tortoises (Testudinidae) have fully adapted to terrestrial life. Long-lived and slow-growing, turtles and tortoises lay leathery shelled eggs that are usually deposited and buried within the ground and left to incubate without parental care. Western pond turtles (*Actinemys* spp.; Emydidae Figure 4.5) and the Mohave Desert Tortoise (*Gopherus agassizii*; Testudinidae; Figure 4.6) are the most widespread taxa in our area. Western pond turtles rarely exceed about 21 cm in length and are omnivorous aquatic animals that inhabit sloughs, lakes, estuaries, marshes, streams, and irrigation ditches. They are often seen basking on logs in these settings. They do, however, also utilize upland terrestrial habitats typically for breeding, oviposition, and overwintering. Individuals occupying northern portions of the range will hibernate either under leaf litter in terrestrial contexts or in the soft mud of a streambed. The herbivorous Mohave Desert Tortoise, one of the most beloved animals of western deserts, frequents desert oases, riverbanks, washes, sand dunes, and rocky slopes. The Mohave Desert Tortoise spends 95% of its life in underground burrows where it can avoid surface ground temperatures that can exceed 140°F. Federally listed as a threatened species, populations have declined dramatically in recent decades owing to urban and military expansion, exotic weed intrusion, increasing densities of predators (especially Common Raven, *Corvus corax*), and competition with grazing livestock.

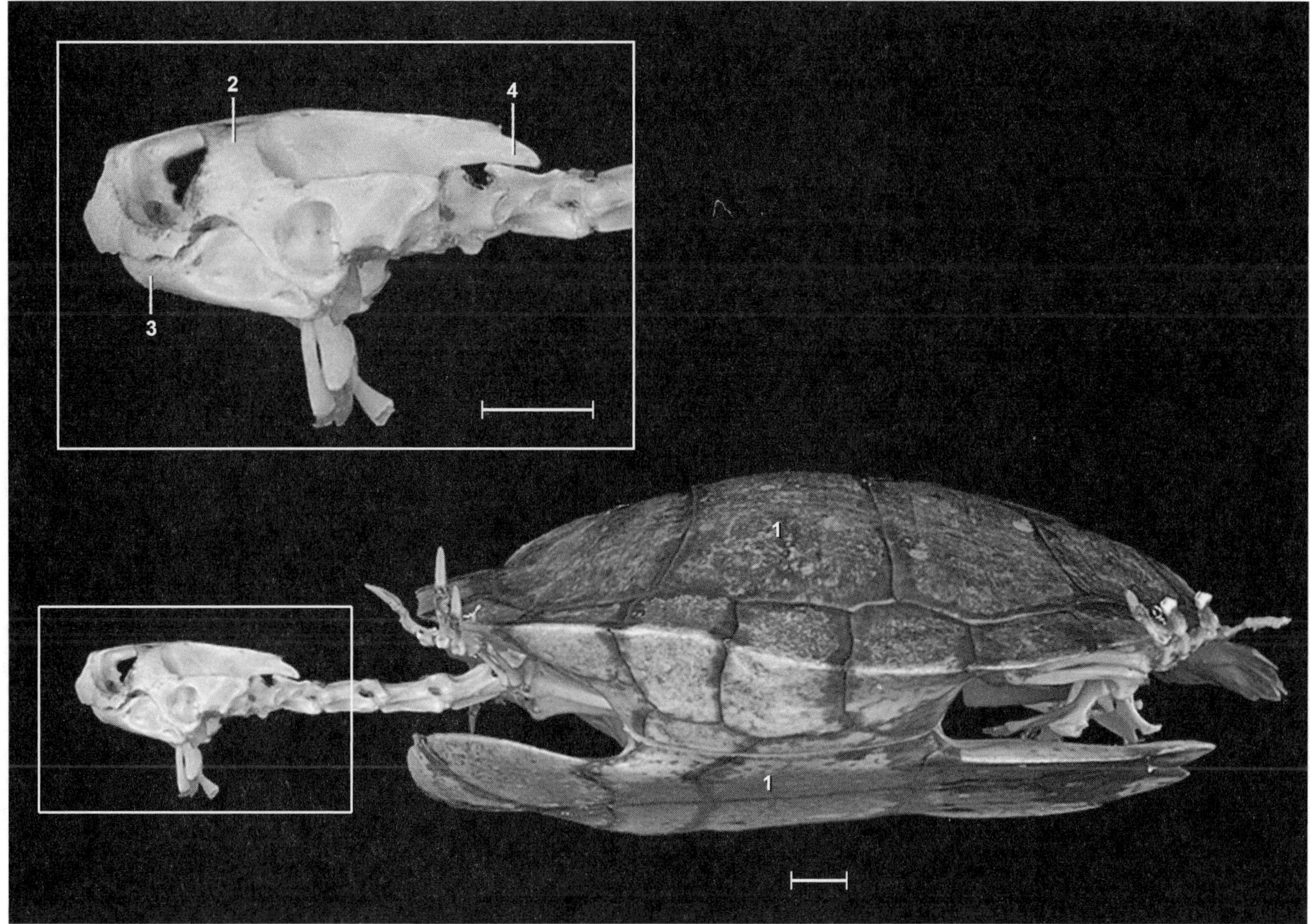

FIGURE 4.7. Skeleton of painted turtle (*Chrysemys* sp.), lateral view. Osteological features (Testudines): 1, prominent dorsal carapace and ventral plastron form bony shell; 2, cranium edentulous; 3, mandible broad and deep, with dorsal projection or hook at symphysis in most taxa (e.g., *Actinemys*); 4, supraoccipital bears posteriorly projecting sagittal crest.

ORDER Squamata (Lizards and Snakes)

The Squamata (L., squamatus = scaly) represent surviving lepidosaurs derived from the diapsid branch of early reptiles. Lizards and snakes have scale-covered bodies and bifid tongues that they extend to detect and retrieve chemical stimuli. They process such stimuli with distinctive sets of sensory cells in their mouths called Jacobson's organs. Many taxa swallow large prey whole with little or no chewing, facilitated by multiple, semiflexible skull joints. These joints are more elaborated in snakes than lizards. Male snakes and lizards have paired copulatory organs, the hemipenes, which are retracted within the vent (cloaca) except when mating. Hemipenes are often visible as two bulges behind the cloaca at the base of the tail. While lizards blink, snakes do not. Their constantly open eyes are protected by a clear covering called the spectacle. Lizards have an external ear within an otic notch behind the quadrate that supports a prominent tympanum. Airborne vibrations are transmitted through the tympanum to the tiny stapes and into an inner ear. In snakes, the elongated columella attaches to the quadrate and lower jaw and transmits vibrations from the ground.

As noted, while both lizards and snakes are derived from diapsid stock, the paired temporal fenestrae have been substantially modified in both these modern groups. In lizards, the arch of bone ventral to the lower opening (quadratojugal) is lost, which gives the appearance of a single large opening separated by a single bony bar—the postorbital-squamosal arch. Even this bar is lost in snakes with the squamosal reduced to a thin plate attached to the posterior braincase and thus no fenestra is evident. With the quadratojugal lost in all squamates, the quadrate becomes free and is associated with substantial mobility and flexibility of the jaws, as noted. This is referred to as a streptostylic jaw articulation.

SUBORDER Sauria or Lacertilia (Lizards)

In the west there are 10 lizard families and 21 genera (Figure 4.8; Color Plates 12–14). These lizards are mostly insectivorous and diurnal, but a few larger local species are herbivorous (e.g., Desert Iguana [*Dipsosaurus dorsalis*] and Chuckwalla [*Sauromalus ater*]). Horned lizards (*Phrynosoma*; Figure 4.8; Color Plate 12), widely known for ejecting blood from the eyes when threatened by canid predators, are specialized ant eaters. The only venomous lizard, the enormous Gila Monster (*Heloderma suspectum*), is mostly carnivorous, feeding on small mammals and nestlings of ground-nesting birds. Osteologically (Figures 4.1–4.3), lizards are characterized by: 1) a generalized appendicular skeleton with four limbs (excepting the legless lizards [Anniellidae]); 2) less flexible skull joints and more limited cranial kinesis compared to snakes; 3) homodont dentition comprised of simple conical teeth with pleurodont attachments; and 4) prominent postorbital-squamosal arches—features lost in snakes.

SUBORDER Serpentes (Snakes)

Six snake families and 29 genera occur in the west. Fossil evidence suggests that snakes evolved from burrowing lizards, and although snakes have lost functional appendages some species (e.g., Boidae, *Charina*) have retained vestiges of the pelvic girdle. These structures form external pelvic spurs on each side of the vent that are used in courtship, mating, and competitive interactions between males. All snakes are carnivorous and can swallow very large prey whole, a practice allowed by highly elaborated cranial kinesis. Specifically, flexible joints occur between the following element pairs: maxilla-prefrontal, prefrontal-frontal, nasal-frontal, premaxilla-vomer, quadrate-squamosal, quadrate-articular, and angular-dentary. In addition, elastic ligaments connect opposing dentaries allowing the lower jaws to separate widely during feeding. Adjusting to the streamlined body form, the paired organs of snakes (e.g., kidneys) are oriented antero-posteriorly instead of side by side, also, most possess a single functional lung. Osteological differences between snakes and the lizards are summarized in the preceding section. Although six snake families occur in the west, the vast majority of species fall into two of them, Crotalidae (formerly Viperidae) and Colubridae.

FIGURE 4.8. Greater Short-horned Lizard (*Phrynosoma hernandesi*).

FAMILY Colubridae (Harmless Egg-laying Snakes)

Colubridae is a large, catchall family that contains about three-quarters of the west's snake species, including such familiar species as racers and coachwhips (*Coluber*), kingsnakes (*Lampropeltis*), and Gophersnakes (*Pituophis catenifer*; Figures 4.9–4.10; Color Plate 15). Several taxa have evolved physiological and behavioral adaptations resulting from a long history of association with rattlesnakes. For instance, Gophersnakes mimic rattlesnakes morphologically and behaviorally to deter predators—the two taxa not only have very similar scale colors and markings, but when threatened, Gophersnakes will flatten their heads to form a more triangular, rattlesnake look and rapidly vibrate their tails to mimic a rattlesnake's rattle.

Another group, the kingsnakes (*Lampropeltis*), prey on a wide range of small vertebrates including

FIGURE 4.9. Gophersnake (*Pituophis catenifer*).

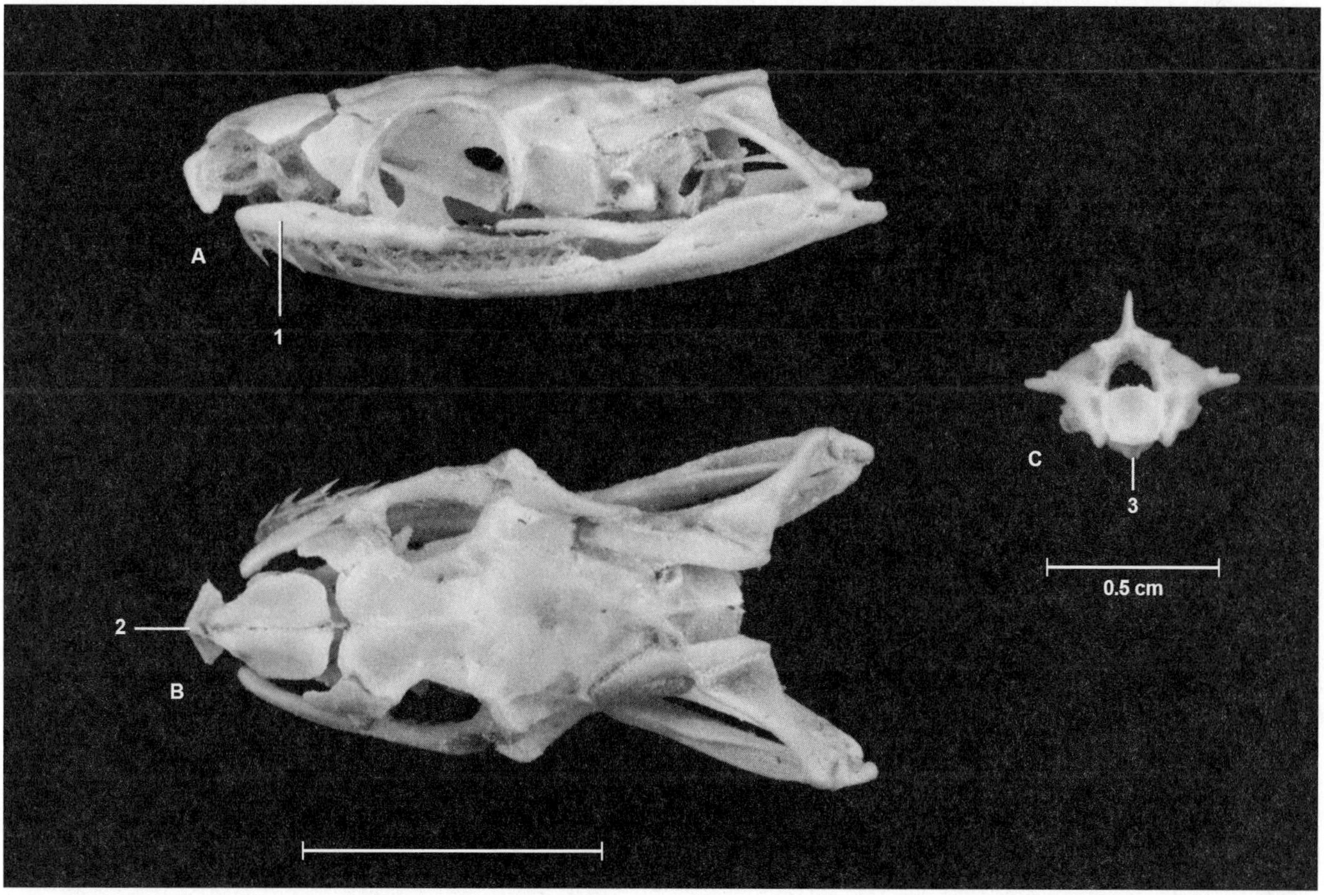

FIGURE 4.10. Gophersnake (*Pituophis catenifer*) cranium, lateral (A) and dorsal (B) views, and vertebra, anterior view (C). Osteological features (Colubridae): 1, maxilla narrow, elongate with uniform-sized teeth; 2, anterior cranium (snout), pointed, elongate; 3, haemal spines on vertebrae less prominent, restricted to anterior section of column.

FIGURE 4.11. Western Rattlesnake (*Crotalus oreganus*).

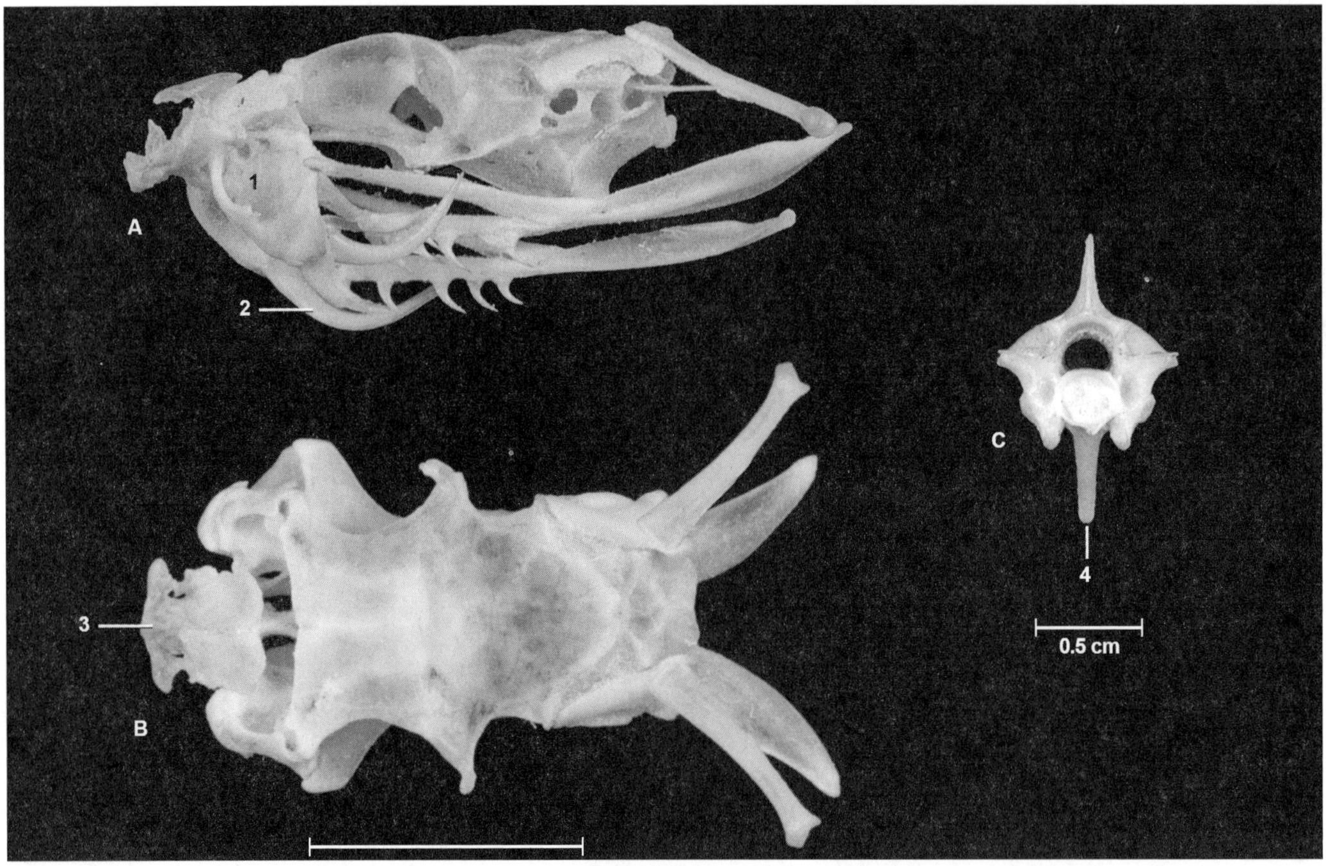

FIGURE 4.12. Western Rattlesnake (*Crotalus oreganus*) cranium, lateral (A) and dorsal (B) views, and vertebra, anterior view (C). Osteological features (Crotalidae): 1, maxilla irregular in shape with deep circular fossa (for loreal pit); 2, maxilla with prominent hollow fangs—replacement teeth located posteriorly; 3, anterior cranium (snout) blunt, foreshortened; 4, vertebrae with prominent, sharply pointed, postero-ventrally projecting haemal spines.

other snakes and, famously, even venomous ones such as rattlesnakes. Kingsnakes possess an enzyme that quickly breaks down the toxins used by venomous snakes that both minimizes harm from possible bites and allows for the digestion of bodies that contain venom. As is typical for many colubrids, both Gophersnakes and kingsnakes kill small vertebrate prey by constriction.

FAMILY Crotalidae (Pitvipers)

This family includes rattlesnakes (*Crotalus* and *Sistrurus*) and cottonmouths and copperheads (*Agkistrodon* spp.), but only rattlesnakes reach our area (Figures 4.11–4.12; Color Plate 16). Pitvipers have the most highly developed venom-injection systems of all snakes. These involve large, paired, hollow, and moveable fangs projecting off the anterior maxillae. When at rest, the fangs fold parallel to the palate, but in biting they swing forward to stab prey or would-be predators to inject a complex cocktail of toxins. Depending on the species, these may include hemotoxins, which destroy tissues, anticoagulants, and neurotoxins that can cause circulatory arrest and respiratory failure. The amount of venom injected in a bite can vary widely. Any bite from a rattlesnake should be regarded as a life-threatening medical emergency. Pitvipers also have distinctive heat sensory organs, or loreal pits, located behind each nostril and embedded within cavities in the maxillae that aid in locating and securing prey at night. Rattlesnakes have heavy bodies; slender necks; broad, triangular heads; and, of course, the prominent, segmented "rattles" at the end of their tails. Formed from modified scales, the rattle serves as a warning device that is deployed when the animal is alarmed. Rattlesnakes are not aggressive and are typically more afraid of people than we are of them. They occupy a wide range of habitats from sea level to 3,350 meters.

Notes

Our presentation of reptilian characteristics, origins, and osteology is adapted from Liem et al. (2001), Hildebrand (1995), and King and Custance (1982). We follow Olsen (1968) for osteological nomenclature, and Liem et al. (2001) for more detail on the cranium. See Bever et al. (2015) for more on the diapsid origin of turtles. The taxonomic nomenclature for reptiles follows Collins and Taggart (2015). Olsen (1968) discusses identification criteria for reptilian remains in archaeological faunas. Our treatment of the natural history and ecology of reptiles draws heavily from Stebbins (2003)—the best source for amphibian and reptile natural history, but the taxonomy is in need of revision.

Mammals

I. General Osteology of Mammals

Characteristics

Mammals are endothermic ("warm-blooded"), air-breathing vertebrates in which females possess mammary glands to produce milk to feed young, and both sexes have sweat glands, hair or fur, three middle ear bones, and an expanded neocortex region of the brain. Mammals are active, agile vertebrates that have complex social, feeding, and reproductive behaviors. Other distinguishing features include determinate growth, the possession of a single bone in the lower jaw—the dentary—and a dentary-squamosal jaw articulation. Most have well-developed and differentiated teeth that are typically replaced as a set once during an individual's life. Most mammals are viviparous, but there are a few oviparous taxa. Worldwide, a total of about 5,420 mammal species live today.

Origins

Early mammals evolved from synapsid reptilian ancestors during the Triassic period of the early Mesozoic era about 220 million years ago. Most of the characteristics just described that can be diagnosed osteologically are present in a group known as the cynodonts, which includes both derived reptiles and primitive mammals. For example, cynodonts share with mammals but not other reptiles: 1) well-differentiated teeth with cheekteeth that bear prominent cusps; 2) the loss of ribs on the lumbar vertebrae; and 3) an enlarged dentary with corresponding reduction or loss of the other bones of the lower jaw. Early mammals were small and were most likely nocturnal insectivores, judging from their small size (< 15 cm) and aspects of the dentition. Most modern mammalian orders evolved during the early Cenozoic (Paleocene, Eocene, or Oligocene) between about 65 to 40 mya.

Two major groups of mammals include the metatherians, or marsupials, and the eutherians, or placental mammals; the latter term is misleading, however, since both groups have placentas. Reproduction in marsupials is characterized by a very short intrauterine life. After birth, relatively underdeveloped young attach to nipples usually located under a flap of skin called a marsupium where they complete their development. Marsupials are mostly found in the Southern Hemisphere and include such familiar animals as opossums and kangaroos. Eutherian mammals have a much longer intrauterine gestation, and young are born at a more advanced stage of development. With the exception of the Virginia Opossum (*Didelphis virginiana*), all North American mammals are eutherians.

Osteology

As endotherms, early mammals had a high metabolism and thus increased energy demands. Keen sense organs and an enlarged, complex brain evolved to support a lifestyle that required a greater, more efficiently assimilated food supply relative to ectothermic amphibians and reptiles. Part of this also involved different feeding methods and greater mastication to improve digestibility and nutrient assimilation; this influenced not only the differentiation and elaboration of teeth but also the evolution of a strong akinetic skull with powerful jaws, jaw muscles, and related bony architecture (Figures 5.1–5.9). For example, a large outward-bowing zygomatic arch developed to accommodate enlarged jaw muscles, specifically the temporalis and masseter. In most mammals, the temporal fenestra is evident as a large gap between the zygomatic arch and the side of the braincase and is continuous with the eye orbit.

As the braincase enlarged and jaws became more powerful, the entire complex of mechanical stresses was altered. As a result, some bones that were separate in earlier vertebrates fused to become compound bones in mammals. For example, the prootic and opisthotic found on the lateral wall of the chondrocranium in early tetrapods fused to form the petrosal bone that encloses the middle and inner regions of the ear. The petrosal, in turn, united with the squamosal, to form the compound temporal, a bone unique to mammals (Figures 5.1, 5.4, 5.9). In addition, the exoccipital, basioccipital, and supraoccipital, fuse to form the single occipital bone in mammals and only a single element, the dentary, forms the lower jaw or mandible (Figures 5.5–5.6).

Other noteworthy changes are related to the development of more acute hearing and reorganization of the structure of the middle ear. As noted, mammals have three, tiny, middle-ear bones or ossicles: the malleus and incus, that evolved from the articular and quadrate, respectively; and the stapes, originally derived from the hyomandibular-columella, named for its stirrup shape. Recall that reptiles have but a single inner ear bone, the columella. The crania of mammals exhibit several other features not found in reptiles, including paired occipital condyles, a fully enclosed bony palate, and a jaw joint formed by the articulation of the dentary and temporal bones. Again, teeth are highly differentiated with respect to function, and the different main types and numbers of teeth in mammals are described by dental formulae (Figures 5.1–5.9).

The vertebral column in mammals is clearly divisible into sections and reflects more specialized adaptations to terrestrial locomotion, compared to those in reptiles and amphibians. The vertebral column acts as a beam to support the total body weight and is thus dorsally arched, but it also bears a number of muscles and ligaments connecting with the appendages that function in locomotion. The anterior and posterior bodies or vertebral centra are mostly flat, or acoelous (Figure 5.10). Connections between adjacent vertebrae are, however, further strengthened in mammals, with tightly interlocking zygapophyses (Figure 5.10). Starting anteriorly, the neck or cervical vertebrae, with rare exceptions, include seven elements; the first two, the atlas and axis, are modified to permit rotation of the head. All mammalian cervical vertebrae are distinguished by the prominent foramina located within each transverse process, allowing passage of the vertebral arteries and veins.

Moving posteriorly, the next section—the thoracic vertebrae—bear ribs. This section is typically represented by 12 to 13 elements. (Note that in reptiles ribs occur on all vertebrae anterior to the tail.) Thoracic vertebrae can thus be identified by articulating facets for ribs (costal facets), but they also tend to have the longest and sharpest spinous processes and smaller centra or bodies than the vertebrae that lie posteriorly, namely the lumbar vertebrae.

In the lumbar section, there are typically five vertebrae. These have larger centra; smaller, more squared-off spinous processes; and prominent anteriorly projecting transverse processes.

Last, there are also five sacral vertebrae that decline in width and size posteriorly down the column. Individual sacral vertebrae often fuse in adult animals to form a single structure, the sacrum, which provides a secure connection between the vertebral column and the pelvic girdle. A highly variable number of caudal vertebrae extend posteriorly from the sacrum and form a bony foundation for the tail. They are thus very narrow, elongate, cylindrical structures that lack the prominent articulating processes and projections of all the other vertebrae.

Mammalian appendicular skeletons vary greatly in relation to specific locomotor adaptations but are generally more well developed and located more directly ventral to the trunk, compared to reptilian ancestors (Figures 5.11–5.20). Orientation of the limbs to allow for a more upright gait requires adjustments to the musculature and bony muscle attachment sites. Major changes in the pectoral girdle result from enlargement of the spinatus muscles that pull the limbs forward. The scapula is substantially expanded in mammals and is equipped with a pronounced spine. In the pelvic girdle, a range of muscles were reorganized reflecting the shift from sprawling, laterally projecting, limbs to more ventrally oriented appendages, enabling greater speed and efficient movement. To accommodate these changes, the ilium is enlarged and oriented more dorsally and anteriorly while the pubis and ischium, attachment sites for postural muscles in reptiles, are reduced in size and fuse together.

In reptiles and amphibians, distal appendages are almost uniformly represented by four or five digits, but there is considerable variation in digit counts for mammals. Highly cursorial forms (adapted for running) have evolved long limbs and reductions in the number of digits. The limbs of horses, for instance, terminate in a single digit. Artiodactyls (artio = even, dactyl = toes) such as elk, moose, and bison stand on two toes per foot. Other mammals have highly dexterous hands and feet (e.g., primates) and typically retain five digits per limb. Different mammalian groups are also distinguished by the posture of their feet or how they contact the ground during locomotion. Many mammals, including humans, bears, opossums, and raccoons, walk on the soles of their feet, where the heel is on the ground as they stand and strikes first during each stride. These animals are said to be plantigrade (= sole + walking; Figure 5.20). Effective leg length and hence stride length is increased in other mammals, such as in carnivores, that are digitigrade (= finger + walking) and walk on what amount to the balls of their feet—only the phalanges touch the ground during locomotion. In the most cursorial mammalian groups—including the perissodactlys (horses) and artiodactyls (deer, sheep, antelope, etc.)—only the toe tips touch the ground, like a ballet dancer. This is referred to as unguligrade (= hoof + walking) and is the source of the term "ungulate," which refers to the hoofed mammals that all exhibit this type of foot posture.

Bone growth also differs in mammals compared to all other vertebrate classes. In fishes, amphibians, and reptiles, bone growth (whether dermal or endochondral) typically occurs on the margins of bones, and immature bones are capped by cartilage. Mammals, by contrast, have true bony epiphyses. These are separated from the shafts by cartilaginous growth plates that are ultimately invaded by bone when growth in length is completed (see Figures 1.2, 1.5). This "epiphyseal fusion," as it is called, occurs at different ages for different elements and species. It gives mammals secure bony articulations even as the bone grows. The presence or absence of fused epiphyses can also provide information regarding an animal's age at death.

Remarks

The mammal skeleton is easily the most familiar to students since the human skeleton is commonly studied in a variety of disciplines. Mammal bone is also the most commonly encountered in archaeological faunas, as mammals were economically the most important group of vertebrates exploited worldwide by prehistoric peoples. Mammal materials are also the most abundant class in other commonly studied depositional contexts, owl pellets, for example. Highly specialized analyses of mammal bone occur in the context of animal domestication, especially in the Old World, where mammals were the most important group of domesticated animals. Finally, detailed microscopic analyses of mammalian growth, age, and development can be conducted from incremental structures (such as annuli) that occur within tooth cementum.

Of all vertebrates, more care needs to be taken when handling live or recently killed mammals. Most have ticks and fleas that can occasionally carry a range of diseases (Lyme disease, bubonic plague, Rocky Mountain spotted fever, tularemia, relapsing fever, etc.).

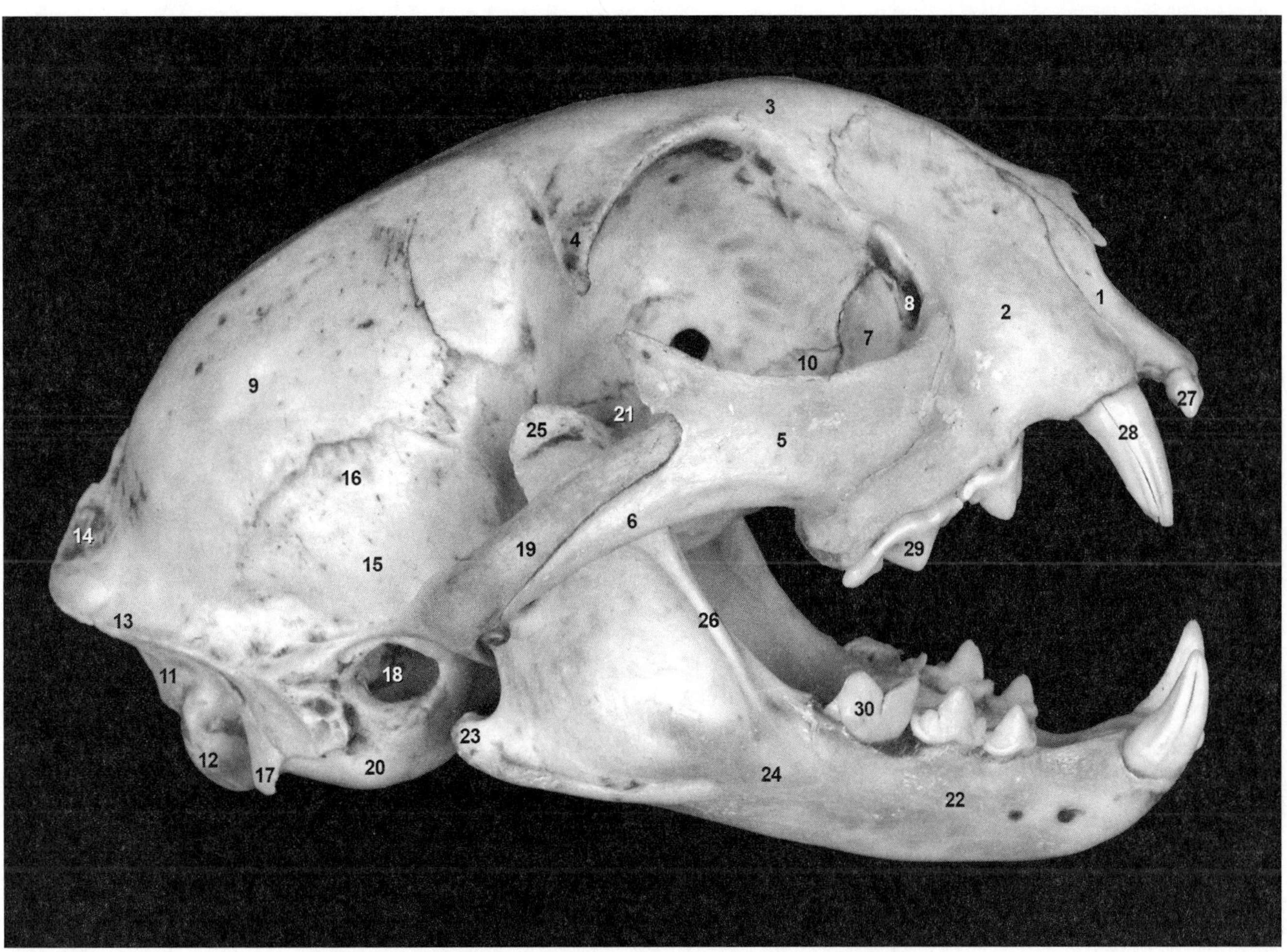

FIGURE 5.1. Cranium of Bobcat (*Lynx rufus*), lateral view.

1. Premaxilla
2. Maxilla
3. Frontal
4. Postorbital process
5. Zygomatic
6. Temporal process
7. Lacrimal
8. Nasolacrimal canal
9. Parietal
10. Ethmoid
11. Occipital
12. Occipital condyle
13. Nuchal crest
14. Interparietal
15. Temporal
16. Squamous portion
17. Mastoid process
18. External auditory meatus
19. Zygomatic process
20. Tympanic bulla
21. Sphenoid
22. Mandible
23. Angular process
24. Mandibular body
25. Coronoid process
26. Mandibular ramus
27. Incisor
28. Canine
29. Premolar
30. Molar

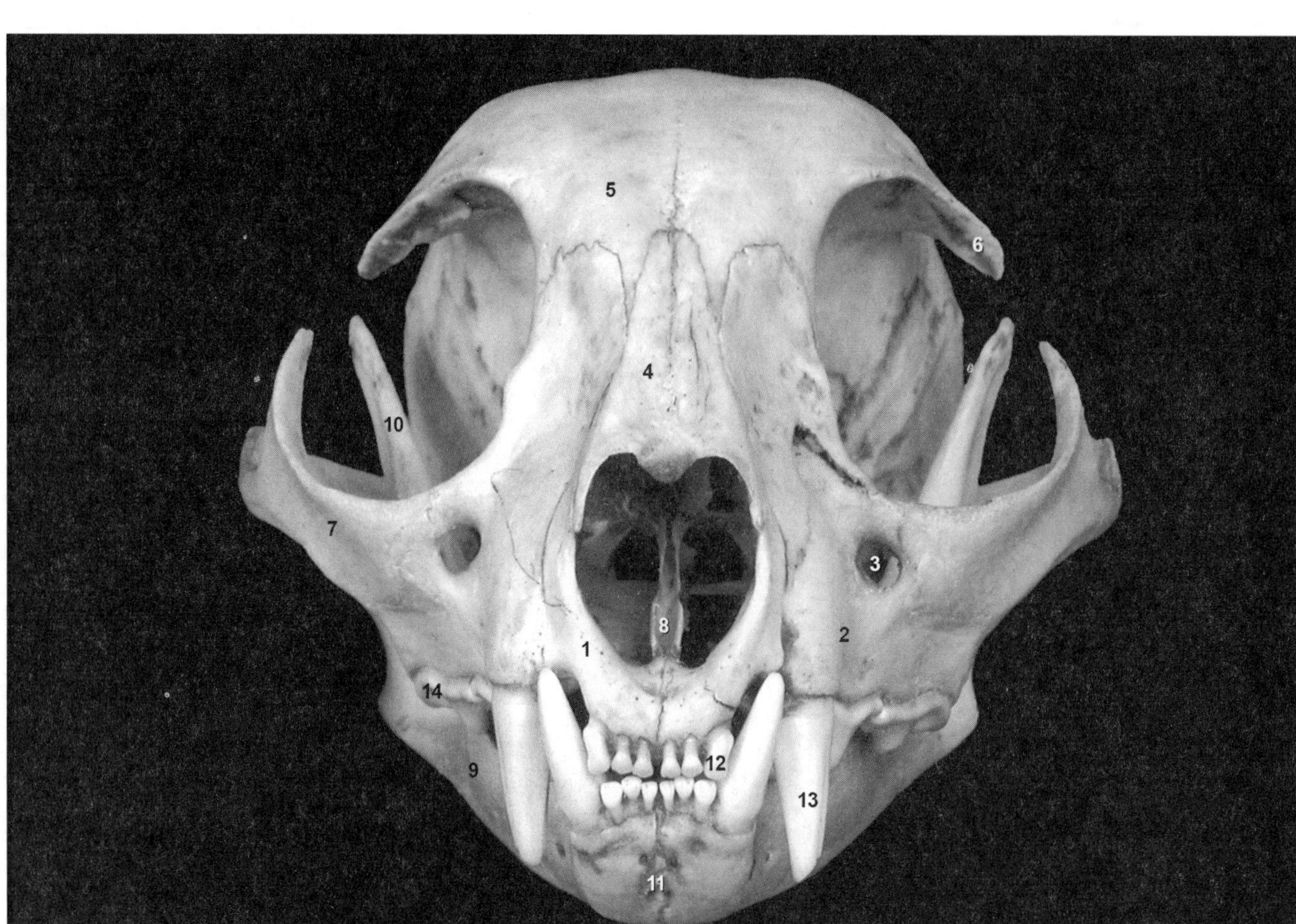

FIGURE 5.2. Cranium of Bobcat, anterior view.

1. Premaxilla
2. Maxilla
3. Infraorbital foramen
4. Nasal
5. Frontal
6. Postorbital process
7. Zygomatic
8. Vomer
9. Mandible
10. Coronoid process
11. Mandibular symphysis
12. Incisor
13. Canine
14. Premolar

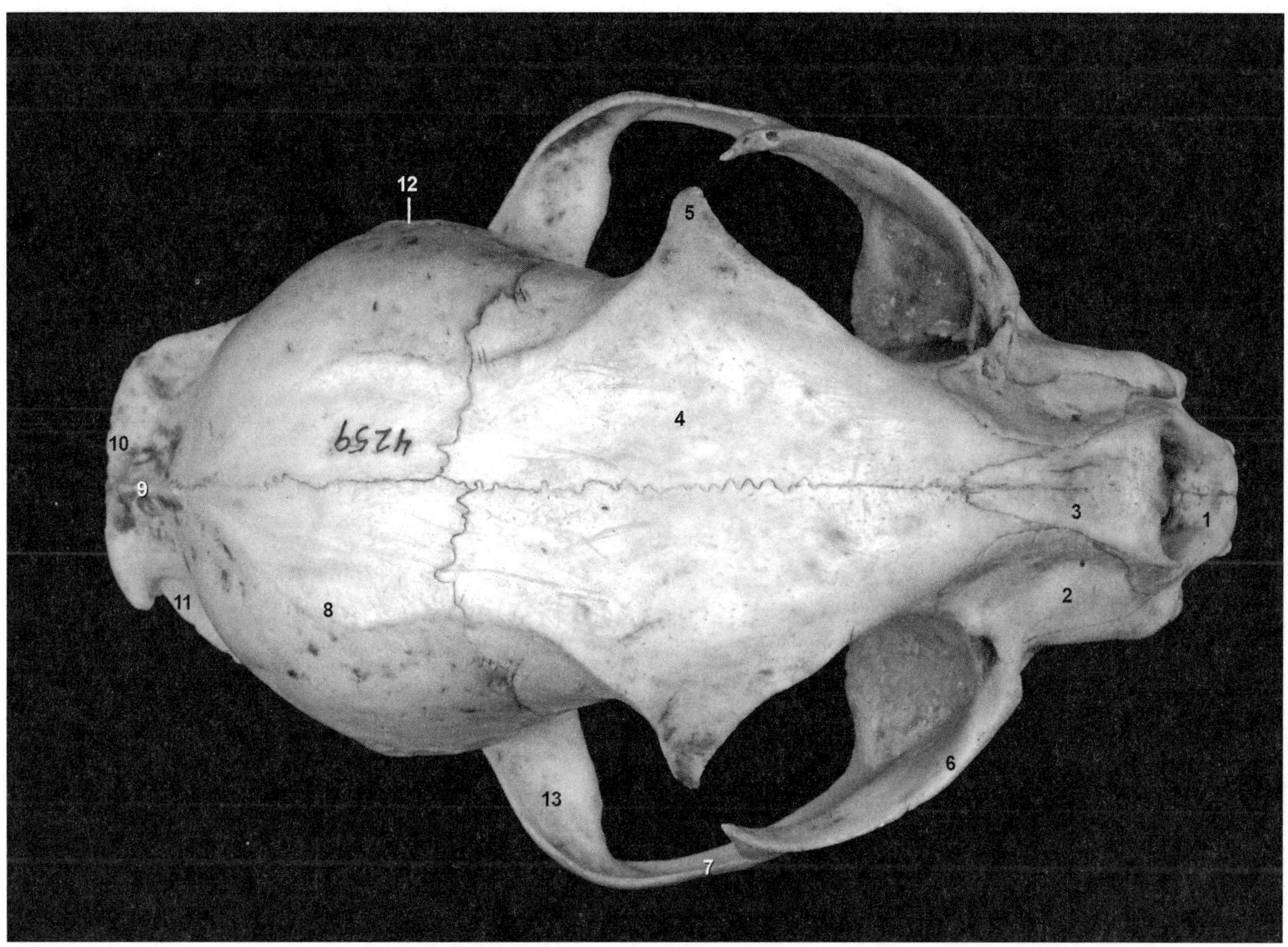

FIGURE 5.3. Cranium of Bobcat, dorsal view.

1. Premaxilla
2. Maxilla
3. Nasal
4. Frontal
5. Postorbital process
6. Zygomatic
7. Temporal process
8. Parietal
9. Sagittal crest
10. Occipital
11. Nuchal crest
12. Temporal
13. Zygomatic process

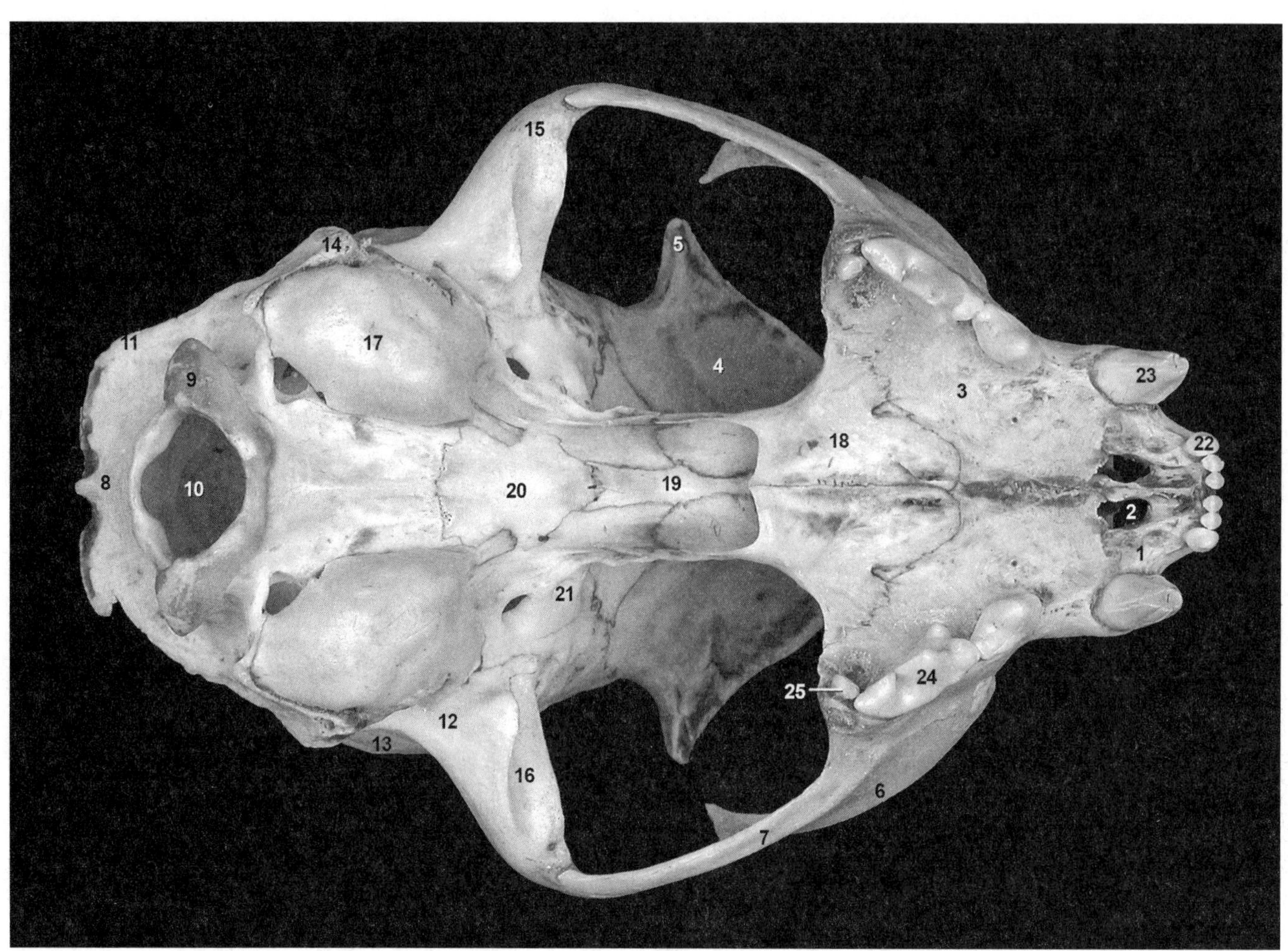

FIGURE 5.4. Cranium of Bobcat, ventral view.

1. Premaxilla
2. Incisive foramen
3. Maxilla
4. Frontal
5. Postorbital process
6. Zygomatic
7. Temporal process
8. Occipital
9. Occipital condyle
10. Foramen magnum
11. Nuchal crest
12. Temporal
13. Squamous portion
14. Mastoid process
15. Zygomatic process
16. Mandibular fossa
17. Tympanic bulla
18. Palatine
19. Presphenoid
20. Basisphenoid
21. Sphenoid
22. Incisor
23. Canine
24. Premolar
25. Molar

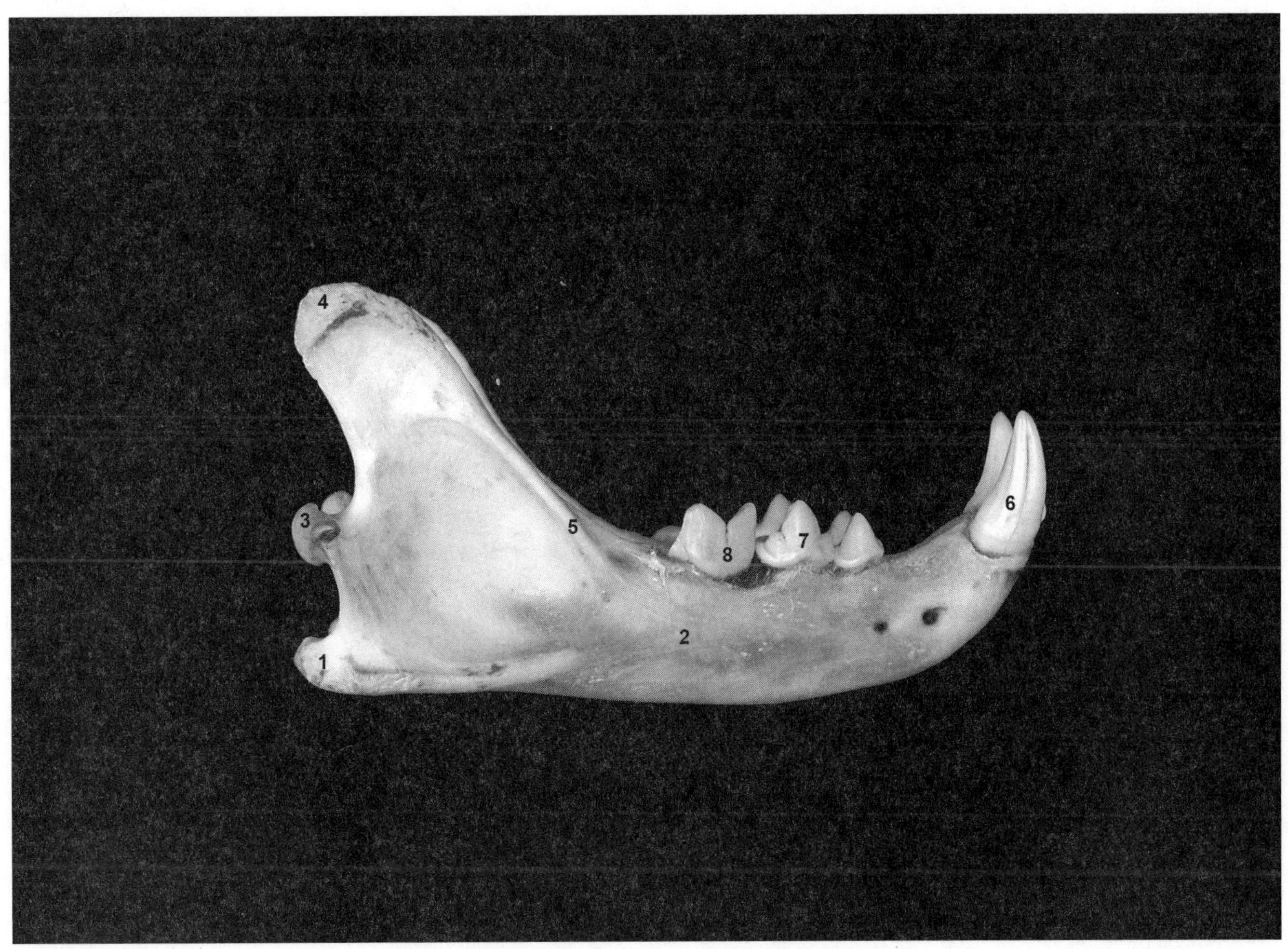

FIGURE 5.5. Right mandible of Bobcat, lateral view.

1. Angular process
2. Mandibular body
3. Condyloid process
4. Coronoid process
5. Mandibular ramus
6. Canine
7. Premolar
8. Molar

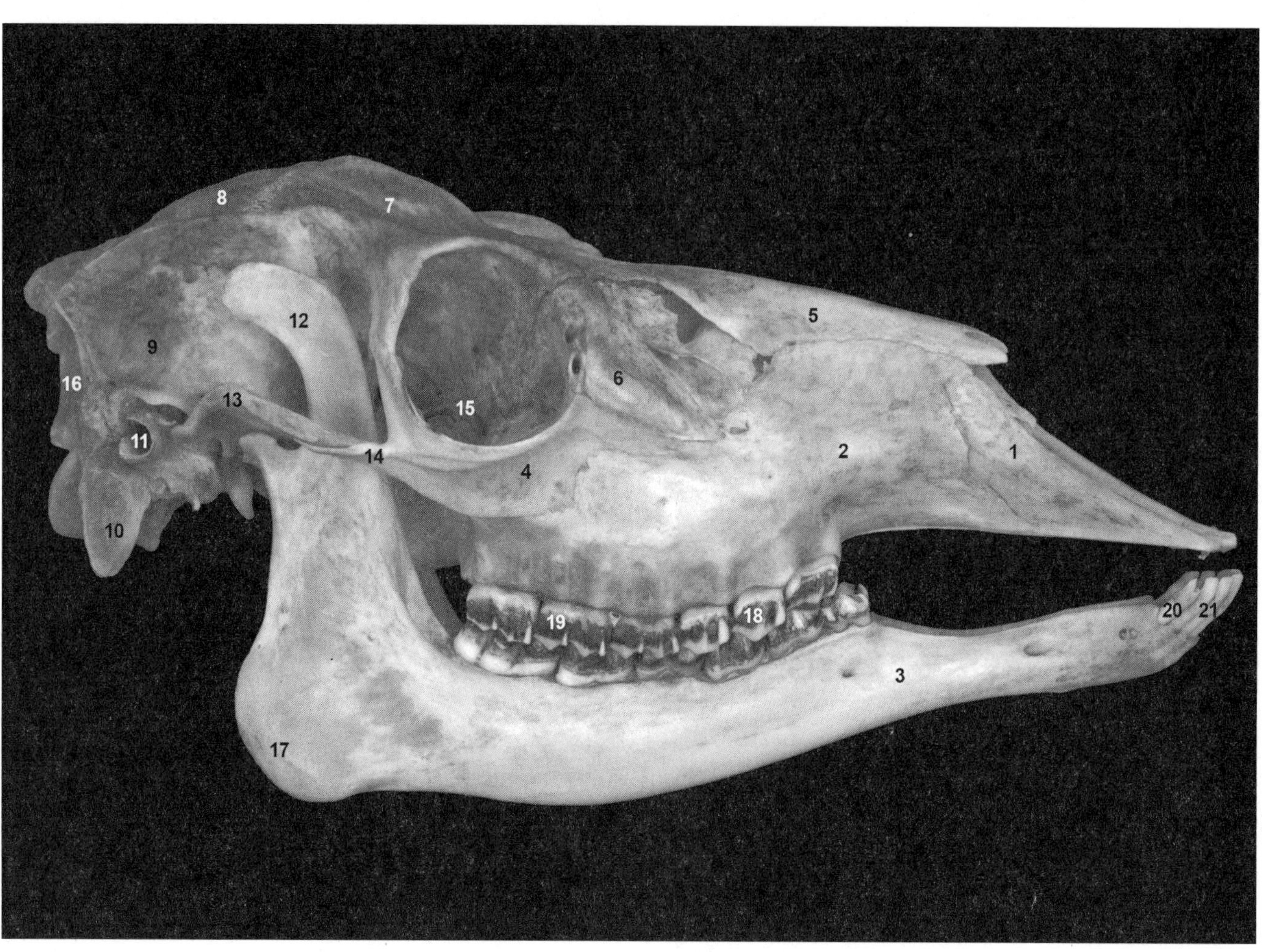

FIGURE 5.6. Cranium of Mule Deer (*Odocoileus hemionus*), lateral view.

1. Premaxilla
2. Maxilla
3. Mandible
4. Zygomatic
5. Nasal
6. Lacrimal
7. Frontal
8. Parietal
9. Temporal
10. Mastoid process
11. External auditory meatus
12. Coronoid process
13. Zygomatic process
14. Temporal process
15. Ethmoid
16. Occipital
17. Angular process
18. Premolar
19. Molar
20. Canine
21. Incisor

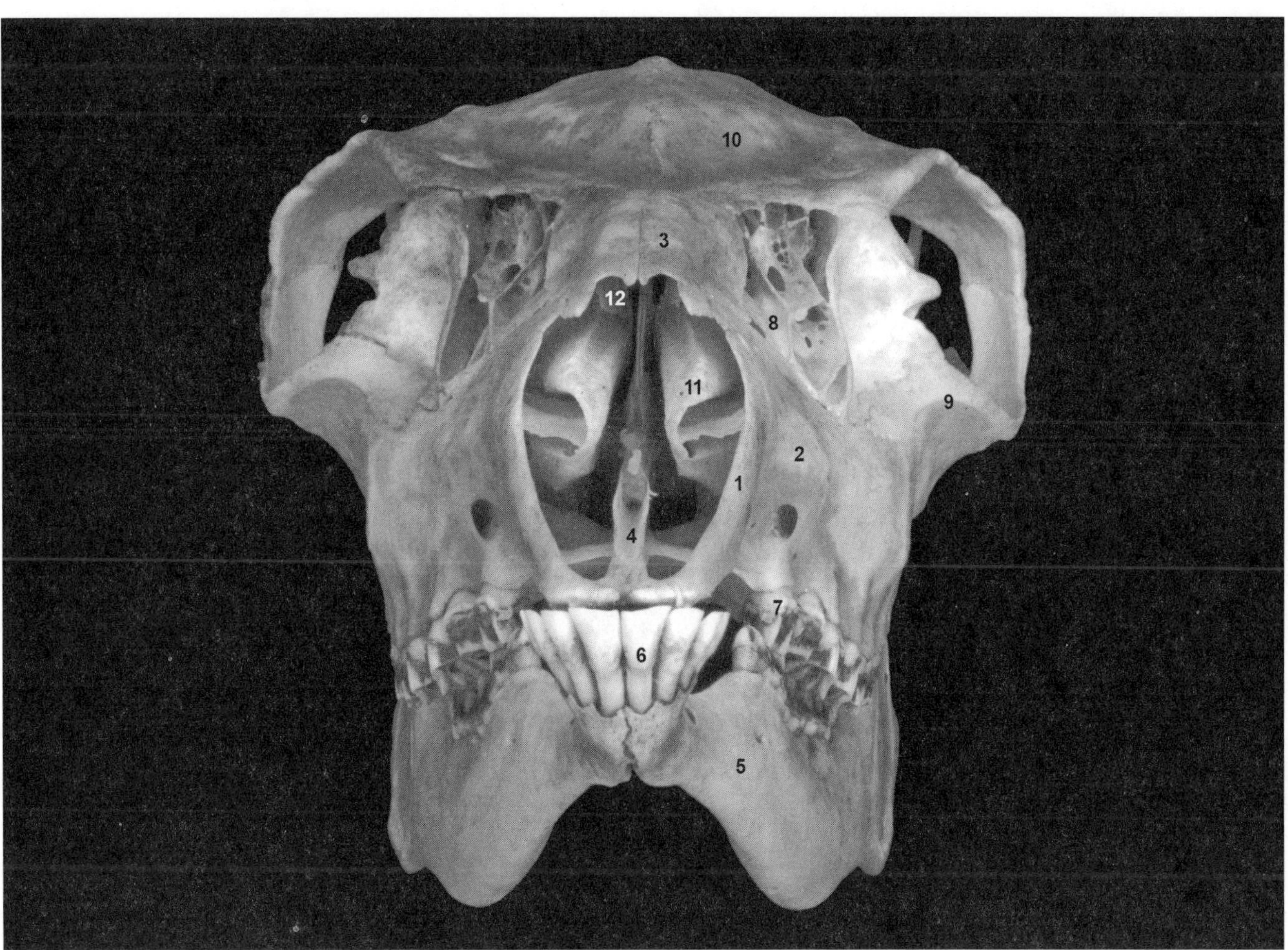

FIGURE 5.7. Cranium of Mule Deer, anterior view.

1. Premaxilla
2. Maxilla
3. Nasal
4. Vomer
5. Mandible
6. Incisor
7. Premolar
8. Lacrimal
9. Zygomatic
10. Frontal
11. Middle nasal conchae
12. Superior nasal conchae (ethmoid)

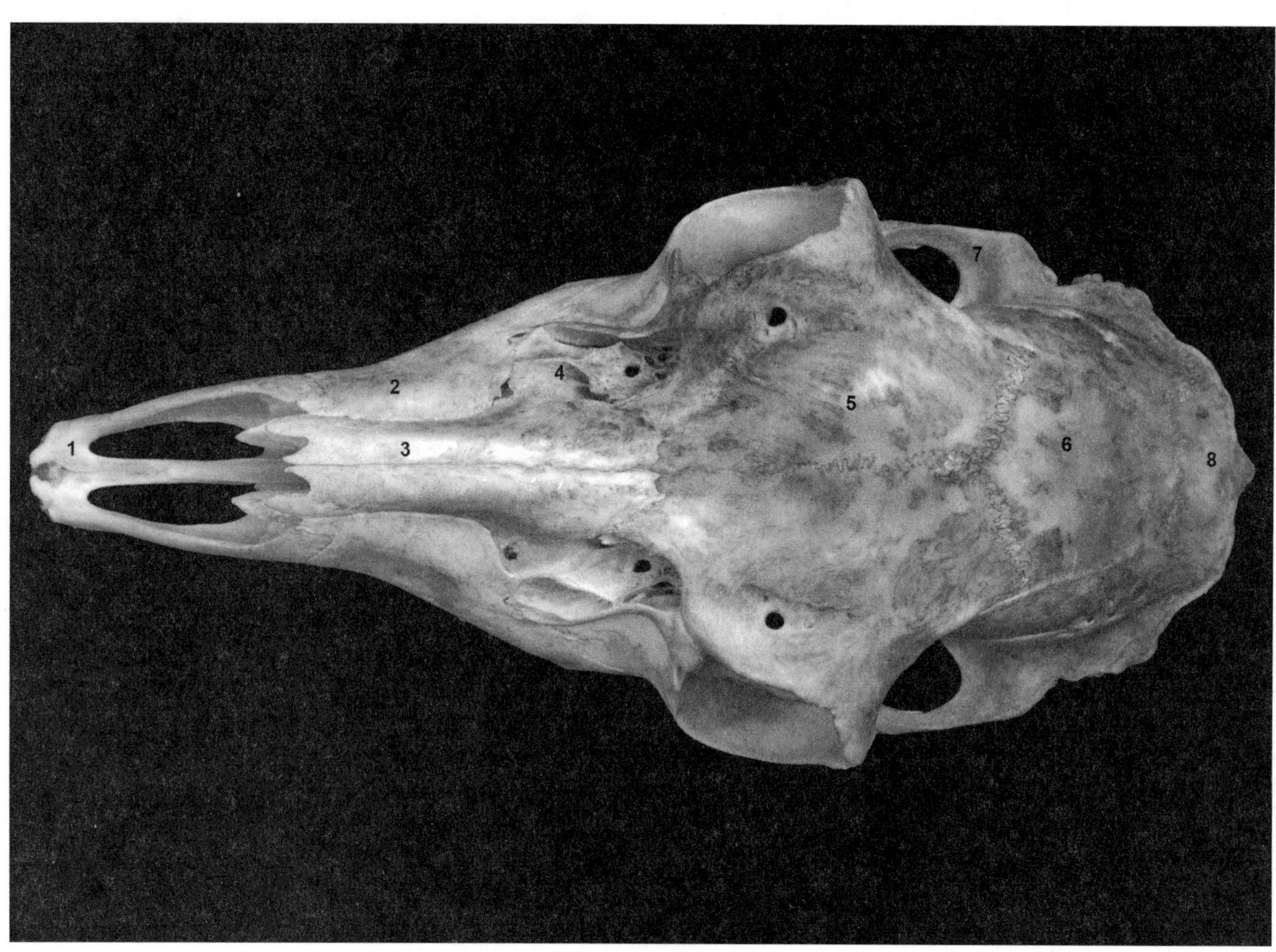

FIGURE 5.8. Cranium of Mule Deer, dorsal view.

1. Premaxilla
2. Maxilla
3. Nasal
4. Lacrimal
5. Frontal
6. Parietal
7. Zygomatic process
8. Occipital

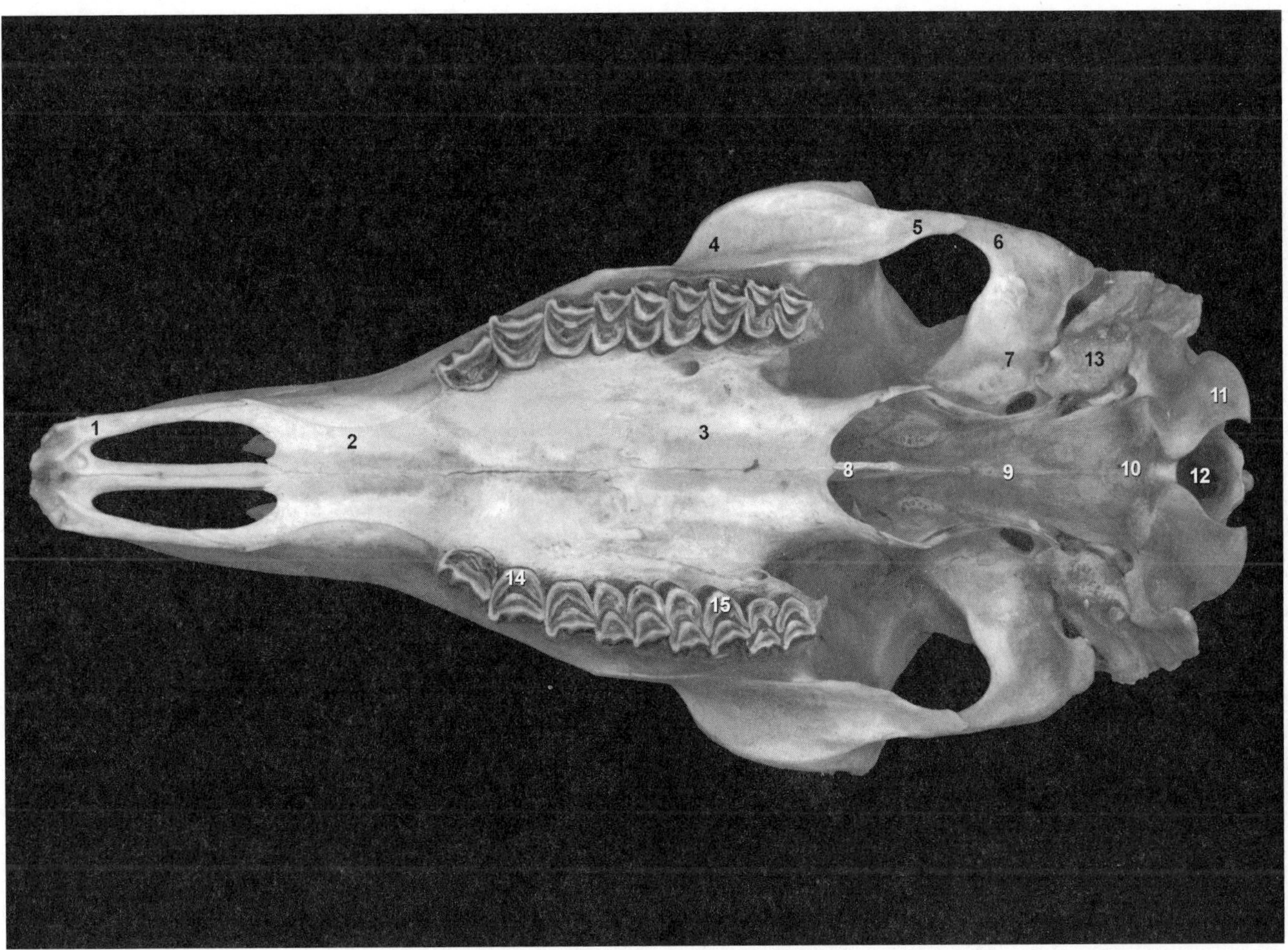

FIGURE 5.9. Cranium of Mule Deer, ventral view.

1. Premaxilla
2. Maxilla
3. Palatine
4. Zygomatic
5. Temporal process (of zygomatic)
6. Zygomatic process (of temporal)
7. Temporal
8. Vomer
9. Basisphenoid
10. Occipital
11. Occipital condyle
12. Foramen magnum
13. Tympanic bulla
14. Premolar
15. Molar

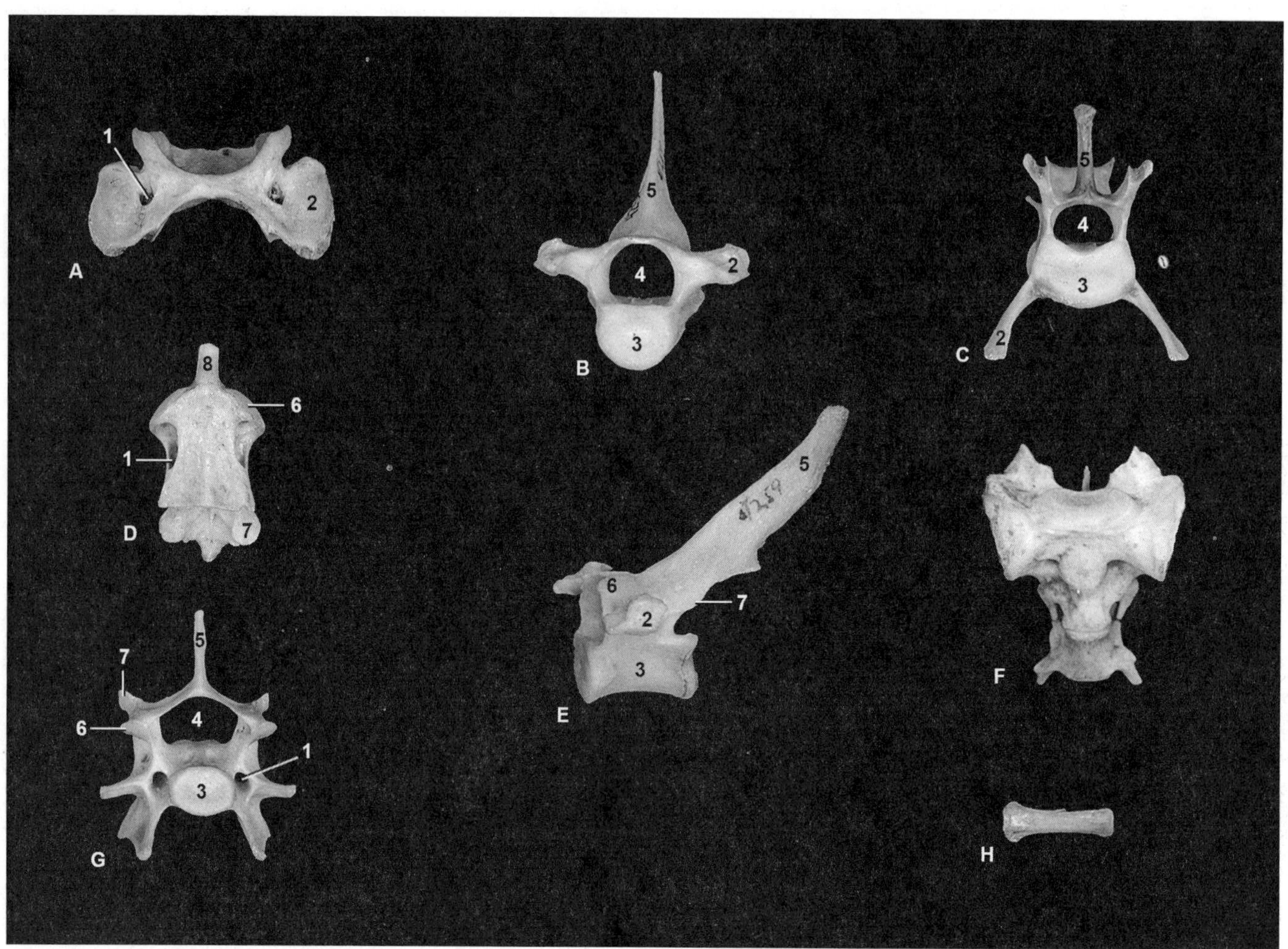

FIGURE 5.10. Vertebrae of Bobcat: atlas, ventral view (A), thoracic vertebra, anterior view (B), lumbar vertebra, anterior view (C), axis, ventral view (D), thoracic vertebra, lateral view (E), sacrum, ventral view (F), cervical vertebra, anterior view (G), caudal vertebra, ventral view (H).

1. Transverse foramen
2. Transverse process
3. Vertebral body
4. Vertebral foramen
5. Spinous process
6. Prezygopophysis
7. Postzygopophysis
8. Odontoid (or dens) process

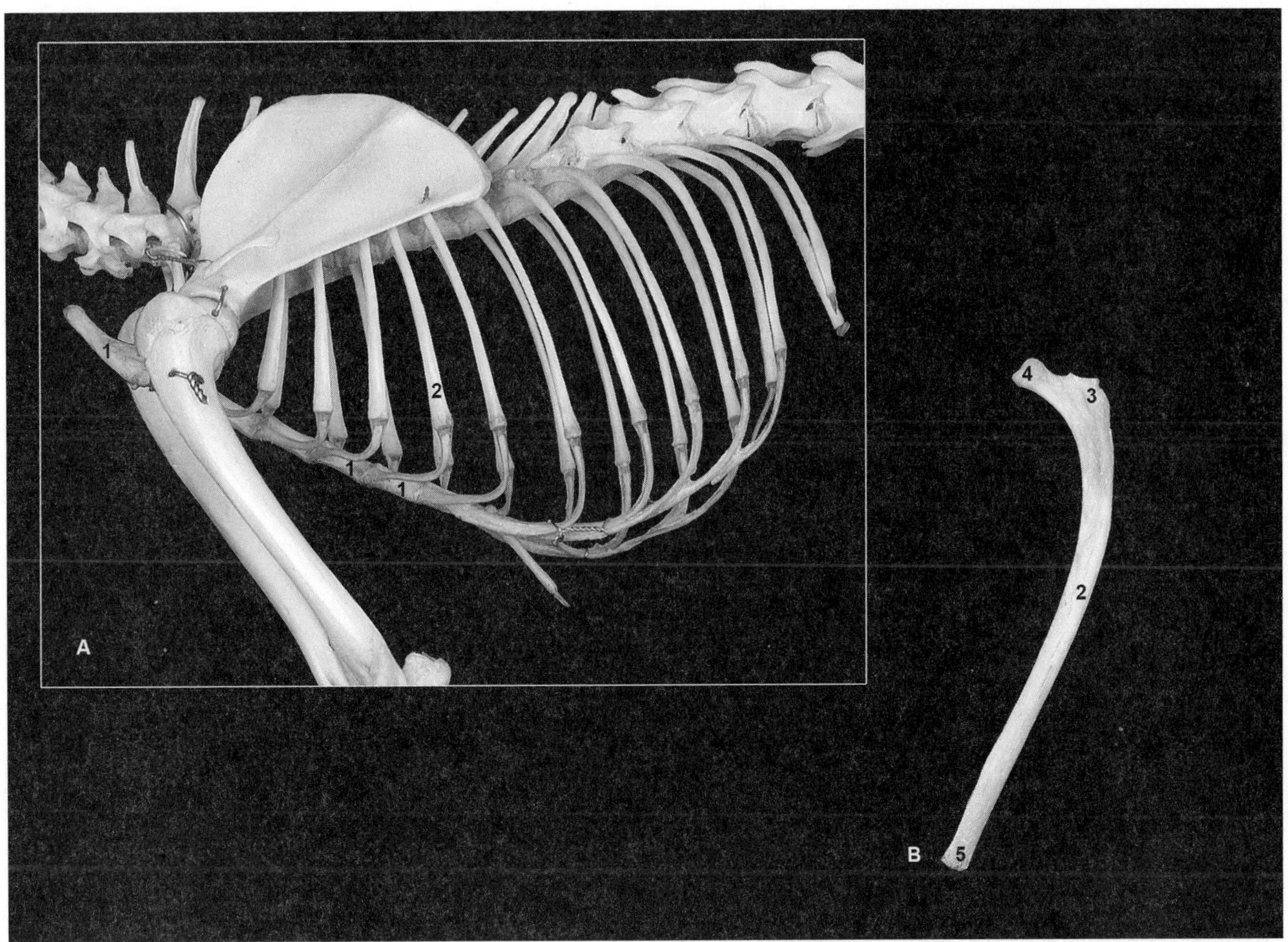

FIGURE 5.11. Thoracic region and l. pectoral girdle (A) of Domestic Cat (*Felis catus*) and l. rib (B) of Bobcat.

1. Sternabrae
2. Ribs
3. Tubercle
4. Costal head
5. Sternal end

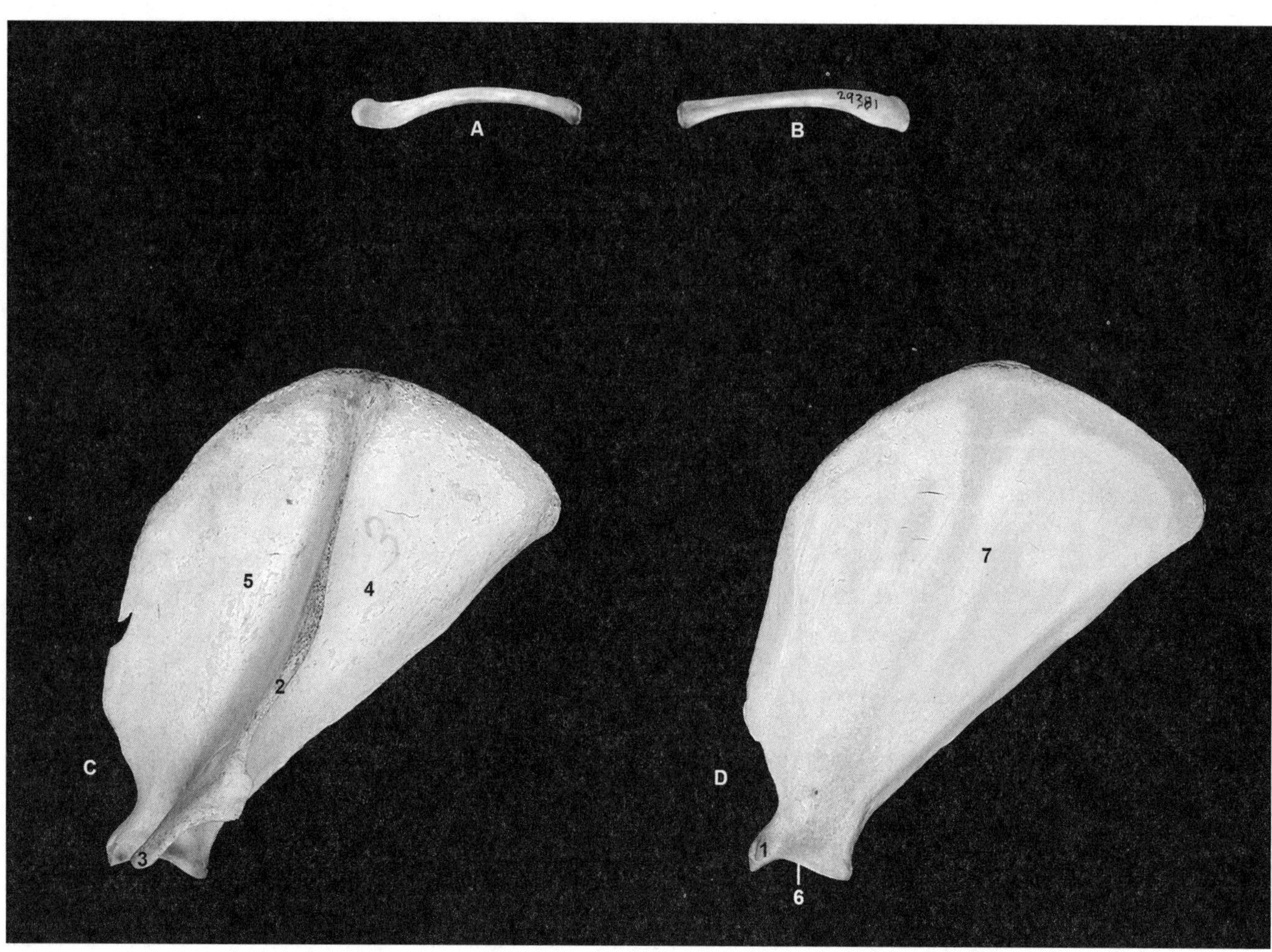

FIGURE 5.12. Yellow-bellied Marmot (*Marmota flaviventris*) right clavicles, posterior (A), and anterior (B) views. Bobcat scapulae, l. scapula, lateral view (C), and r. scapula, medial view (D).

1. Coracoid process
2. Scapular spine
3. Acromion
4. Infraspinous fossa
5. Supraspinous fossa
6. Glenoid fossa
7. Subscapular fossa

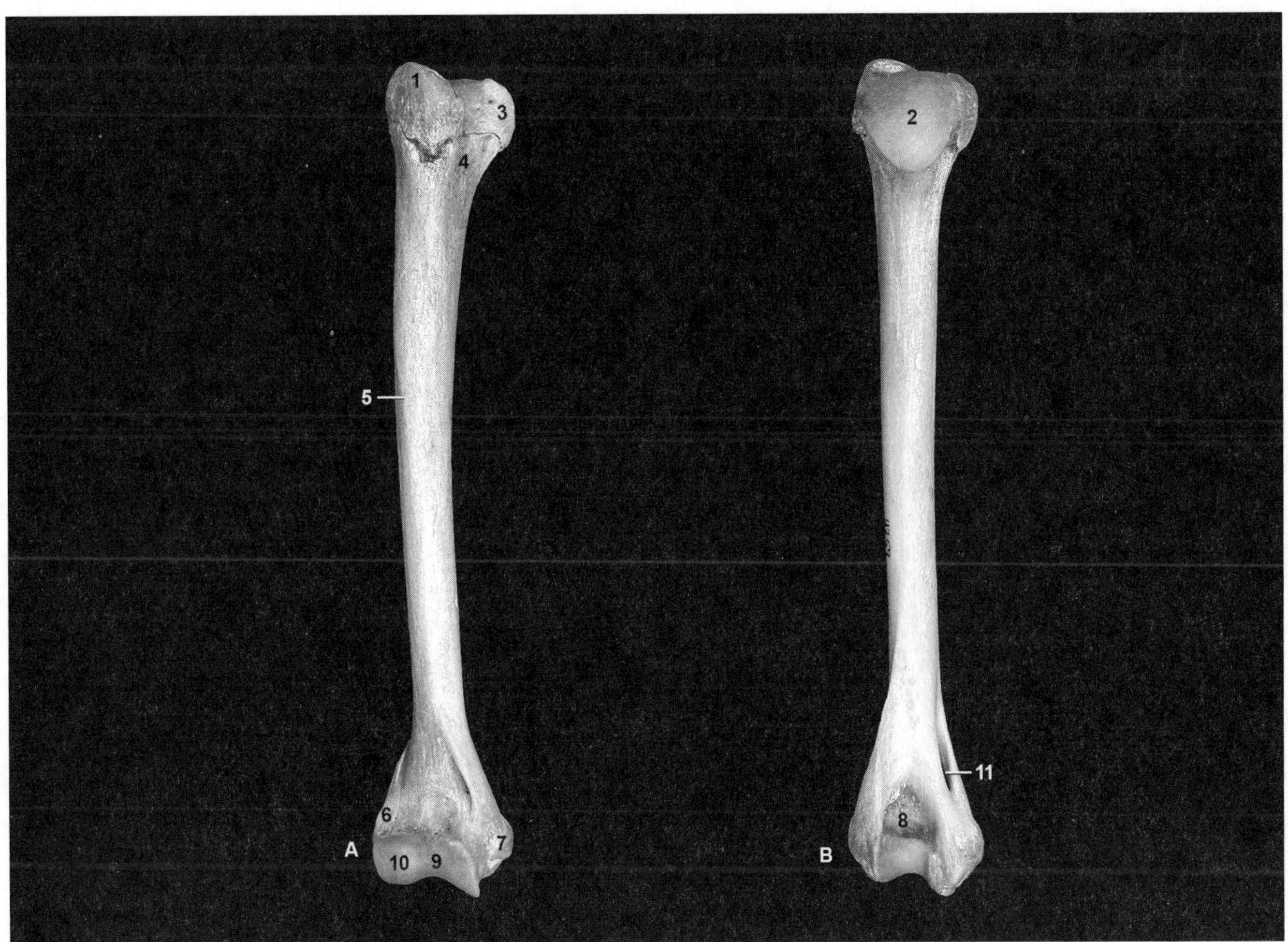

FIGURE 5.13. Bobcat humerus, r. anterior view (A), and l. posterior view (B); proximal is top, distal is bottom.

1. Greater tubercle
2. Humeral head
3. Lesser tubercle
4. Bicipital groove
5. Deltoid ridge
6. Lateral epicondyle
7. Medial epicondyle
8. Olecranon fossa
9. Trochlea
10. Capitulum
11. Supracondyloid foramen

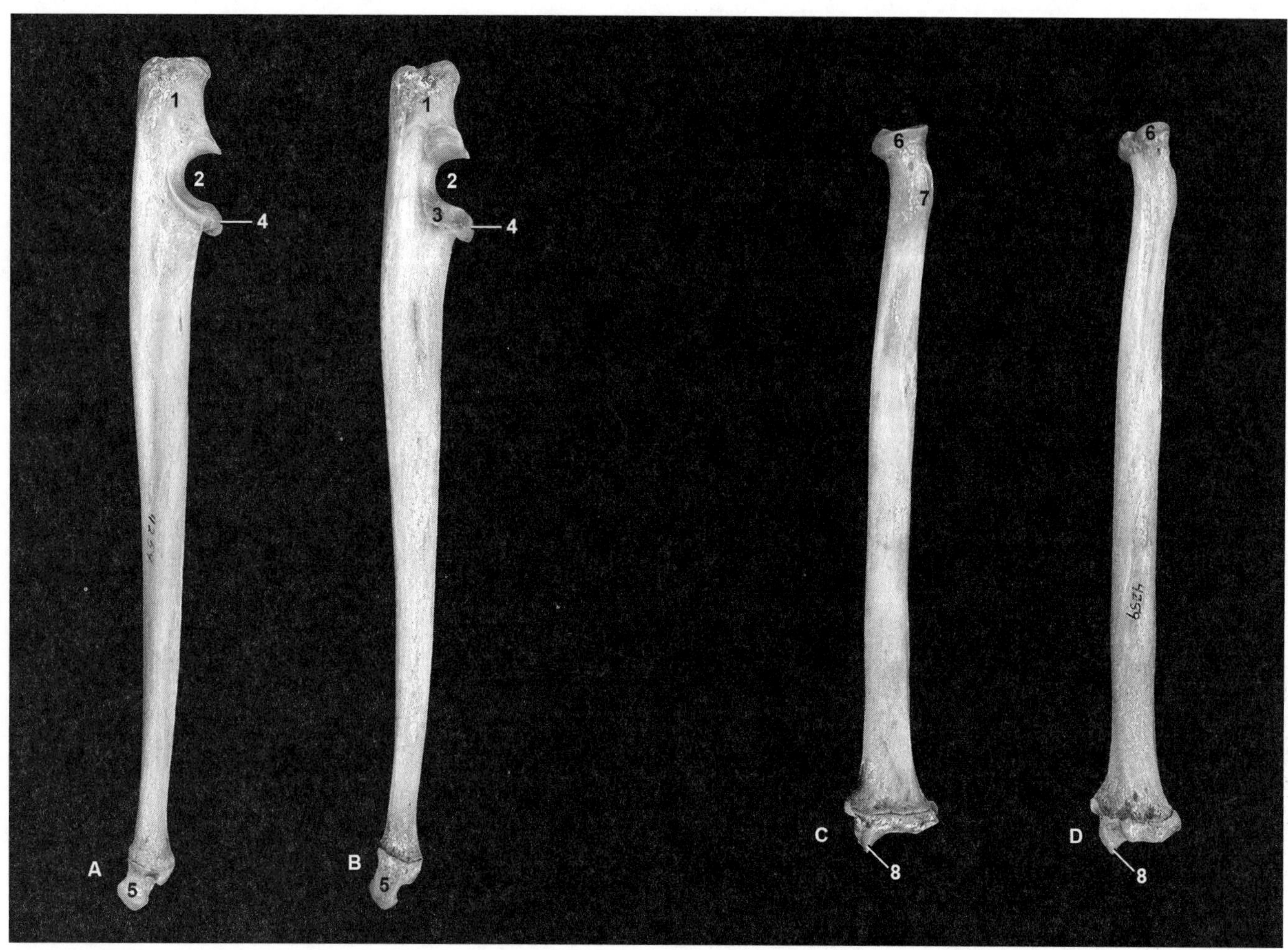

FIGURE 5.14. Bobcat l. ulna, medial view (A); r. ulna, lateral view (B); l. radius, posterior view (C); and r. radius, anterior view (D); proximal is top.

1. Olecranon process
2. Semilunar notch
3. Radial notch
4. Coronoid process
5. Styloid process
6. Radial head
7. Bicipital tuberosity
8. Styloid process

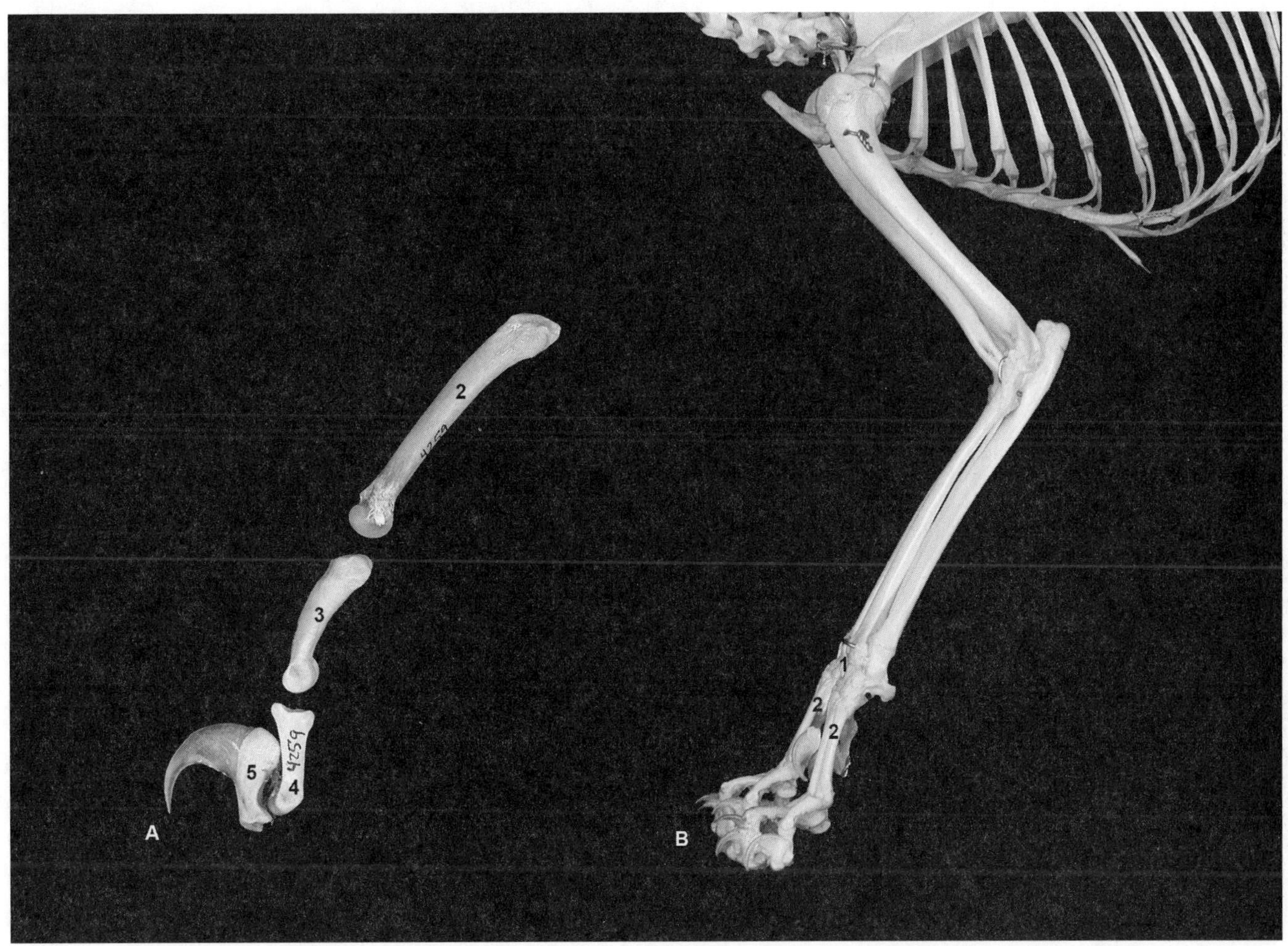

FIGURE 5.15. Lateral views of digit elements of Bobcat (A) and pectoral girdle and forelimb (B) of Domestic Cat.

1. Carpals
2. Metacarpals
3. Proximal (or 1st) phalanx
4. Middle (or 2nd) phalanx
5. Distal (or 3rd) phalanx

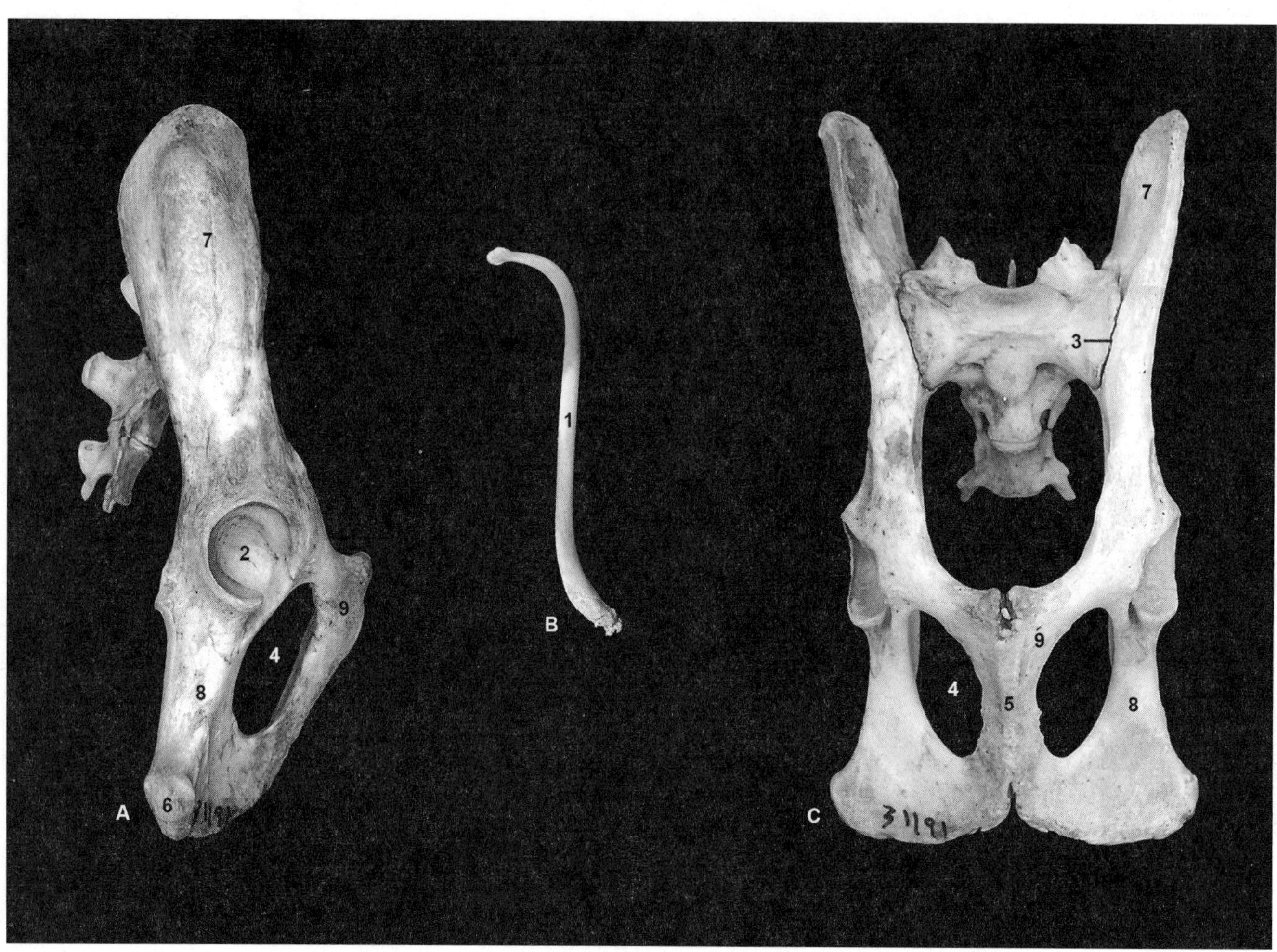

FIGURE 5.16. Lateral views of r. os coxae and sacrum (A) of Bobcat and os baculum (B) of Raccoon (*Procyon lotor*); ventral view of Bobcat pelvis (C); top is anterior.

1. Os baculum
2. Acetabulum
3. Auricular impression
4. Obturator foramen
5. Pubic symphysis
6. Ischial tuberosity
7. Ilium
8. Ischium
9. Pubis

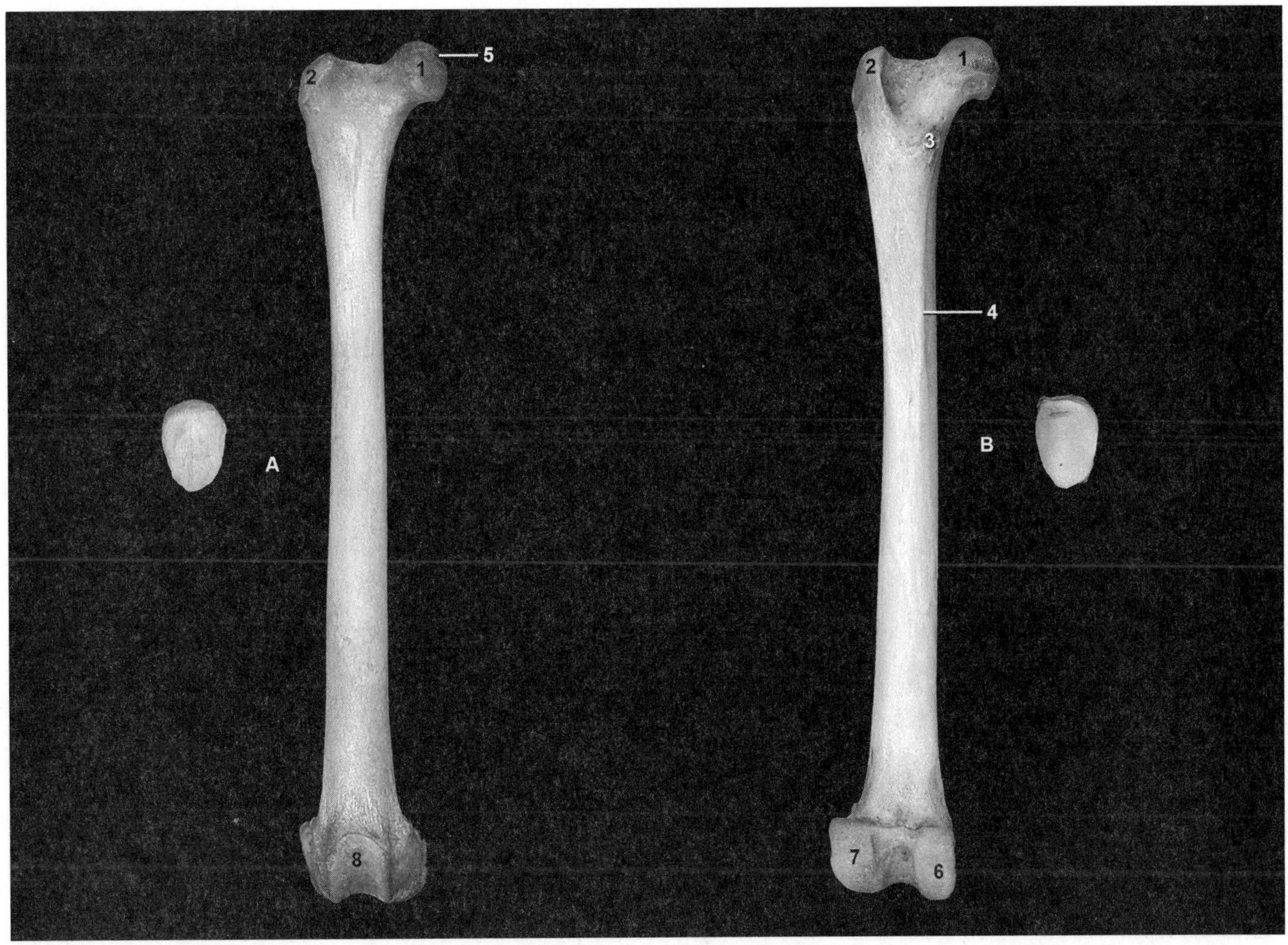

FIGURE 5.17. Bobcat r. femur and patella, anterior views (A), and l. femur and patella, posterior views (B); top is proximal.

1. Femoral head
2. Greater trochanter
3. Lesser trochanter
4. Linea aspera
5. Fovea capitis
6. Medial condyle
7. Lateral condyle
8. Patellar surface

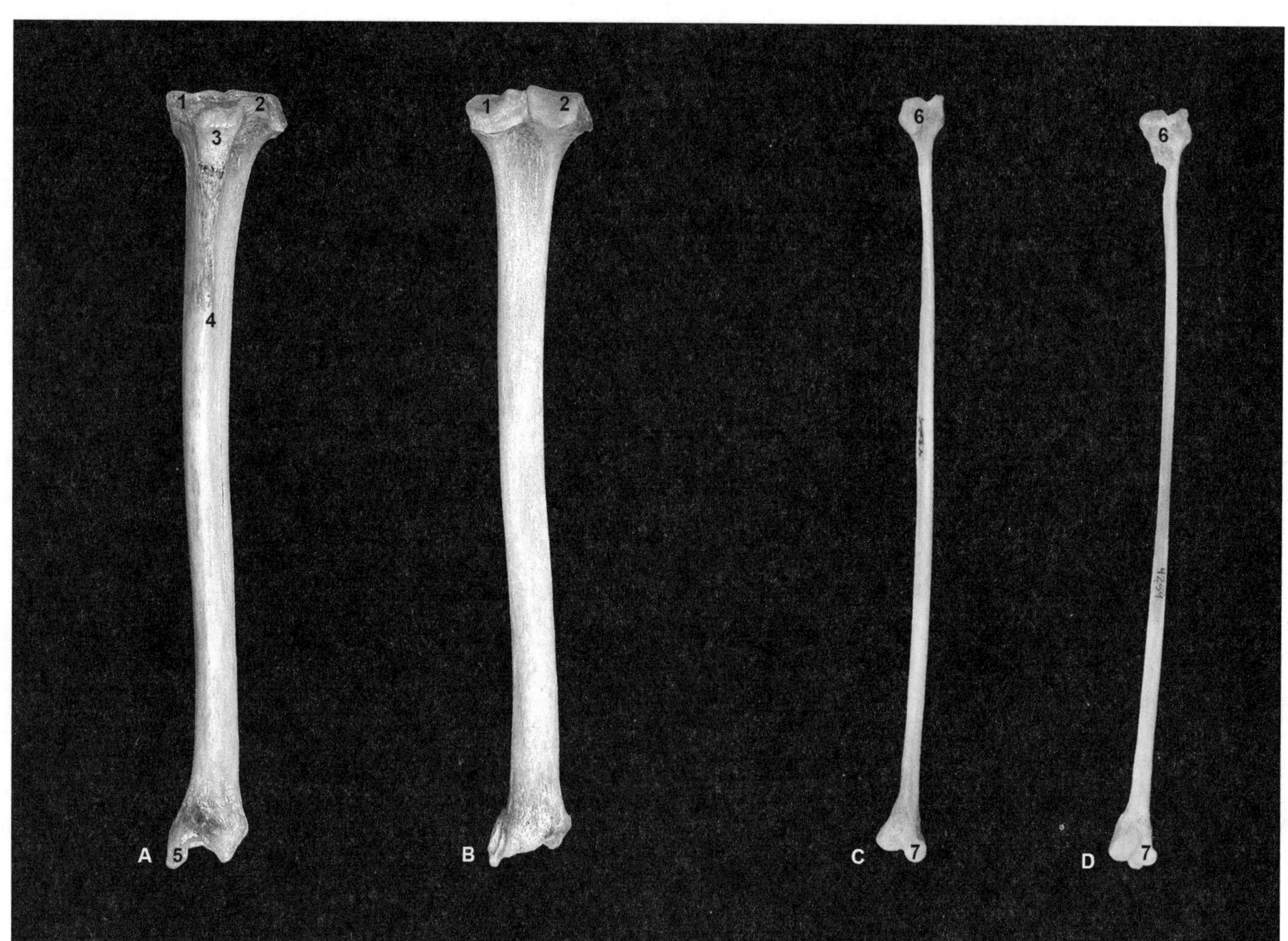

FIGURE 5.18. Bobcat tibia and fibula: l. tibia, anterior view (A); r. tibia, posterior view (B); l. fibula, lateral view (C); and r. fibula medial view (D); proximal is top.

1. Medial condyle
2. Lateral condyle
3. Tibial tuberosity
4. Tibial crest
5. Medial malleolus
6. Head
7. Lateral malleolus

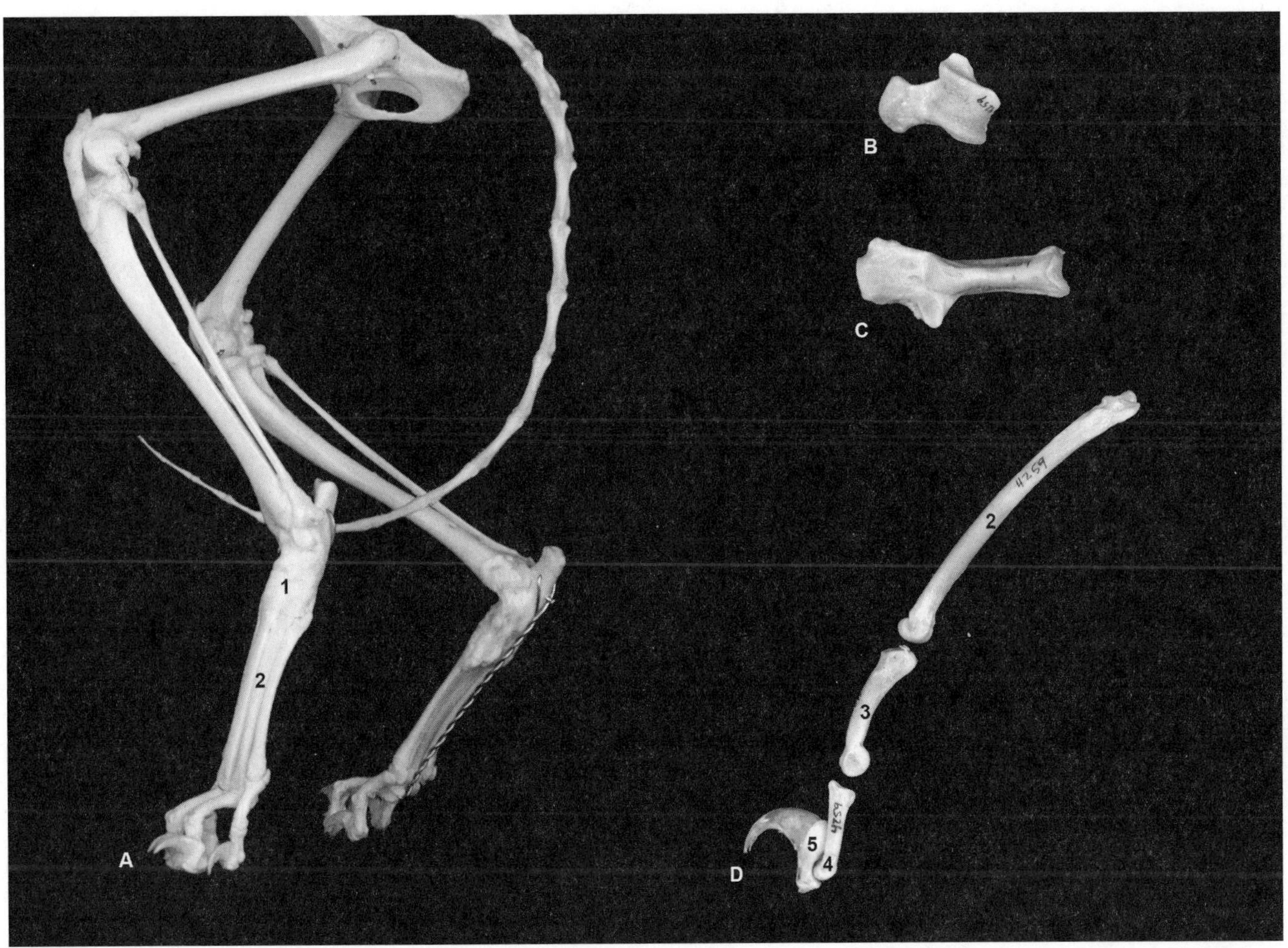

FIGURE 5.19. Hindlimb (A) of Domestic Cat. Bobcat, r. astragalus (B) and calcaneus (C), dorsal views, and digit elements, lateral view (D).

1. Tarsals
2. Metatarsals
3. Proximal (or 1st) phalanx
4. Middle (or 2nd) phalanx
5. Distal (or 3rd) phalanx

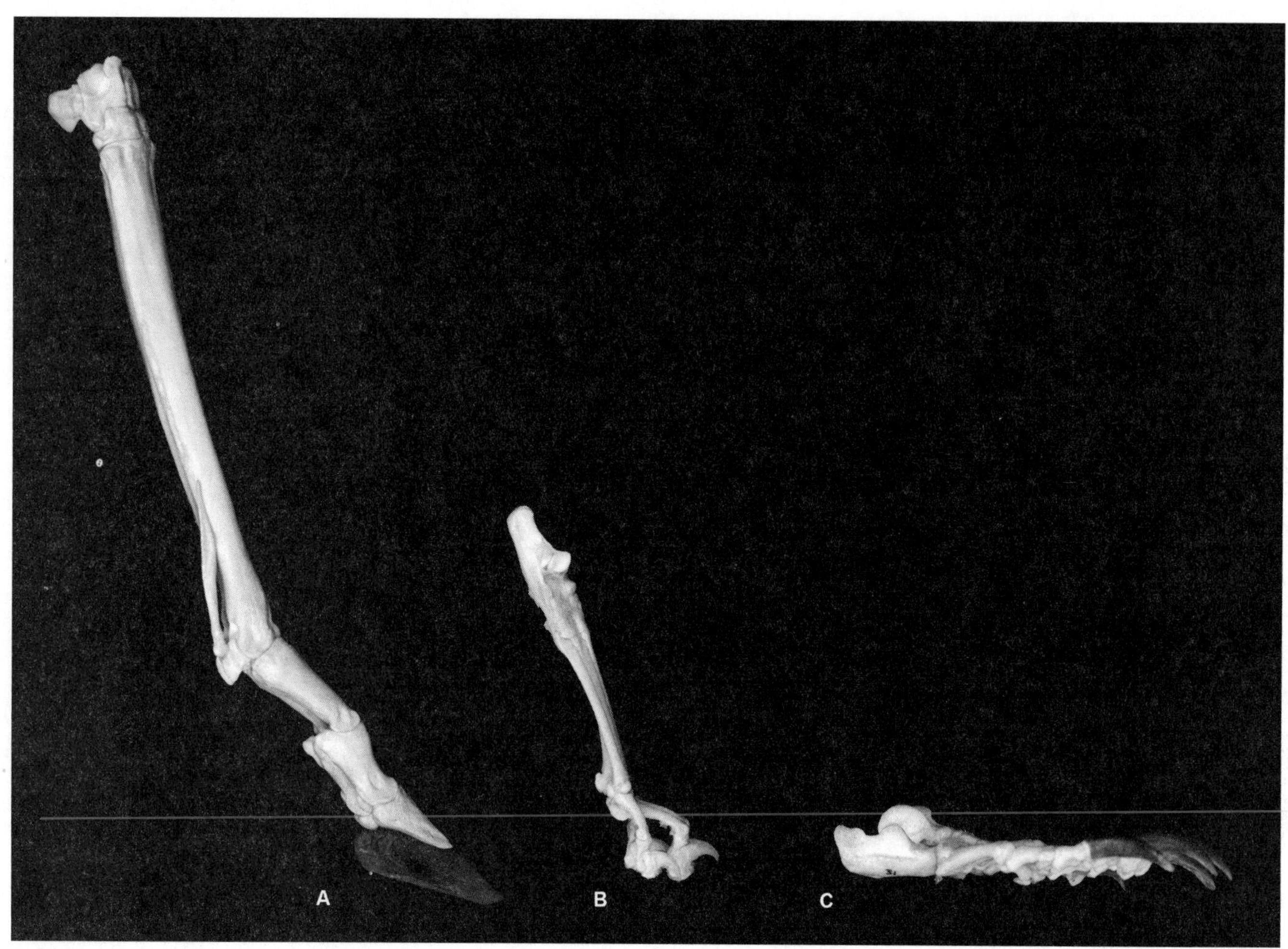

FIGURE 5.20. Unguligrade (deer) (A), digitigrade (cat) (B), and plantigrade (porcupine) (C) foot postures.

II. Taxonomy and Osteological Variation of Western Mammals

CLASS Mammalia (Mammals)
ORDER Rodentia (Rodents)

Rodentia (L., rodere = to gnaw, dens = teeth) is the largest order of mammals in the world and includes over 2,277 species—42% of all mammal species. Rodents also dominate in western North America and are represented by eight families and over 160 species. Next to Passeriformes (perching birds) no other vertebrate order in the region is represented by so many species (Figures 5.21–5.40; Color Plates 17–22). Most rodents are herbivorous, small, and mouse- or rat-like, except for the American Beaver (*Castor canadensis*) and North American Porcupine (*Erethizon dorsatum*). Most have four toes on the front feet and five on the hind feet. Rodents have a unique dentition with two pairs of evergrowing incisors, no canines, and a large diastema separating the incisors from the cheekteeth. Since only the anterior portions of the incisors are enameled (usually orange in color due to the presence of iron), the inner (posterior) portion wears much faster and produces a distinctive beveled edge that is ideal for gnawing.

Most species are nocturnal, with squirrels, chipmunks, and marmots (Sciuridae) being the noteworthy exceptions. Live trapping and night drives are thus essential to observe, study, and admire rodents. Many are active throughout the year, but many others hibernate during the winter; some also estivate to escape the heat of the summer. Some do both (e.g., Piute Ground Squirrel, *Urocitellus mollis*) and are thus dormant for all but a few months during the spring. Both hibernation and estivation involve dramatic reductions in body temperature and metabolism.

Although their small size dictates a generally low return rate to human foragers, rodents were intensively utilized in many prehistoric settings in the west. Indeed, their relative abundance in archaeological faunas compared to large game has been widely used as an index of human foraging efficiency with high proportionate abundances indicating lower overall hunting returns. Since many taxa are fossorial, or adapted to digging, determining the depositional origin of rodent remains in archaeological settings is problematic. Finally, since many rodent species have very narrow environmental tolerances, they are the most widely used vertebrates in reconstructing past environments.

FAMILY Aplodontiidae (Mountain Beaver)

This family is represented by a single species, the most primitive living rodent, the Mountain Beaver or Sewellel (*Aplodontia rufa*; Figures 5.21–5.22). The name Mountain Beaver may be easier to pronounce than Sewellel, but it is poorly suited since the animal is not confined to mountains and it is completely unrelated to the American Beaver (Castoridae). This peculiar rodent has small ears and eyes and is similar in size (about 31 cm in

FIGURE 5.21. Mountain Beaver (*Aplodontia rufa*). Photo by Dale T. Steel

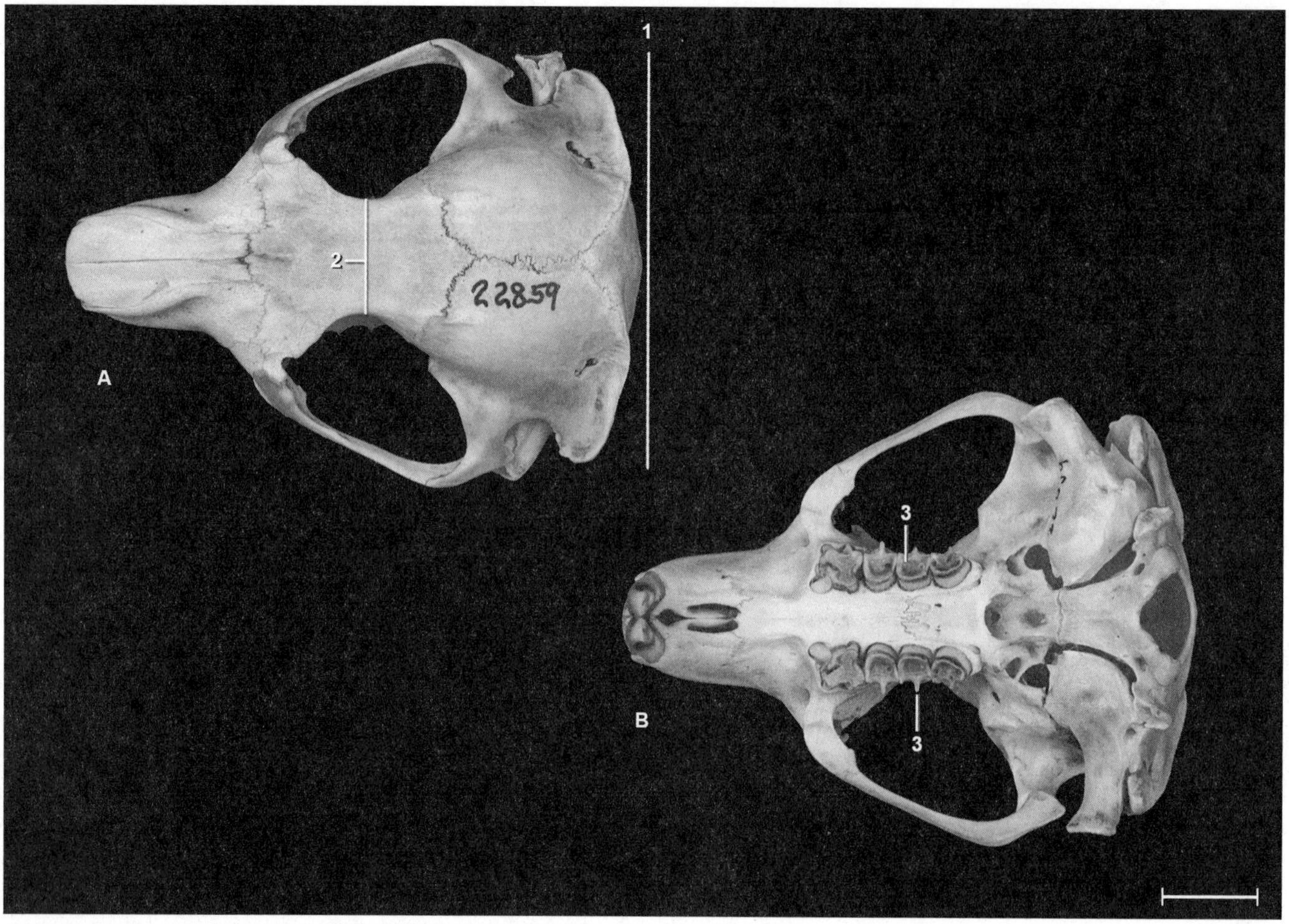

FIGURE 5.22. Cranium of Mountain Beaver (*Aplodontia rufa*), dorsal (A) and ventral (B) views. Osteological features (Aplodontiidae): 1, cranium distinctively wedge-shaped, with broad posterior tapering to front; 2, interorbital breadth narrow; 3, cheekteeth D-shaped, with prominent, rounded, medial margins and spike-like lateral projections.

FIGURE 5.23. Red Squirrel (*Tamiasciurus hudsonicus*).

length) and appearance to the Muskrat (*Ondatra zibethicus*), but it lacks the Muskrat's long tail. Restricted to moist forests of the Pacific Northwest and the northern Sierra Nevada, it specializes on forest plants that are toxic or unpalatable to most other mammals (ferns, firs, rhododendrons). Sewellel's spend much of their time underground in extensive burrow systems.

FAMILY Sciuridae
(Squirrels, Chipmunks, and Allies)

The family name means "shade tail," owing to the large, bushy tails of some squirrels that are held upright over the body (Figures 5.23–5.25). Eight genera occur in the west; all members of the family except the Northern Flying Squirrel (*Glaucomys sabrinus*) are diurnal, thus the sciurids are the most conspicuous, familiar, and beloved rodents.

Many hibernate; some also estivate. Sciurids are typically quite vocal, using a variety of barking or chattering alarm calls. Most ground squirrels (*Otospermophilus*, *Urocitellus*, *Ammospermophilus*) and prairie dogs (*Cynomys*) are colonial burrowers.

FAMILY Castoridae (Beavers)

Reaching weights up to 32 kg, the American Beaver (*Castor canadensis*) is the largest rodent and the only living member of the family in North America (5.26–5.27). Beavers are named for paired "castor" glands located between the pelvis and the base of the tail; they are present in both sexes and produce an oily, highly odoriferous substance called castoreum that is used along with products from anal scent glands in scent communication. The American Beaver is well known for

FIGURE 5.24. Yellow-bellied Marmot (*Marmota flaviventris*).

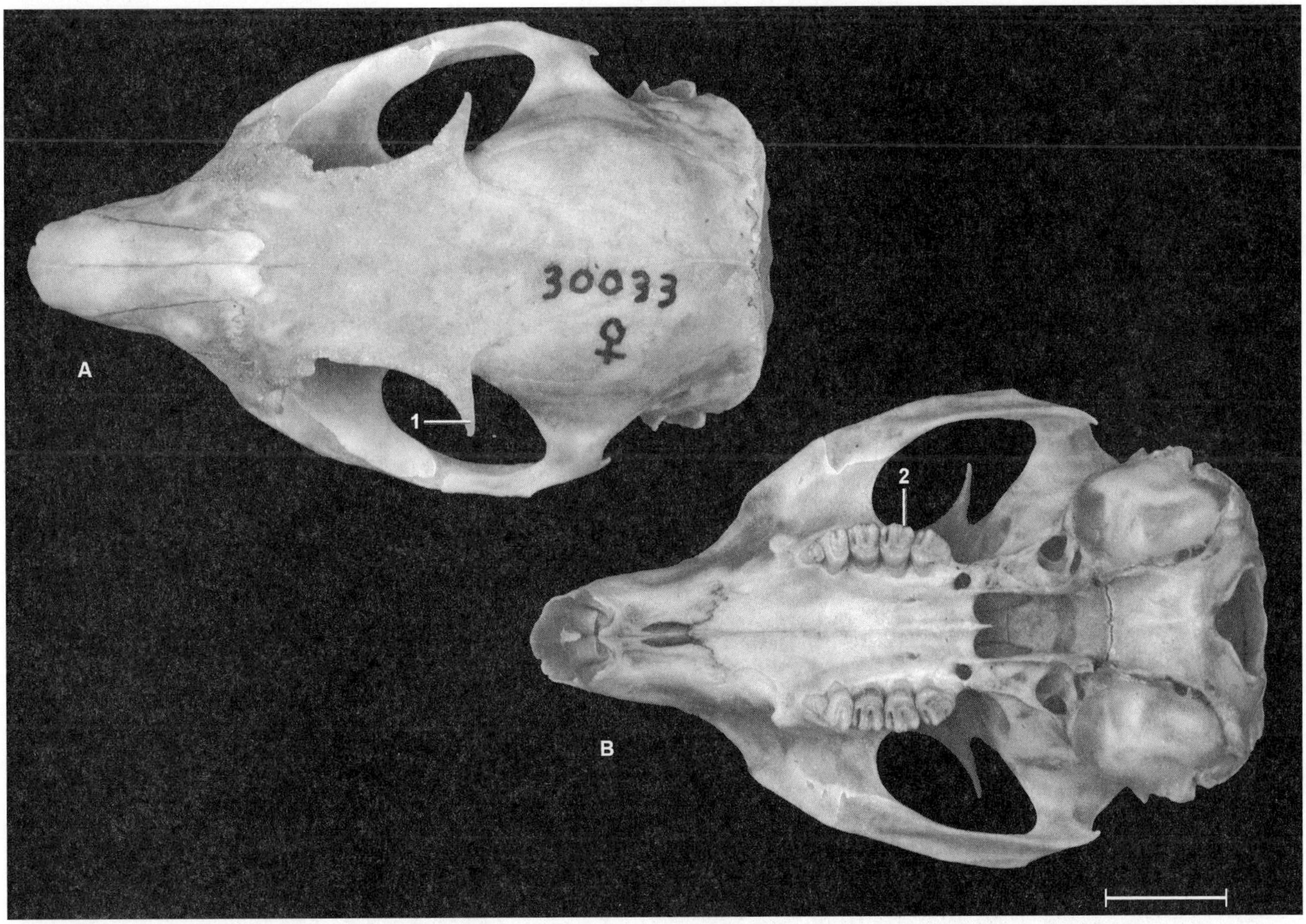

FIGURE 5.25. Cranium of Rock Squirrel (*Otospermophilus variegatus*), dorsal (A) and ventral (B) views. Osteological features (Sciuridae): 1, prominent postorbital processes; 2, upper cheekteeth with distinctive rounded medial borders and laterally extending, finger-like cusps—appearance resembles (three-fingered) baseball glove; 3, cheekteeth multiple-rooted—mandibles and maxillae with lost teeth show multiple, circular alveoli (not shown).

FIGURE 5.26. American Beaver (*Castor canadensis*). Photo by D. G. E. Robertson

its ability to fell trees and alter the landscape by building dams with sticks, branches, and mud across streams; nests are constructed within these structures or in burrows along river banks. Beavers have many unique adaptations related to their aquatic environments. These include a large, naked, paddle-shaped tail that is used as a rudder; short limbs; fully webbed hind feet; small eyes and external ears; and valved nostrils. The large tail is used to slap the surface of the water to warn others of danger. They live in small family groups, mate in winter, and give birth to litters of usually four to five kits in the spring. Giant, black bear-sized beavers (*Castoroides*) evolved 10 to 14 mya and occupied North America until sometime toward the end of the Pleistocene. Trappers seeking beaver pelts originally explored much of western North America.

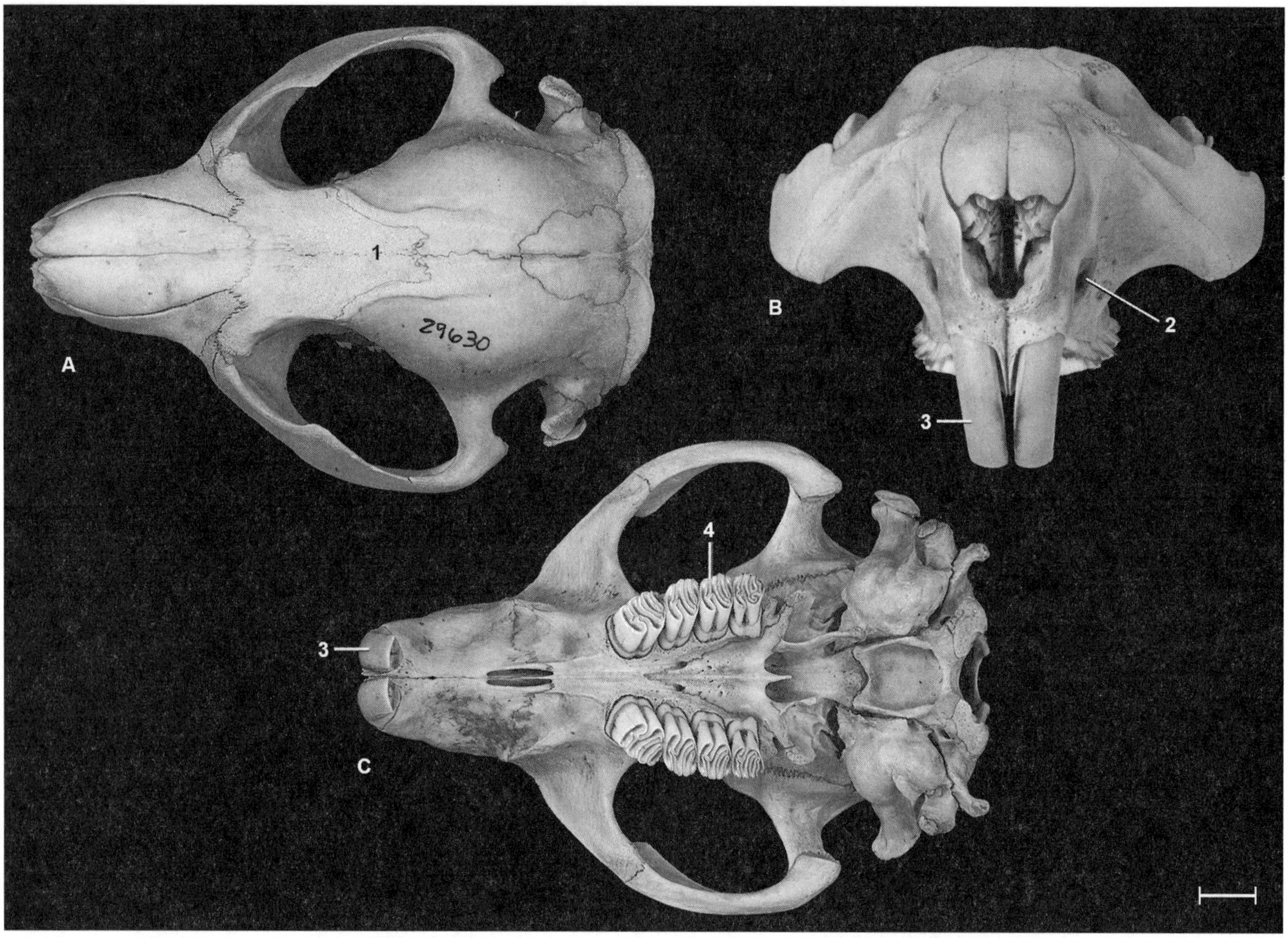

FIGURE 5.27. Cranium of American Beaver (*Castor canadensis*), dorsal (A), anterior (B), and ventral (C) views. Osteological features (Castoridae): 1, large size distinguishes beaver from all other rodents—confusion possible only with North American Porcupine (*Erethizon dorsatum*); 2, unlike porcupine, cranium has small, slit-like infraorbital foramen; 3, incisors are massive; 4, cheekteeth lophodont.

FAMILY Heteromyidae (Kangaroo Rats and Mice, and Pocket Mice)

The heteromyids (Figures 5.28–5.30) fall within the suborder Castorimorpha that also includes beavers (Castoridae) and pocket gophers (Geomyidae) and are thus more closely related to the latter than they are to other rats or mice (Cricetidae). Kangaroo rats (*Dipodomys*; Figure 5.28) and kangaroo mice (*Microdipodops*) are aptly named for their enormous hind legs, long tails, and tiny forelimbs. Kangaroo mice are miniature versions of kangaroo rats but have much thicker tails. Pocket mice (*Chaetodipus*, *Liomys*, *Perognathus*; Figure 5.29) are nondescript mice with equal-sized hind- and forelimbs that have smaller ears than true mice (e.g., *Peromyscus*, *Reithrodontomys*). Almost all taxa are nocturnal granivores (seed-eaters), seeds are stored in fur-lined cheek pouches until deposited in underground larders. The Chisel-toothed Kangaroo Rat (*Dipodomys microps*) is exceptional in that it eats shadscale (*Atriplex*) leaves. Kangaroo rats can survive on the metabolic water produced from seed digestion, although they will drink when water is available. Heteromyids are solitary, territorial, and build underground tunnels with chambers for sleeping, storage, and nesting. Kangaroo rats can be readily captured on summer nights with a flashlight and

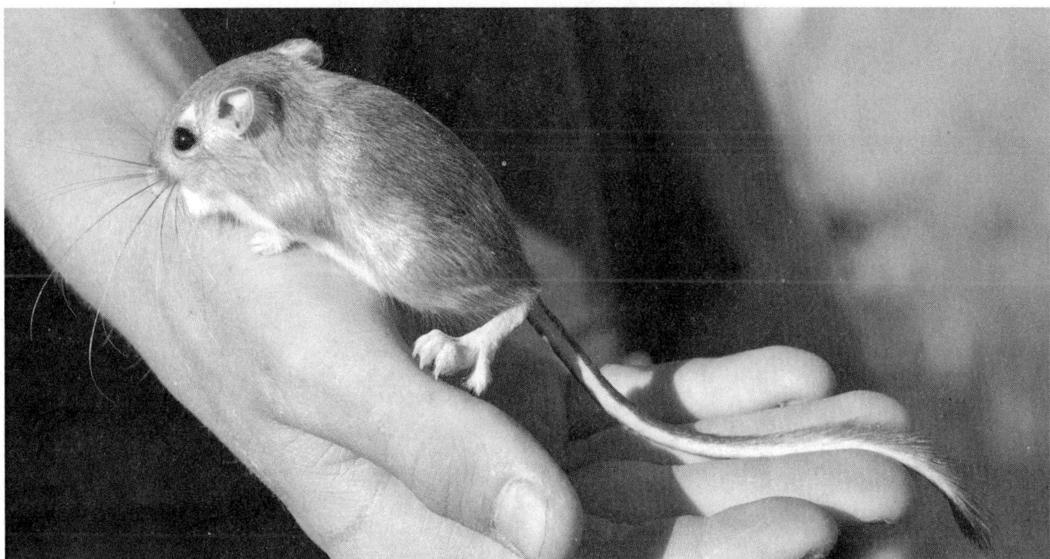

FIGURE 5.28. Ord's Kangaroo Rat (*Dipodomys ordii*).

FIGURE 5.29. Great Basin Pocket Mouse (*Perognathus parvus*).

butterfly net. They are more active on warm, overcast, or moonless nights. Most do not hibernate. Kangaroo rats are extremely docile and endearing; they almost never bite—unlike most rodents—and have even been known to eat seeds from the hand! Heteromyids are generally adapted to arid, open environments, but the degree of xeric adaptation varies by species.

FAMILY Geomyidae (Pocket Gophers)

Only a single genus of pocket gopher, *Thomomys*, is widespread in the west (Figures 5.31–5.32). (The ranges of two other genera, *Geomys* and *Cratogeomys*, extend from the east into southern and eastern New Mexico). Well adapted to subterranean life, pocket gophers have stocky cylindrical bodies; large heads; no obvious neck; short, naked tails; small ears and eyes; and powerful forelimbs. They spend virtually their entire lives underground within a series of tunnels and function-specific chambers. The "pockets" refer to the fur-lined cheek pouches in which they transport vegetation for food and nesting material. The lips close behind the incisors, allowing them to gnaw underground without ingesting dirt. Their orange incisors are thus always on prominent display. Seldom venturing above ground, their presence is nonetheless revealed by rounded mounds of dirt on the surface. When they are seen, it is almost always the head and shoulders of an animal pushing dirt from a tunnel opening. Pocket gophers are solitary vegetarians that feed on a range of plant materials, especially roots and tubers. They are active both day and night and do not hibernate.

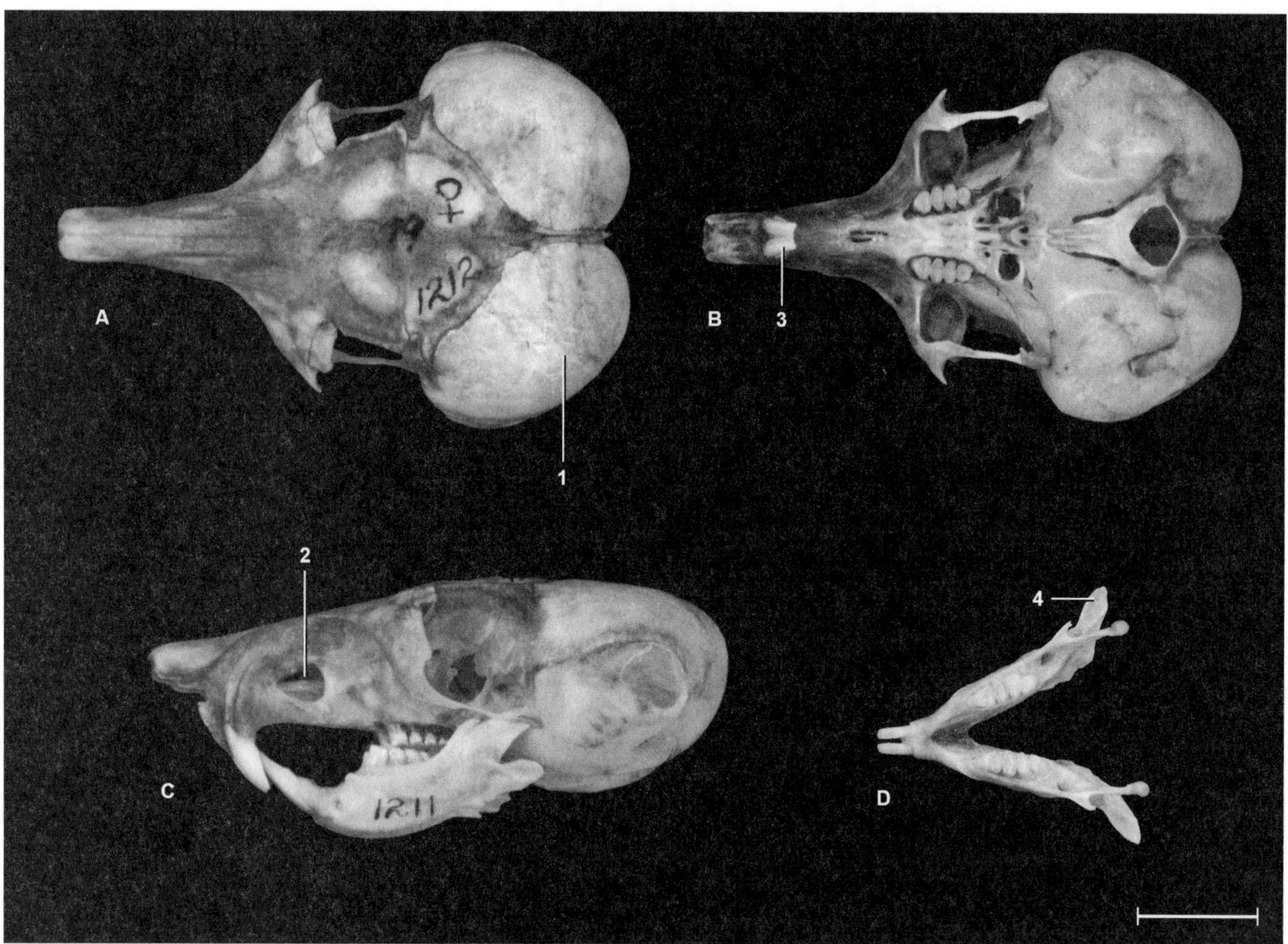

FIGURE 5.30. Desert Kangaroo Rat (*Dipodomys deserti*) cranium, dorsal (A), ventral (B), and lateral (C) views and mandible, dorsal view (D). Osteological features (Heteromyidae): 1, extremely large auditory bulla; 2, infraorbital foramina pass through rostrum (bony snout); 3, grooved upper incisors (except in Mexican Spiny Pocket Mouse (*Liomys irroratus*); 4, angular process of mandible flares laterally (except in Mexican Spiny Pocket Mouse).

FIGURE 5.31. Botta's Pocket Gopher (*Thomomys bottae*). Photo by Leonardo Weiss

Pocket gophers are well known as garden pests and for disturbing and size-sorting archaeological sediments.

FAMILY Dipodidae (Jumping Mice)

Readily identified by their extremely long tails and hindfeet, jumping mice belong to a family of principally Old World desert rodents; two genera occur in North America (*Napaeozapus* and *Zapus*), but only three *Zapus* species occur in the west (Figures 5.33–5.34). These mice are capable of incredible jumps, as their common name suggests, sometimes as far as 3 m in a single bound. Their tails are used for balance; individuals with damaged tails somersault and rarely land on all fours. The most widespread western species are Pacific Jumping Mouse (*Zapus trinotatus*) and Western

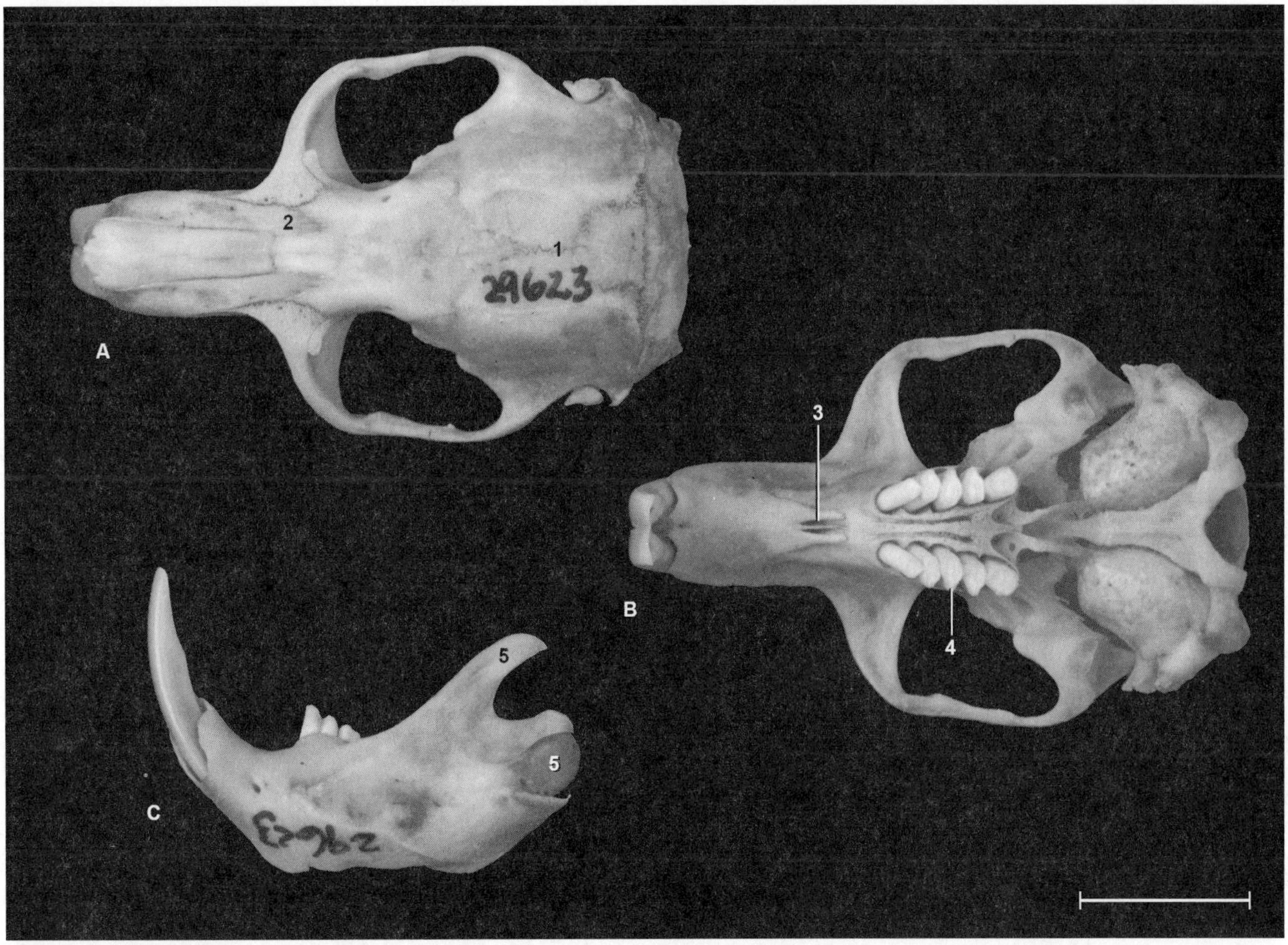

FIGURE 5.32. Botta's Pocket Gopher (*Thomomys bottae*) cranium, dorsal (A), and ventral (B) views, and mandible, lateral (C) view. Osteological features (*Thomomys*): 1, large squamosal portions of temporal connect midsagittally and comprise much of braincase; 2, premaxilla extends posteriorly well past nasals; 3, tiny incisive foramen; 4) cheekteeth oriented strongly posterior; 5, mandible robust with prominent coronoid process and bulbous condyloid and angular processes.

FIGURE 5.33 Woodland Jumping Mouse (*Napaeozapus insignis*). Photo by D. G. E. Robertson.

Jumping Mouse (*Zapus princeps*). Both inhabit mesic settings, near streams, wetlands, and meadows within coniferous forests or alder-willow associations. Both taxa enter a lengthy hibernation in winter (lasting six months or more in northern areas), sustained by substantial fat deposits that accumulate by fall.

FAMILY Cricetidae (Cricetid Rats and Mice)
Represented by 14 genera and 54 species, Cricetidae is the single largest family of mammals in western North America. Crecetidae is comprised of two subfamilies: Arvicolinae, which includes the voles (e.g., *Microtus*, *Clethrionomys*, *Phenacomys*), lemmings (*Synaptomys*, *Dicrostonyx*), and Muskrat (*Ondatra zibethicus*); and Neotominae, which includes the deermice (*Peromyscus*), grasshopper mice (*Onychomys*), harvest mice (*Reithrodontomys*), woodrats (*Neotoma*), and al-

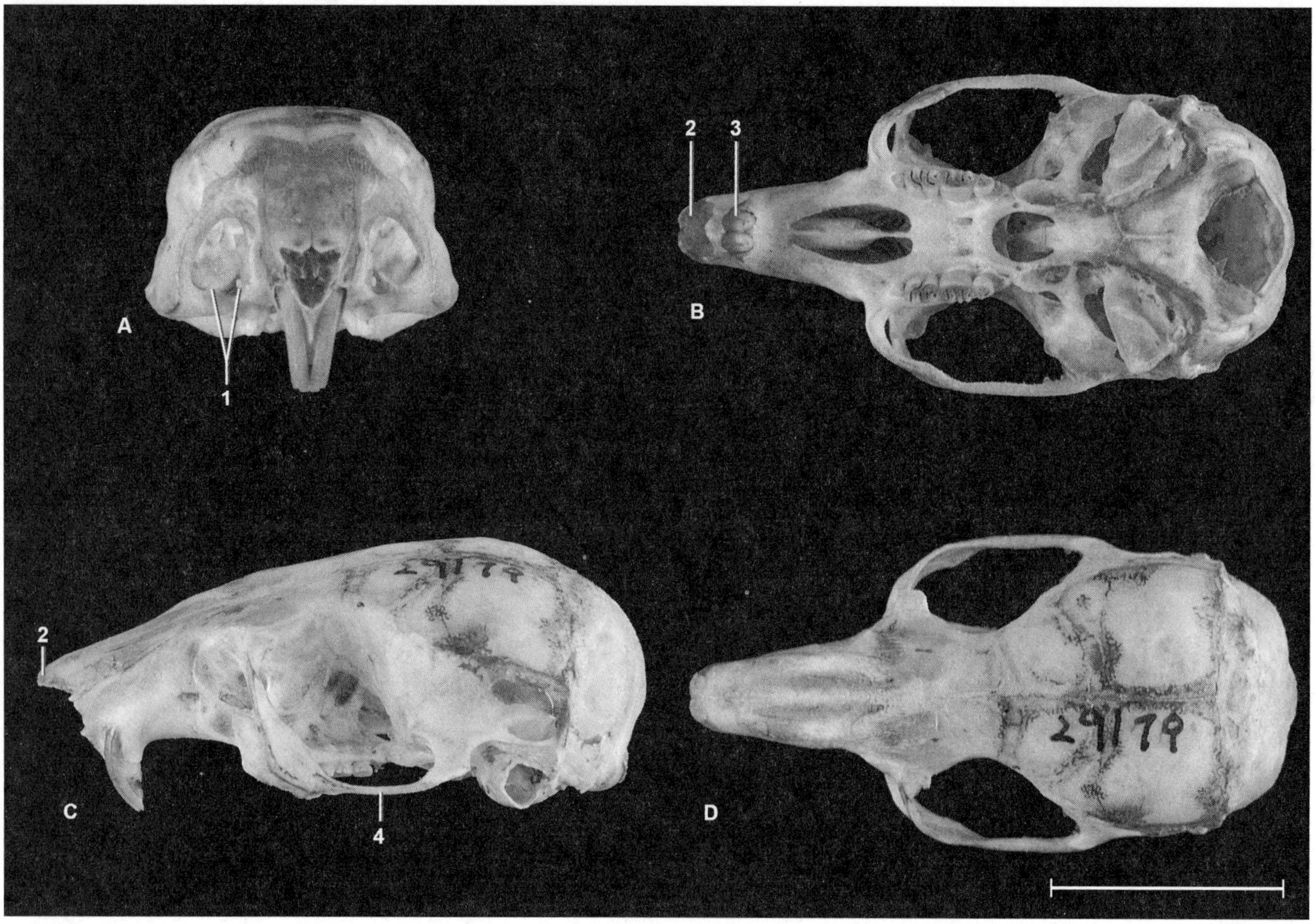

FIGURE 5.34. Cranium of Western Jumping Mouse (*Zapus princeps*), anterior (A), ventral (B), lateral (C), and dorsal (D) views. Osteological features (Dipodidae): 1, paired infraorbital foramina on each side (largest is dorsal); 2, nasals extend beyond premaxilla, incisors; 3, grooved maxillary incisors; 4, zygomatic arch extends ventrally below level of palate.

lies. They vary substantially in morphology, habitat preferences, and habits, but most use burrows and none hibernate. They share several cranial features, such as a broad zygomatic plate and narrow infraorbital foramina; they also lack the postorbital processes that are common in other rodents (e.g., Sciuridae).

Woodrats or packrats are famous for the large, aboveground nests they make from various plant materials. These nests can form large "middens" of urine-encased plant parts that accumulate over generations of use. Packrat middens can survive for tens of thousands of years in protected settings such as rock crevices, caves, or rockshelters. Analyses of the plant macrofossils recovered from these middens have provided detailed records of past vegetation change over much of the arid west. Two widespread species in the west include the lower-elevation, more xeric-oriented Desert Woodrat (*Neotoma lepida*; Figure 5.35), and the more mesic, higher-elevation, Bushy-tailed Woodrat (*Neotoma cinerea*). The occlusal surfaces of the cheekteeth are distinctive, appearing as a zipper-like set of interdigitated triangles.

Nine genera of voles (subfamily Arvicolinae) occur in western North America. These ubiquitous animals are mouse-like but have shorter tails and more gopher-like bodies. They are the most important prey for a variety of predators including Coyote (*Canis latrans*) and Barn Owl (*Tyto alba*); hence, they are routinely observed in carnivore scats and owl pellets. The Muskrat (*Ondatra zibethicus*) is a large, semi-aquatic vole with silky, brown fur and a scaly, laterally compressed tail (Figure 5.36). They use cut vegetation to build rounded houses and may also den in holes dug into the shore. Their cheekteeth are like larger versions of *Neotoma* and *Microtus*.

Deer mice (*Peromyscus*) are small mice with big eyes and large, naked ears, and internal cheek pouches. They are represented by 12 species in the west and have adapted to a wide range of habitats. They are the most common small mammal to enter mountain cabins and are the primary carriers of Hantavirus pulmonary syndrome, an infectious respiratory disease. People can contract this rare disease when they come into contact with the infected animals or inhale the aerosolized virus derived from their urine and fecal pellets. Given the diversity of cranial features of the family, we describe distinguishing features of two of the more widespread and abundant taxa in the region: *Neotoma* (Figure 5.37), representing Neotominae, and Muskrat (Figure 5.38), representing Arvicolinae.

FIGURE 5.35. Desert Woodrat (*Neotoma lepida*).

FIGURE 5.36. Common Muskrat (*Ondatra zibethicus*) Photo by Kira Broughton-Klubnikin

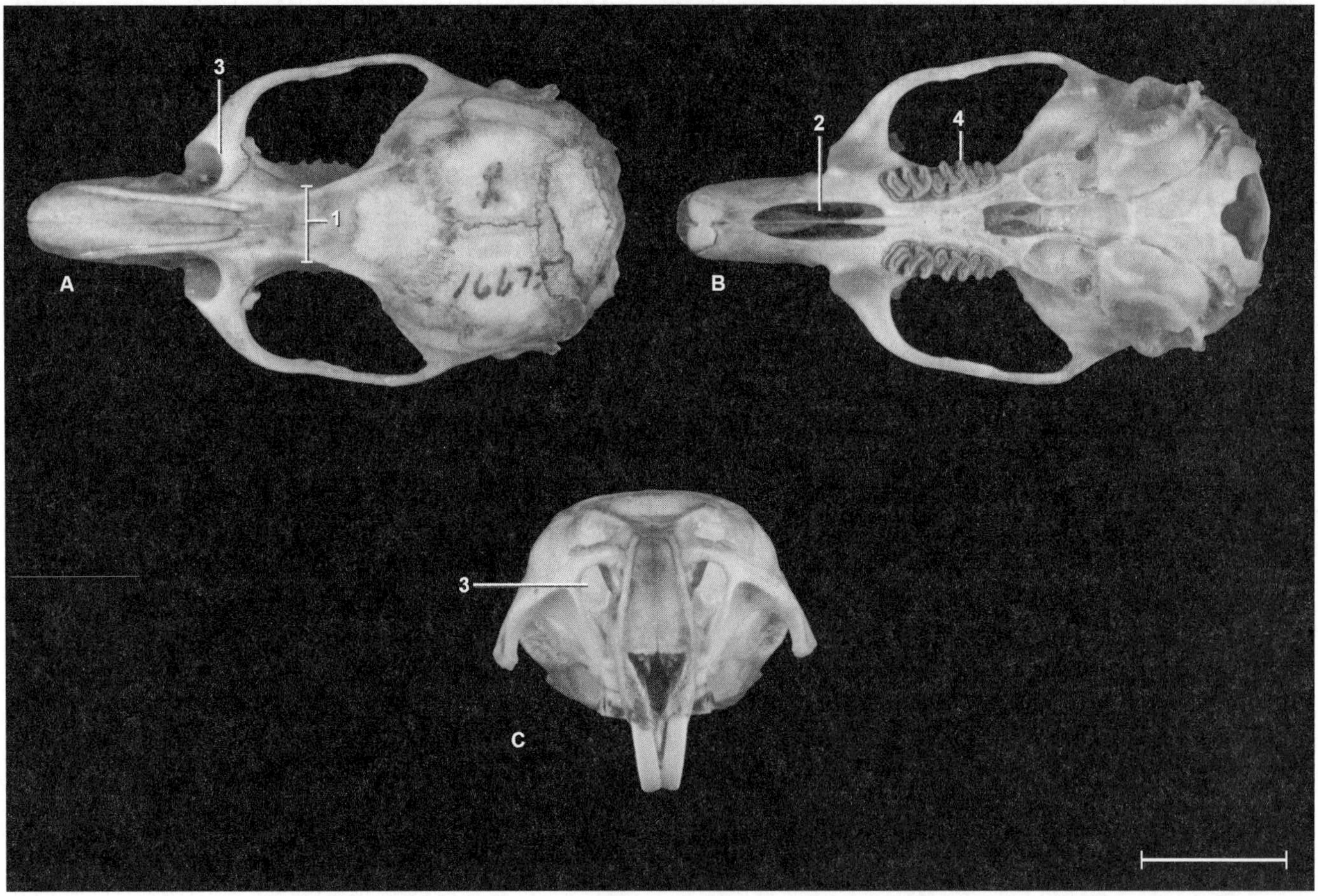

FIGURE 5.37. Cranium of Desert Woodrat (*Neotoma lepida*), dorsal (A), ventral (B), and anterior (C) views. Osteological features (*Neotoma*): 1, cranium narrow interorbitally, maxilla is visible from dorsal view; 2, incisive foramen long, closely approaching level of M^1; 3, large infraorbital foramen forms prominent gap between zygomatic and base of rostrum; 4, three, flat-crowned, hypsodont cheekteeth per quadrant, each with three (two in M_3) connected enamel loops surrounding "puddles of dentin."

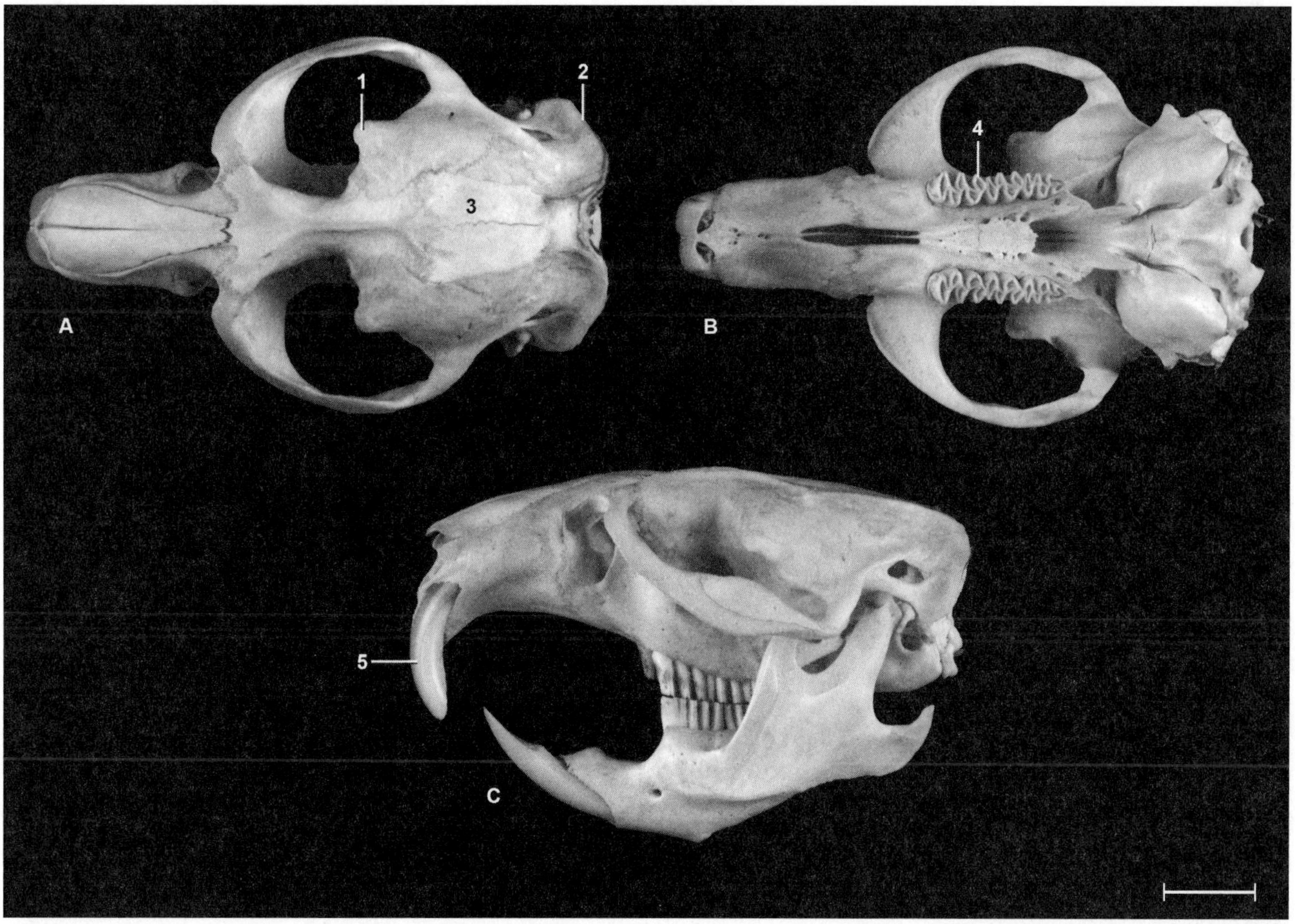

FIGURE 5.38. Cranium of Common Muskrat (*Ondatra zibethicus*), dorsal (A), ventral (B), and lateral (C) views. Osteological features (Arvicolinae): 1, temporal with strong angular projection at dorso-anterior edge; 2, nuchal crests well developed; 3, features 1 and 2 give squared appearance to braincase; 4, cheekteeth distinctive with series of alternating triangles and loops; 5, incisors extend beyond (anteriorly) nasals.

FAMILY Erethizontidae (New World Porcupines)

This South American family is represented by only a single species in North America: the North American Porcupine (*Erethizon dorsatum*; Figures 5.39–5.40). Among North American rodents, the porcupine is second only in size to the American Beaver. It is widely known for the long, sharp quills (modified hairs) that cover most of the animal's dorsal surface and are especially large and numerous on the rump and tail. These quills have barbs on the tips that cause them to work their way deep into the skin. The quills are not "thrown," but previously loosened ones can be propelled if the animal whips its tail. Porcupines are mostly nocturnal and active year round; they move slowly and are easily approached. They eat a variety of plant materials, including tree buds, nuts, grass, berries, and leaves in spring and summer

FIGURE 5.39. North American Porcupine (*Erethizon dorsatum*).

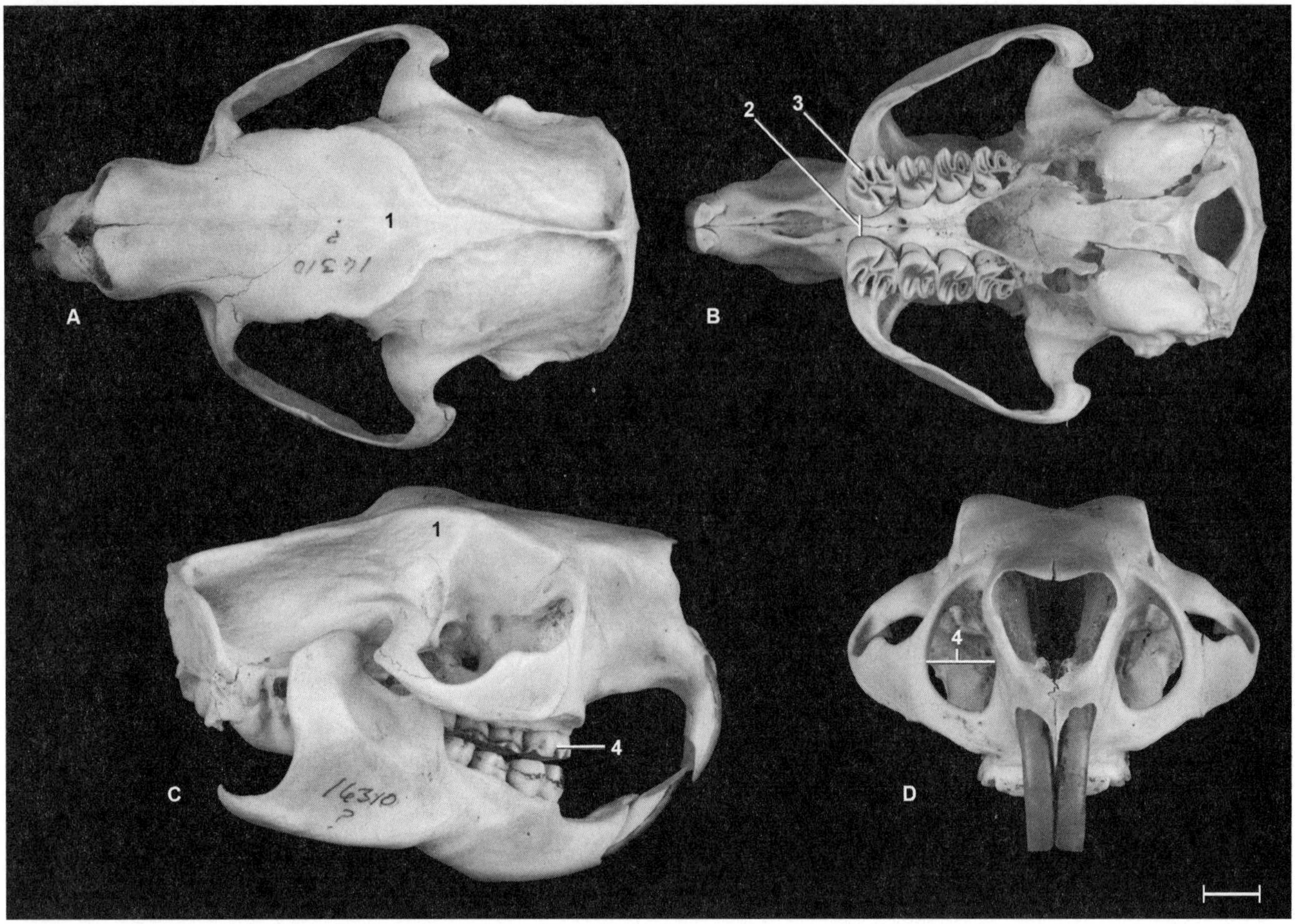

FIGURE 5.40. Cranium of North American Porcupine (*Erethizon dorsatum*), dorsal (A), ventral (B), lateral (C), and anterior (D) views. Osteological features (*Erethizon dorsatum*): 1, size of skull exceeded only by American Beaver; 2, cheekteeth rows converge anteriorly; 3, P^1 largest tooth in row; 4, extremely large infraorbital foramen (through which masseter muscle passes).

FIGURE 5.41. American Pika (*Ochotona princeps*).

and the cambium layer of conifers in winter. They are well known for their salt craving and often frequent roadsides for residue from winter salting.

ORDER Lagomorpha (Rabbits, Hares, and Pikas)

Two lagomorph families (Gr., lago = hare, morphe = form), Leporidae (rabbits and hares) and Ochotonidae (pikas), and three genera occur in the west (Figures 5.41–5.44; Color Plates 23–25). Lagomorphs are born with three incisors on each side of the premaxilla, but the outer pair is quickly shed. What remain are a large, prominent, functional anterior pair and a small, vestigial, posterior pair. Rodents have only one incisor per side in the upper jaw. Although completely herbivorous, lagomorphs are also coprophagous. They produce two types of spherical fecal pellets: dry pellets and wet pellets. The dry ones are discarded but the wet ones are promptly reingested. This allows nutrients that are broken down by bacterial action in the cecum during the first pass to be absorbed by the intestines during the second trip through.

The less familiar pikas (*Ochotona*, Figure 5.41) have very small ears, no apparent tail, and forelimbs and hindlimbs of similar size; they are similar in overall shape to guinea pigs. Signs include drying grass piles in rocks adjacent to nests in talus slopes and white urine stains on rocks. Unlike the leporids, pikas are highly vocal and live in small colonies. They communicate with a range of loud, chirping calls and nasal bleats.

Leporids have giant ears, enlarged hindlegs for running and jumping, and fluffy tails (usually white on at least one side). The two genera of leporids, *Lepus* (hares or "jackrabbits"; Figure 5.42) and *Sylvilagus* (cottontails; Figure 5.43) are distinguished by body size and relative ear size. *Sylvilagus* are smaller in size with relatively smaller

FIGURE 5.42. Black-tailed Jackrabbit (*Lepus californicus*).

FIGURE 5.43. Desert Cottontail (*Sylvilagus audubonii*).

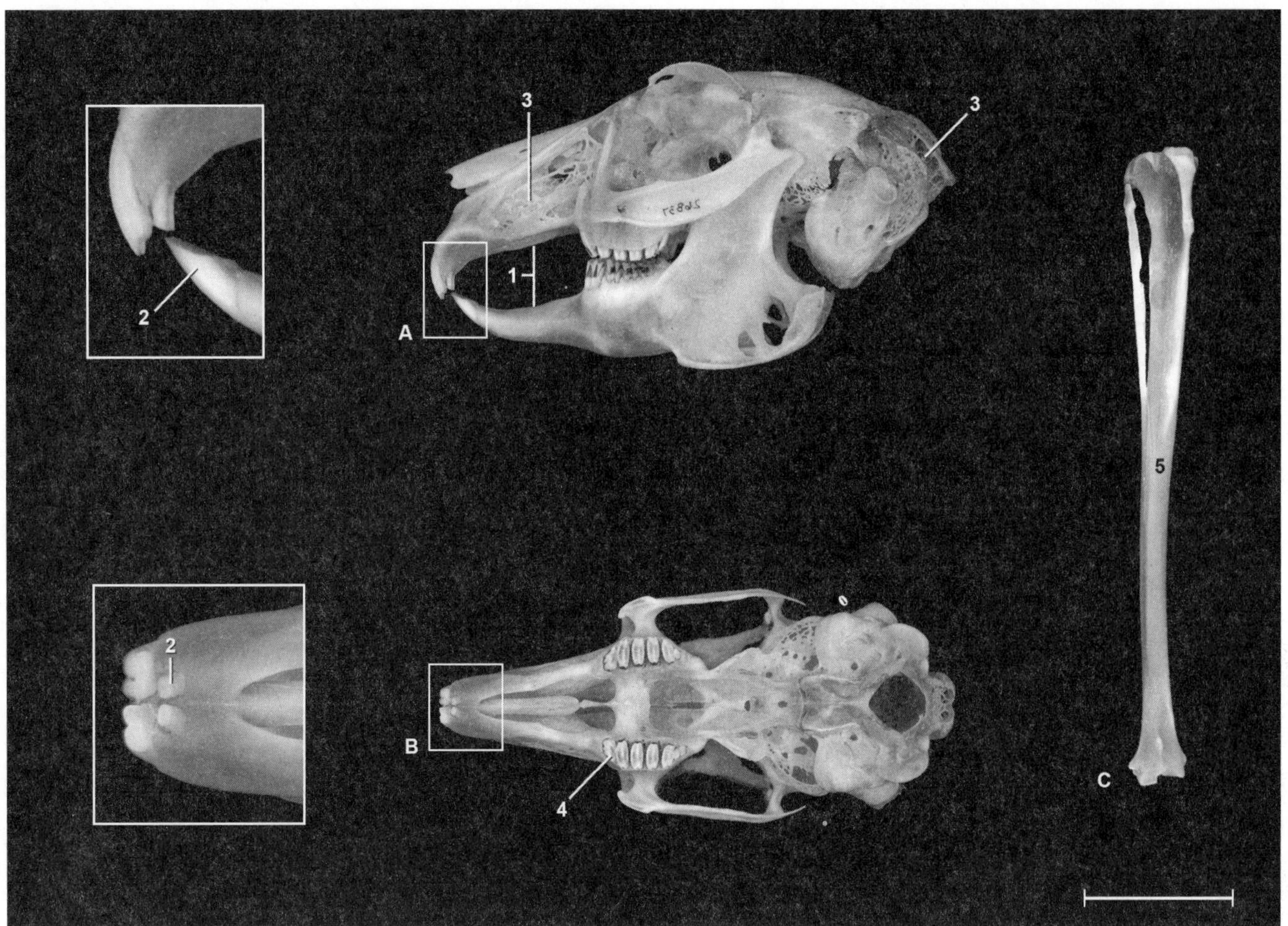

FIGURE 5.44. Black-tailed Jackrabbit (*Lepus californicus*), cranium, lateral (A) and ventral (B) views; and tibia, anterior view (C). Osteological features (Lagomorpha): 1, dentition with pronounced diastema; 2, incisors (2/1), large, white; paired in premaxilla with anterior set large, functional, and grooved (labially); posterior set small, vestigial, peg-like; 3, cranium highly fenestrated, especially on maxilla and occipital; 4, teeth evergrowing, lack roots; 5, limb elements long, thin, bird-like. Cranium of pikas (Ochotonidae) similar to leporids, but with five upper cheekteeth per side (rather than six).

ears. Hares do not make maternity nests and give birth to relatively precocial, fully furred, open-eyed young. Rabbits make maternity nests and give birth to more altricial, naked, closed-eyed bunnies.

Activity patterns for the lagomorphs vary by species (e.g., pikas = diurnal; cottontails = usually crepuscular; hares = more nocturnal). In general, *Sylvilagus* is more indicative of denser vegetative cover while *Lepus* is more common in open settings. Genus-level identifications are relatively easy for all elements; species-level identifications are possible only with cranial material (Figure 5.44) and are extremely challenging. Indigenous peoples in many settings intensively exploited lagomorphs across western North America.

ORDER Soricomorpha (Shrews and Moles)

The west is represented by two soricomorph families (L., sorico = shrew): Soricidae (shrews) and Talpidae (moles). Shrews and moles are small animals with short, dense, velvety fur; tiny eyes and ears; a small braincase relative to the elongate rostral and nasal portions of the skull; five, clawed digits on fore- and hindfeet; and sectorial cheekteeth—those that have sharp or bladelike cusps. Most are fossorial and nocturnal. Shrews and moles have long been included in the order Insectivora with hedgehogs, tenrecs, and other allies, but recent evidence suggests that the former group is paraphyletic and recent taxonomies place the shrews and moles in a separate order Soricomorpha. Reflecting the old name for the order,

FIGURE 5.45. Masked Shrew (*Sorex cinereus*).

insects and other invertebrates are the primary foods of all species.

FAMILY Soricidae (Shrews)

The family Soricidae is represented by two genera (*Sorex* and *Notiosorex*) in the west (Figures 5.45–5.46). Superficially similar to mice, shrews have long, pointed, flexible snouts and tiny eyes and ears. Shrews have a very high metabolism and are exceptionally active and nervous: heart rates can exceed 1,200 beats per minute when agitated. They must feed about every three hours, day or night, and can routinely consume twice their body weight in invertebrate prey per day. Shrews will also eat plants, fungi, and small vertebrates, alive or as carrion. Young are born naked and blind in litters of two to ten. When alarmed, young will grasp the base of a sibling's tail with their mouths, forming a long chain, with the mother leading a caravan that at first glance can resemble a snake. In many species, females deliver only a single litter during their lifetimes; after weaning, adults die so that by late summer populations are represented only by young of the year. Different shrew species can be readily identified by variation in the number, shapes, and sizes of their unicuspid (premolar) teeth.

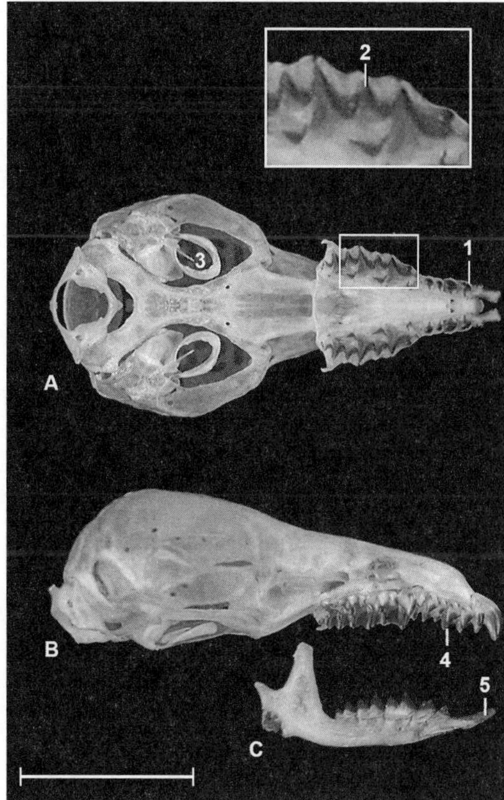

FIGURE 5.46. Dusky Shrew (*Sorex monticolus*), cranium, ventral (A) and lateral (B) views; and mandible, lateral view (C). Osteological features (Soricidae): enamel in all western species embedded with iron (reduces wear), producing wine-red pigmentation; 1, cranium lacks diastema, zygomatic arch; 2, premolars and molars with W-shaped sectorial cusp pattern; 3, incomplete auditory bullae form "tympanic rings"; 4, lateral incisors, canines, and some premolars are undifferentiated, single-cusped teeth called unicuspids; 5, central incisors well developed, projecting anteriorly in mandible, pincer-like in maxilla.

FAMILY Talpidae (Moles)

Moles, represented by two genera (*Scapanus* and *Neurotrichus*) in western North America, are usually larger than shrews, with proportionately shorter tails and broader forepaws; they also lack

FIGURE 5.47. Broad-footed Mole (*Scapanus latimanus*). Photo by Sarah Murray

pigmentation on their teeth (Figures 5.47–5.48). Moles are the most fossorial of all mammals and tunnel for insects and worms with enlarged, laterally projected forelimbs that enable them to virtually "swim" through soft sediments. Sex determination is difficult owing to the penis-like urinary papilla of females. The Shrew-mole (*Neurotrichus gibbsii*) of the Pacific Northwest is aptly named because it is shrew sized and lacks the enlarged forelimbs typical of moles but retains other cranial features of the Talpidae.

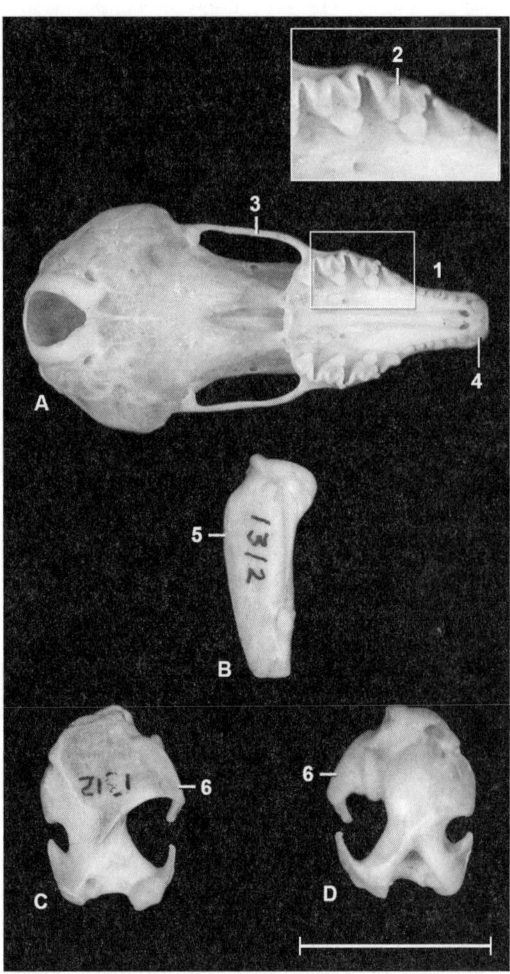

FIGURE 5.48. Broad-footed Mole (*Scapanus latimanus*) cranium, ventral view (A); sternum, lateral view (B); and r. humerus, anterior (C) and posterior (D) views. Osteological features (Talpidae): 1, cranium lacks diastema; 2, W-shaped sectorial cusp pattern on cheekteeth, as in shrews, but no red staining; 3, zygomatic arch present, unlike shrews; 4, vertically oriented mandibular incisors; 5, keeled sternum, for pectoralis muscle attachment; 6, frontlimb elements (especially the humerus) highly modified with exaggerated muscle attachments.

ORDER Chiroptera (Bats)

Represented by over 1,100 species worldwide, the Chiroptera (L., chiro [from Gr., kheiro] = hand; Gr., pteron = wing) are exceeded only by the rodents in number of described species. The west is represented by four bat families, Mormoopidae, Molossidae, Phyllostomidae, and Vespertilionidae—the latter family, the vesper bats, has 21 western regional species. Bats use caves to nest and roost and are thus commonly encountered in western cave deposits. Bats are the only flying (volant) mammals; wings are formed by a web of tough, elastic skin (the patagium), which stretches over the metacarpals, four highly elongated phalanges, and then over parts of the hindlimb and tail. A small bony structure, the calcar, projects from the hindfeet to support the wing where it attaches to the legs (Figures 5.49–5.50; Color Plate 26). A small, clawed thumb is not enclosed in the wing. All bats are nocturnal and most local species are insectivorous; some feed on fruit, nectar, and pollen.

Many bats emit supersonic (typically inaudible to human ears) echolocation calls, at a rate of 30 to 60 per second, to detect insect prey and to avoid obstacles during night flying. Their large, complex ears are an adaptation for this echolocation. In response, moths of the family Noctuidae have evolved a keen sense of hearing to detect bat echolocation sounds. Bats also emit chirps and squeaks for communication, which people can hear. Bats are not blind, and most also have good vision.

Since all local species are insectivorous, and insect prey is generally unavailable during winter, most species hibernate or are migratory. Most bats mate in the fall; females store sperm until the following spring during which time ovulation occurs. They produce a single litter per year, typically one or two offspring. Many form breeding colonies, comprised of females and young, in which the young may suckle from any female, allowing mothers to forage.

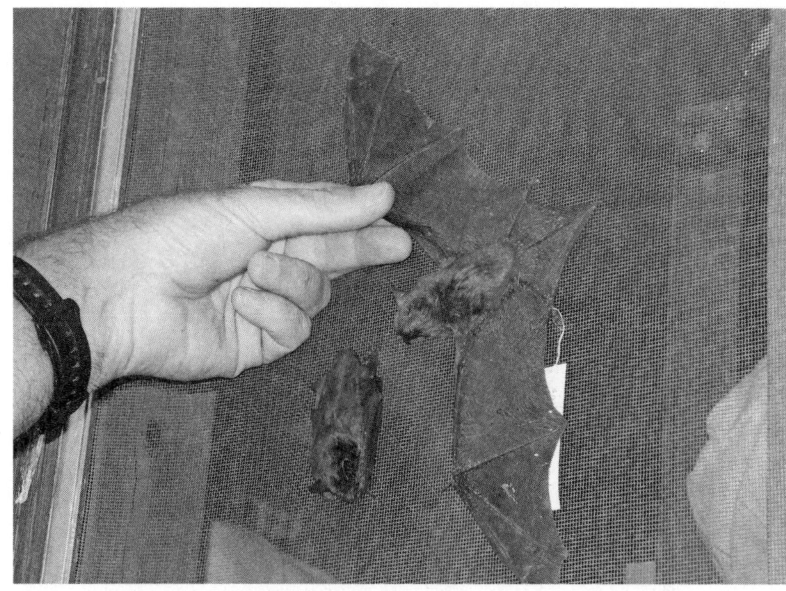

FIGURE 5.49. Big Brown Bat (*Eptesicus fuscus*). Photo by Frank E. Bayham

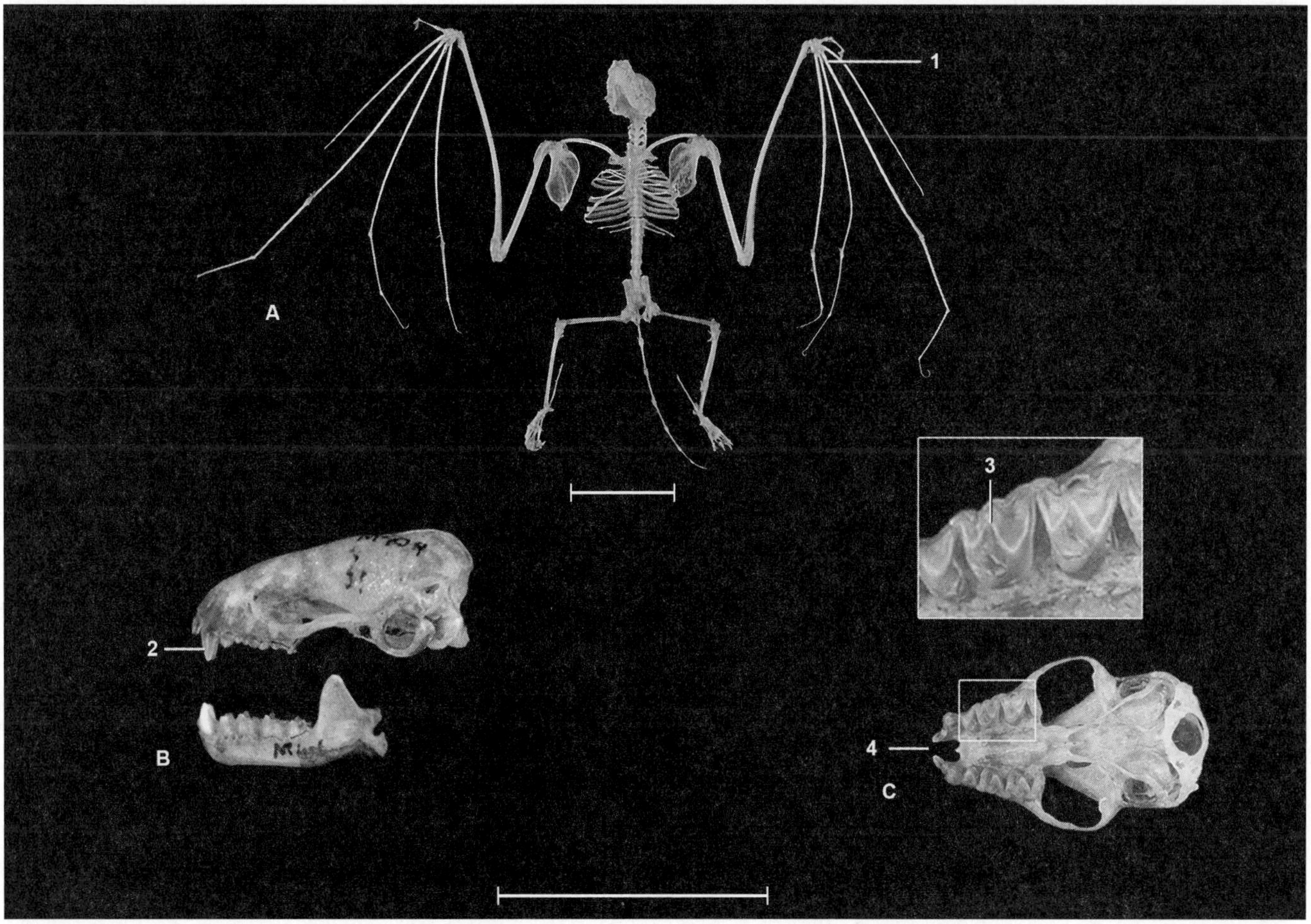

FIGURE 5.50. Complete skeleton of bat (A), and Pallid Bat (*Antrozous pallidus*) cranium, lateral (B) and ventral (C) views. Osteological features (Chiroptera): 1, forelimb elements dramatically elongated forming support for wing (patagium)—metacarpals near equal length of ulna and radius; ulna thin, vestigial; 2, canines well-developed in many species; 3, maxillary cheekteeth sectorial, W-shaped cusp pattern; 4, vesper bats (Vespertilionidae) with distinctive U-shaped gap in nasal region.

Like all mammals, bats can contract rabies. Most recent cases of human rabies in the west are related to bites from bats. If a person is bitten, collect the animal and seek immediate medical attention—rabies is readily curable if treated early. Onset of symptoms may take many weeks, after which no cures exist and death is almost inevitable. However, only about 1 in 1,000 bats are infected with rabies and bat colonies do not have outbreaks of rabies because animals that contract the disease will separate from the colony and fall to the ground. About 10% of bats collected near the ground test positive for rabies, but such animals are not aggressive. Thus, handling isolated bats found on or near the ground is inadvisable.

Different bat species utilize a wide range of habitats from deserts (e.g., *Lasiurus xanthinus*, Western Yellow Bat) to forested settings (e.g, *Lasionycteris noctivagans*, Silver-haired Bat) and can be sensitive indicators of paleoenvironmental conditions.

ORDER Carnivora (Carnivores)

Carnivores (L., carn = flesh, vor = devour) range widely in size from the massive Southern Elephant Seal (*Mirounga leonina*), which can reach over 5,000 kg, to the tiny Least Weasel (*Mustela nivalis*), which weighs little more than 28 gm. Sharing the role as the largest terrestrial mammal in North America, Brown Bears (*Ursus arctos*) and Polar Bears (*Ursus maritimus*) can tip the scales upwards of 816 kg. Exclusive of the marine pinnipeds (seals, sea lions, and walruses), which we do not treat here, six families and 32 carnivore species occur in western North America (Figures 5.51–5.64; Color Plate 27).

As the order name implies, most carnivores are adapted to catch and kill animal prey, and they have a unique dentition well suited to this task. Although all carnivores are not strict meat eaters, and many consume a wide range of vegetal foods, all have three pairs of very small incisors that are used for grooming and pulling and a single pair of large, pointed, conical canines used for piercing and holding animal prey. The cheekteeth (premolars and molars) vary in size and shape with the dietary habits of the particular species. For example, in felids (cats) and canids (dogs) the last upper premolars and first lower molars are adapted for tearing meat: as the jaw opens and closes these teeth slide past one another like the two blades on a pair of scissors and can shear animal tissues. These specialized sectorial teeth are referred to as carnassials (Figure 5.53). Other carnivores, bears for instance, have low-crowned bunodont molars, well suited for an omnivorous diet that includes substantial amounts of fibrous vegetal foods.

Most carnivores, except dogs, are solitary outside of the mating season. Only bears hibernate. The population densities of carnivores are invariably much smaller than the herbivorous mammals that occupy lower trophic levels. Given broad home ranges that crosscut different habitat types, carnivores are typically not sensitive habitat indicators. Nor were they common prey items for prehistoric peoples occupying terrestrial environments; thus they are typically uncommon in archaeological faunas compared to rodents, lagomorphs, and artiodactyls. Carnivores were also exploited for fur in both prehistoric and historic times.

FAMILY Felidae (Cats)

Cats are large or medium-sized carnivores with short faces; small ears; sharp, retractable claws; and digitigrade foot structure (Figures 5.51–5.53). Although several tropical cats reach northern Mexico and southern Arizona (e.g., Margay and Ocelot, *Leopardus*; Jaguarundi, *Puma yagouaroundi*; Jaguar, *Panthera onca*), only three species have broad ranges in the west: Bobcat (*Lynx rufus*; Figure 5.51), Cougar (*Puma concolor*; Figure 5.52), and Canadian Lynx (*Lynx canadensis*). All are secretive, nocturnal, specialized meat-eaters; Canadian Lynx and Bobcat focus on lagomorphs—Canadian Lynx is a specialist hunter of Snowshoe Hare (*Lepus americanus*); cougars hunt deer and medium-sized mammals.

FAMILY Canidae (Dogs, Foxes, and Allies)

Four genera of canids occur in western North America. All species are doglike with digitigrade foot structure (Figure 5.20); long, pointed muzzles; large, erect, triangular ears; long limbs; and long bushy tails. They have five toes on the

FIGURE 5.51. Bobcat (*Lynx rufus*). Photo by Frank E. Bayham

FIGURE 5.52. Cougar (*Puma concolor*).

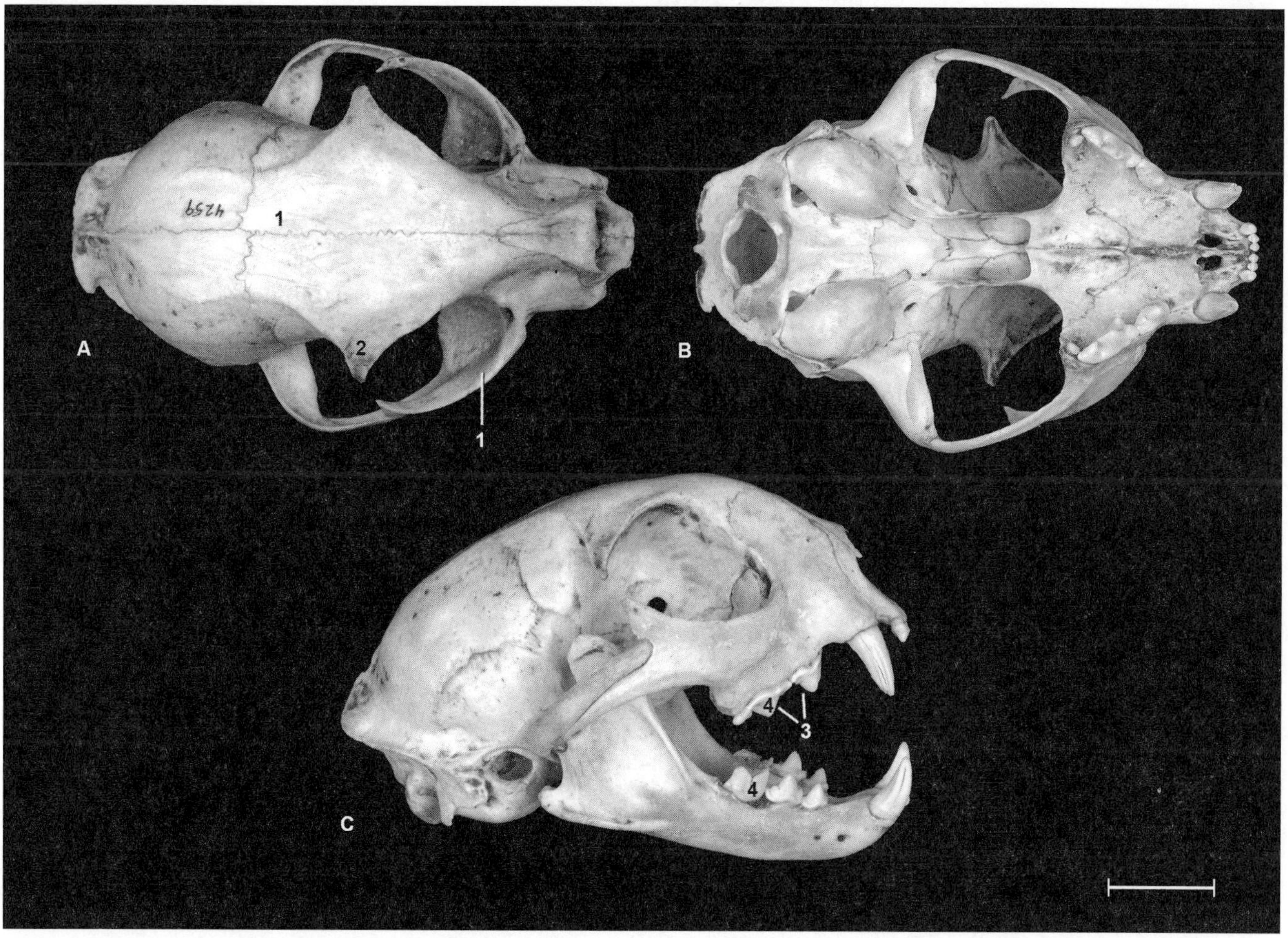

FIGURE 5.53. Cranium of Bobcat (*Lynx rufus*) dorsal (A), ventral (B), and lateral (C) views. Osteological features (Felidae): 1, cranium round and squat with flaring zygomatics and reduced rostrum; 2, well-developed postorbital processes; 3, cheekteeth reduced in number—only a single molar per quadrant (M 1/1, unique to cats); 4, cheekteeth sectorial, carnassials are P^3 and M_1.

FIGURE 5.54. Gray Wolf (*Canis lupus*).

FIGURE 5.55. Gray Fox (*Urocyon cinereoargenteus*).

front feet and four on the back feet; the fifth front toe does not bear weight and, thus, leaves no track. While canids have sharp eyes and acute hearing, their enlarged rostrum supports an exceptionally well-developed sense of smell. This they use to detect and track prey but also to "read" urine and scent posts left by other animals. Gray Wolf (*Canis lupus*; Figure 5.54), Coyote (*Canis latrans*), and Domestic Dog (*Canis lupus familiaris*) can all interbreed. They live in packs and are predominantly nocturnal, although they may be active any time. Foxes (*Vulpes* and *Urocyon*) are also mostly nocturnal but are more solitary hunters of small mammals (or work in pairs); Gray Fox (*Urocyon cinereoargenteus*) is more omnivorous and climbs trees (Figures 5.55–5.56).

Genetic data suggest that the ancestors of domestic dogs split from Gray Wolves as far back as 100,000 years ago in the Old World. However, it is unclear whether humans were involved in that initial divergence: archaeological evidence for a close connection between people and dogs (i.e., dog burials) dates much later to about 14,000 years ago. In North America, dog skeletal material yields dates that rival the age of the oldest human skeletal remains from the region (e.g., 11,000 years ago; Danger Cave, Utah), suggesting that dogs may have tagged along with people on the earliest human migrations into the New World.

FAMILY Ursidae (Bears)

Bears are large, stocky, plantigrade (Figure 5.20) carnivores with small, inconspicuous tails and large, blunt, nonretractable claws (Figure 5.57). Like canids, they have a keen sense of smell. As noted, bears are among the more omnivorous

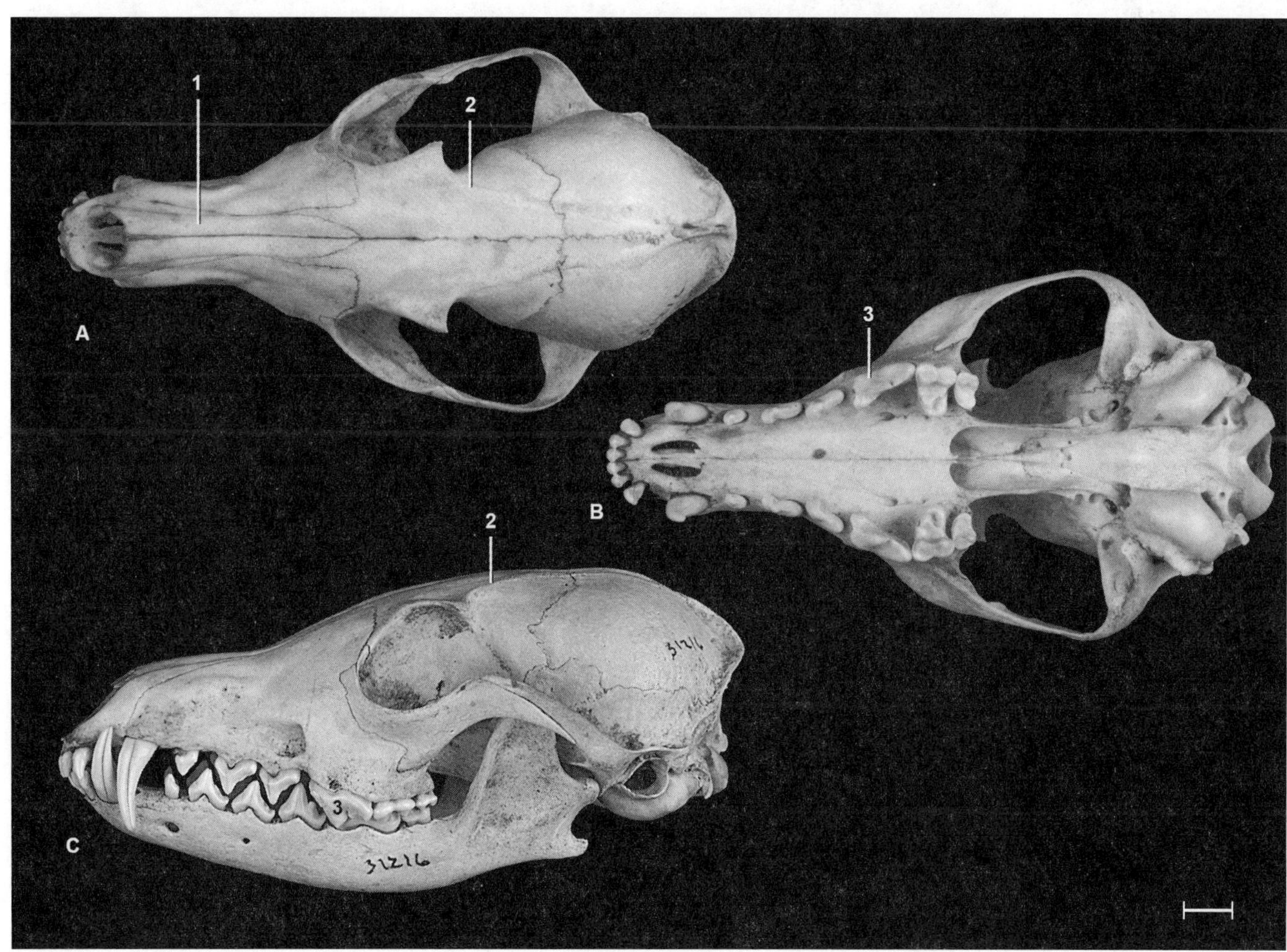

FIGURE 5.56. Cranium of Red Fox (*Vulpes vulpes*), dorsal (A), ventral (B), and lateral (C) views. Osteological features (Canidae): 1, rostrum elongated, pointed; 2, well developed temporal lines (ridges) and/or sagittal crests; 3, well developed sectorial carnassials (P^4 and M_1).

FIGURE 5.57. Brown Bear (*Ursus arctos*). Photo by USFWS/David Menke

carnivore species and have large, impressive canines but broad flat-crowned (bunodont) molars for crushing vegetal food material (Figure 5.58). Bears occupy dens in winter and may hibernate for over six months. Mating takes place in the spring or summer but implantation of the embryos is delayed until the fall. Litters of one to three rat-sized cubs are born in winter in the den; newborn cubs nurse while the mother remains semitorpid. Females produce litters every other year in the Black Bear (*Ursus americanus*) and every two to four years in the Brown Bear (*Ursus arctos*) and Polar Bear (*Ursus maritimus*). Of note in forensic contexts, certain bear elements, especially feet, are superficially similar to those of humans—especially if the terminal phalanges are removed as is often the case when animals are skinned.

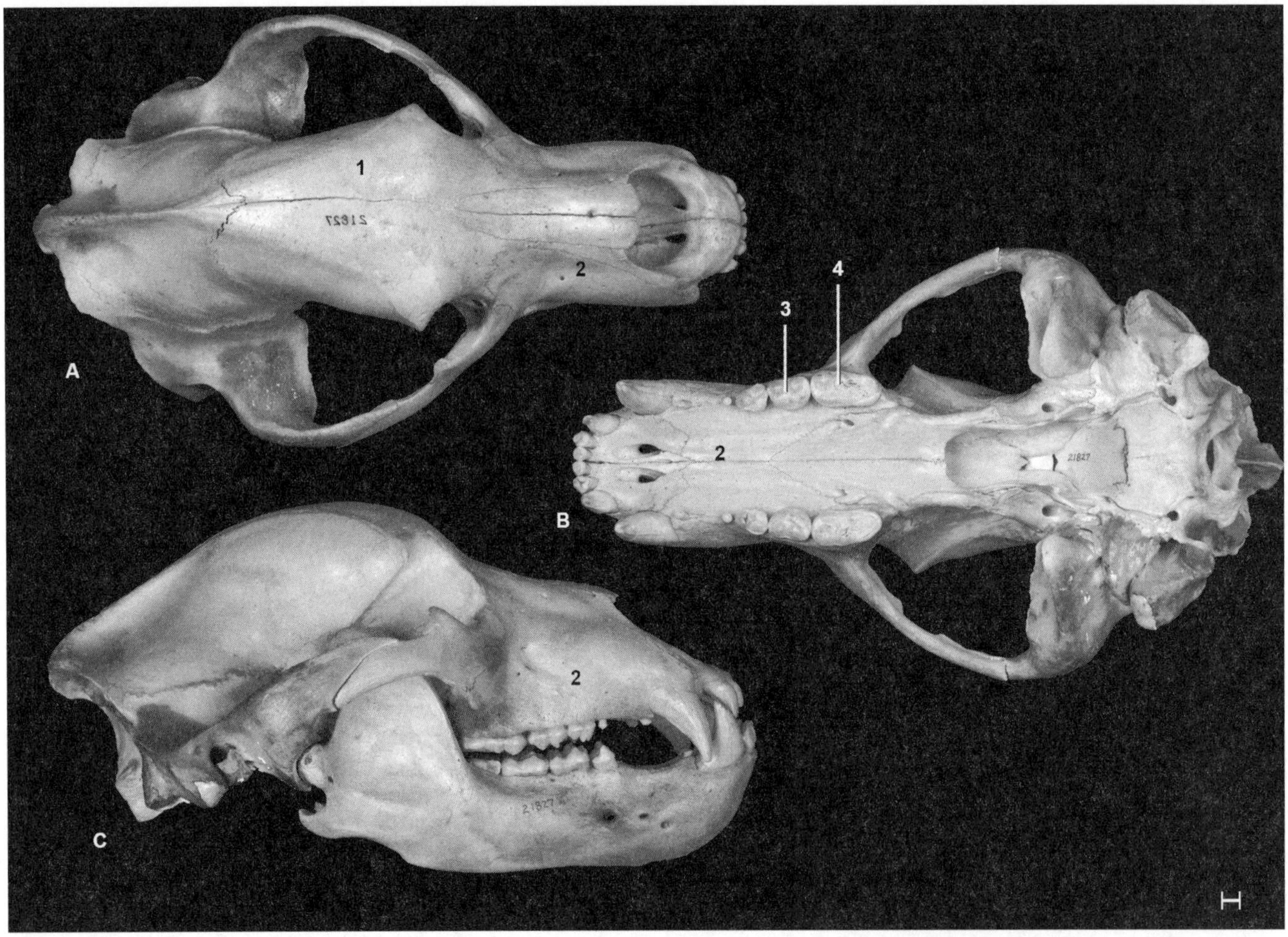

FIGURE 5.58. Cranium of Brown Bear (*Ursus arctos*), dorsal (A), ventral (B), and lateral (C) views. Osteological features (Ursidae): 1, large size of cranium is distinctive; 2, rostrum elongate, but proportionately shorter than canids; 3, molars bunodont, broad and flat; 4, last upper molar (M^2) is largest cheektooth.

FAMILY Mustelidae
(Weasels, Otters, Badgers, and Allies)

Represented by 56 species worldwide, Mustelidae is the largest carnivore family. Although mustelids come in a diversity of shapes and sizes—from the large, lumbering American Badger (*Taxidea taxus*; Figure 5.59) to the nimble, arboreal American Marten (*Martes americana*)—most are relatively small carnivores with long, low-hung bodies, short legs, rounded snouts, short rounded ears, and thick silky coats. Four genera and nine species occur in the west (Figure 5.60). All are solitary, efficient predators, active throughout the year; most are nocturnal. Mustelids have paired anal scent glands from which they emit strong scents to mark territories and for use in social communication.

FIGURE 5.59. American Badger (*Taxidea taxus*). Photo by Nicci Barger

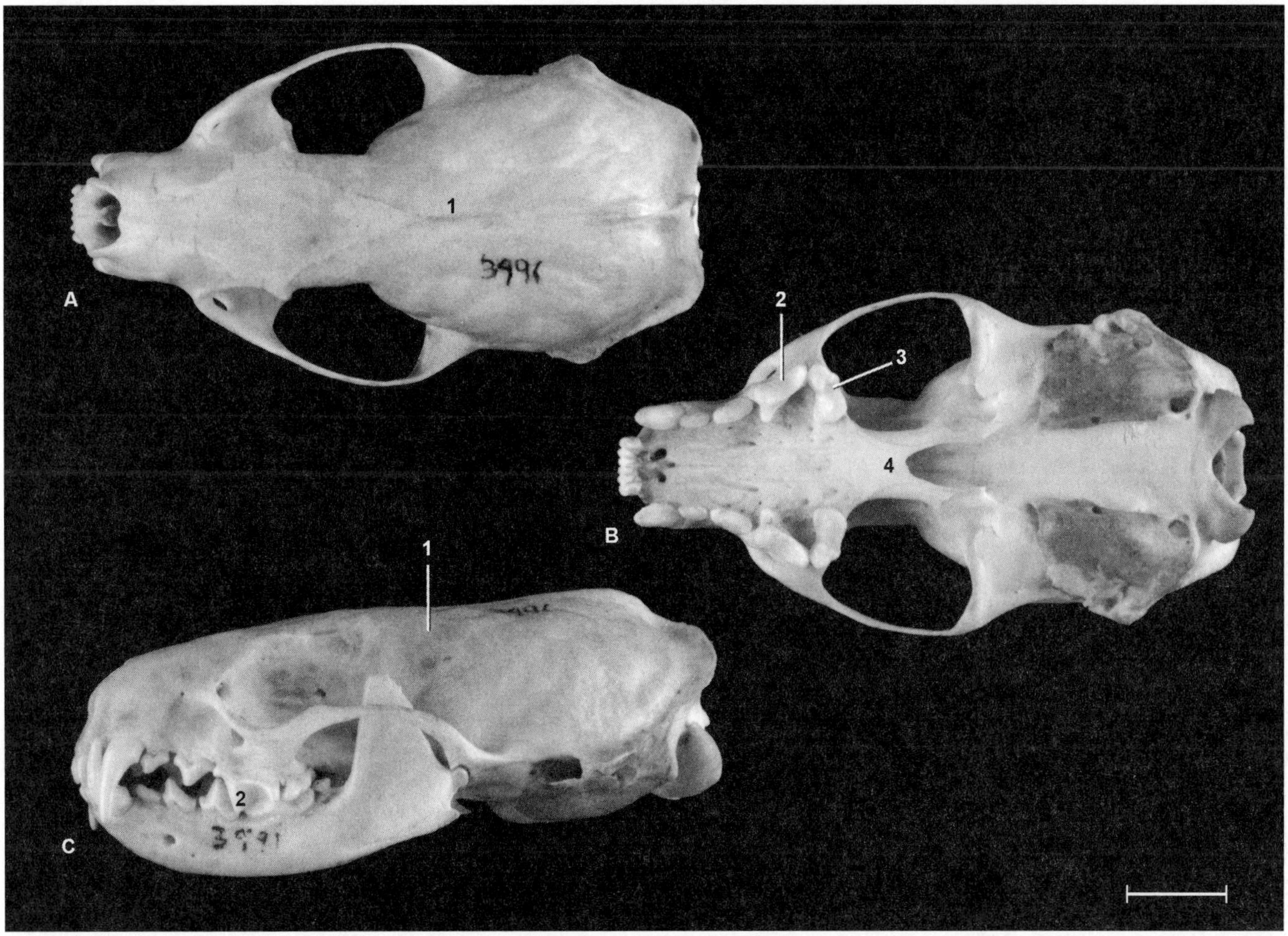

FIGURE 5.60. Cranium of American Mink (*Neovison vison*), dorsal (A), ventral (B), and lateral (C) views. Osteological features (Mustelidae): 1, cranium dorso-ventrally flattened; 2, well developed carnassials (P^3 or P^4 and M_1); 3, single, upper molars wider than long, project medially; 4, bony palate (palatines) extends posteriorly, well past cheekteeth.

FIGURE 5.61. Striped Skunk (*Mephitis mephitis*).

FAMILY Mephitidae (Skunks and Stink Badgers)

Skunks have long been considered members of the family Mustelidae, but recent molecular evidence indicates a more distant relationship. They are now placed in their own family with the stink badgers (*Mydaus*) of Indonesia and the Philippines. Skunks are among the most familiar—and infamous—mammals, with the bold, contrasting black-and-white color and paired anal scent glands from which they spray a powerful musk to deter predators. In Striped Skunks (*Mephitis mephitis*), the spray can be discharged accurately up to about 4.6 m and can cause temporary blindness if it hits the eyes; most animals hold enough scent to deliver five to six shots. It can take up to 10 days to recharge empty scent glands. Thankfully, fair warnings are usually given before they let loose. These include a variety of amusing antics, including handstands, front feet stomping, and hissing. (Although *mephitis* is Latin for "stench,"

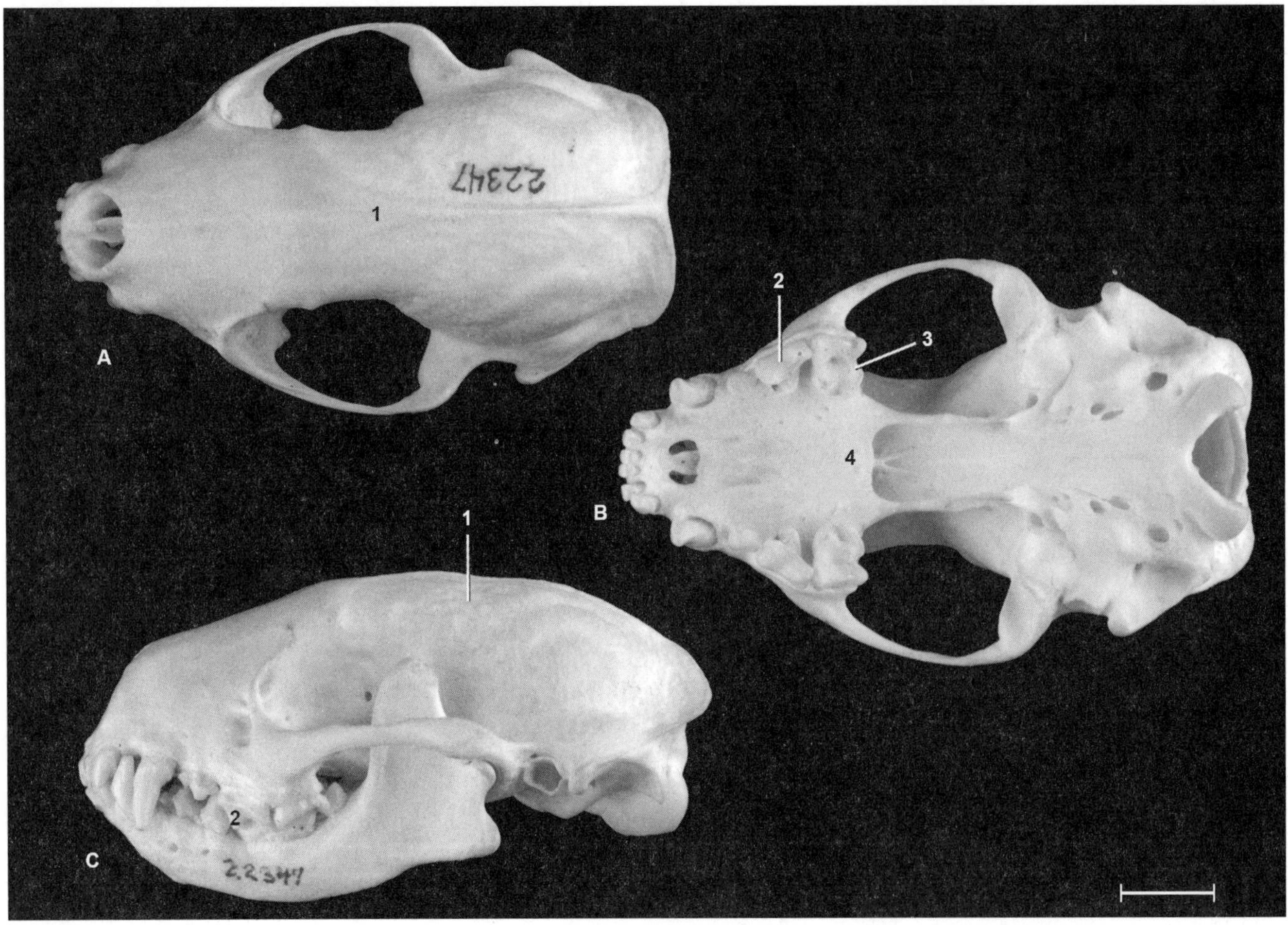

FIGURE 5.62. Cranium of Striped Skunk (*Mephitis mephitis*), dorsal (A), ventral (B), and lateral (C) views. Osteological features (Mephitidae): as in mustelids: 1, cranium dorso-ventrally flattened; 2, well developed carnassials (P^3 or P^4 and M_1); 3, single, upper molars wider than long, project medially; unlike mustelids: 4, bony palate (palatines) extends minimally past cheekteeth.

many find the smell of skunk pleasant in small doses.) With such an effective defense mechanism, skunks have few predators, although Great Horned Owls (*Bubo virginianus*) will take them. Skunks are also very tolerant of people and can be readily approached as they forage in the early evening. Four species of skunk are native to the west: Striped Skunk (Figures 5.61–5.62), Hooded Skunk (*Mephitis macroura*), Western Spotted Skunk (*Spilogale gracilis*), and Western Hog-nosed Skunk (*Conepatus mesoleucus*). They are all plantigrade, crepuscular and nocturnal omnivores, and are active throughout the year.

FAMILY Procyonidae
(Raccoons, Ringtail, Coati, and Allies)

This family is restricted to the New World; most of the 14 species are tropical. Three species occur in western North America: Raccoon (*Procyon lotor*; Figures 5.63–5.64; Color Plate 27), Ringtail

FIGURE 5.63. Raccoon (*Procyon lotor*). Photo by USFWS/Bill Buchanan

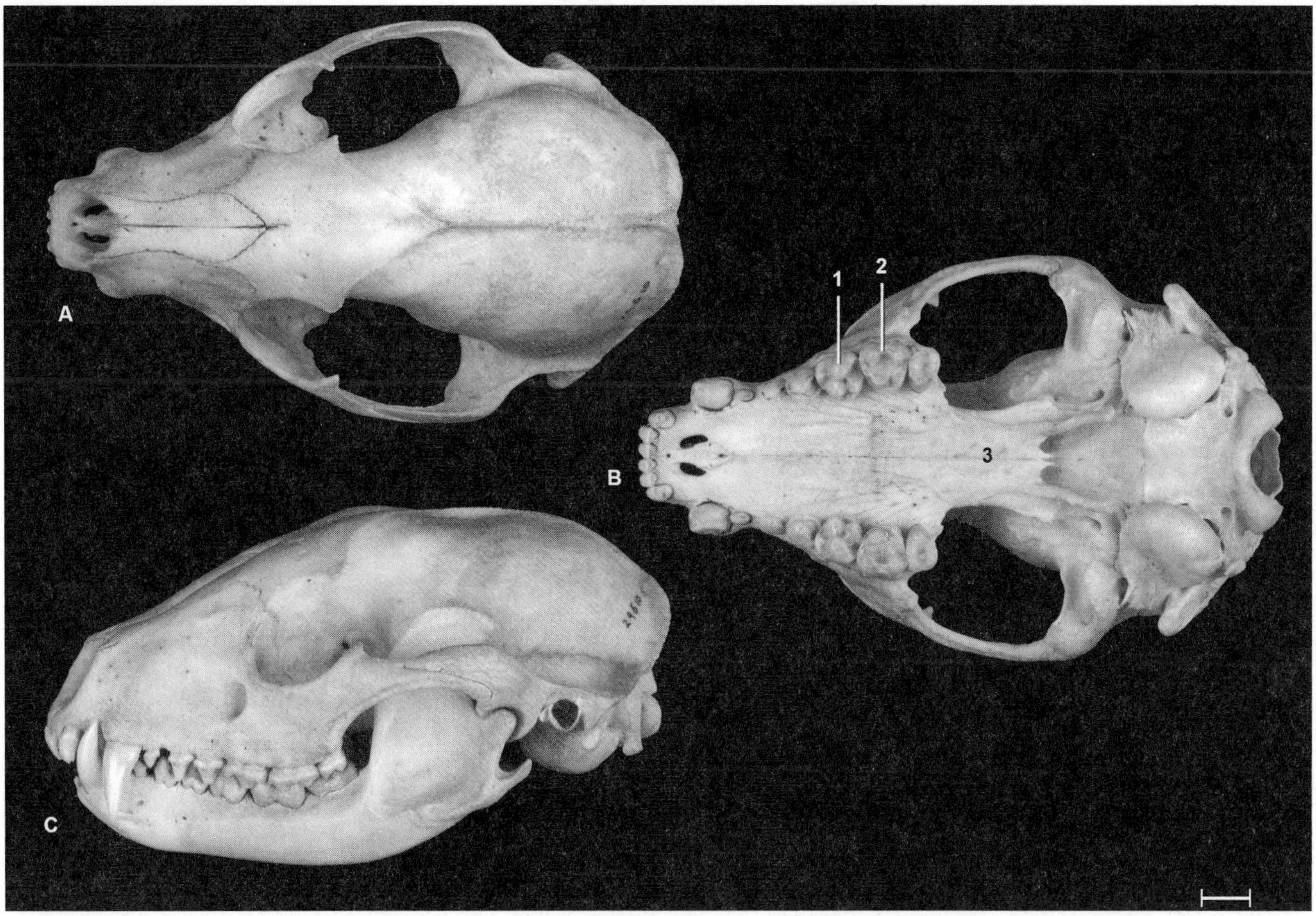

FIGURE 5.64. Cranium of Raccoon (*Procyon lotor*), dorsal (A), ventral (B), and lateral (C) views. Osteological features (Procyonidae): 1, six upper cheekteeth as in canids; *Procyon* and *Nasua* distinguished from canids by: 2, lack of carnassials and broad, flattened molars; 3, bony palate (palatines) extending posteriorly past cheekteeth. Cranium of *Bassariscus* (not pictured) distinguished from canids by smaller size (< 85 mm total length) and short rostrum.

(*Bassariscus astutus*), and White-nosed Coati (*Nasua narica*). The range of the coati is restricted to the extreme southwest; the other two are widely distributed across the west. Procyonids are characterized by long, bushy, and ringed tails (indistinct in the coati); bilobed prominent baculae (Figure 5.16); and five, clawed toes on each foot. Coati and Raccoon are plantigrade; the Ringtail is digitigrade. The familiar Raccoon is an omnivore and semiaquatic forager; it is common in every habitat with water sources. The secretive Ringtail is the size of a small cat, with a distinctive, raccoon-like, ringed tail. Both the Ringtail and Raccoon are nocturnal; the coati is diurnal.

ORDER Artiodactyla (Even-toed Ungulates)

The worldwide order of Artiodactyla (Gr., artio = even, dactylo = finger) includes deer, sheep, goats, bison, pronghorns, pigs, and relatives. Four families occur in the west: Tayassuidae (peccaries), Cervidae (deer), Bovidae (sheep, goats, bison), and Antilocapridae (pronghorns). Surprisingly, artiodactyls are closely related to the Cetacea (whales, porpoises, and dolphins), based on molecular evidence and certain striking osteological similarities, especially in the astragalus. Hence, leaving out the cetaceans from Artiodactyla makes the latter taxon paraphyletic and many authorities suggest a new order, Cetartiodactyla, which would include all artiodactyls and cetaceans.

Most western artiodactyls are large, cursorial mammals that have long, slender legs and elongated and raised foot elements with two functional toes per limb; thus, they are unguligrade and walk and run on their "tip-toes." Only two functional metatarsals and metacarpals (collectively, "metapodials") are retained per limb in most artiodactyls but each of the pairs fuse early in development to form a single element, sometimes referred to as a cannon bone (Figure 5.65). Metatarsals have a linear groove on the anterior shaft—most pronounced in cervids and pronghorn—and a more completely circular proximal end. Metacarpals lack the groove and have more semicircular or D-shaped proximal ends. Males, and in some species both sexes, have prominent horns or antlers.

Except for the omnivorous Collared Peccary (*Pecari tajacu*; Figure 5.66), all are herbivorous and exhibit a high-crowned selenodont dentition (crescent-shaped cusps; Figure 1.8) that lacks incisors in the upper jaw. Most have multichambered stomachs that allow hastily ingested food to be temporarily stored in the largest chamber, the rumen. Food then passes to a second chamber where it is shaped into pellets before being returned to the mouth to be more thoroughly chewed as the animal rests in a more protected setting.

As a group, artiodactyls were the most highly ranked and economically important animals to prehistoric peoples in the west. In many cases, artiodactyl remains dominate archaeological faunas;

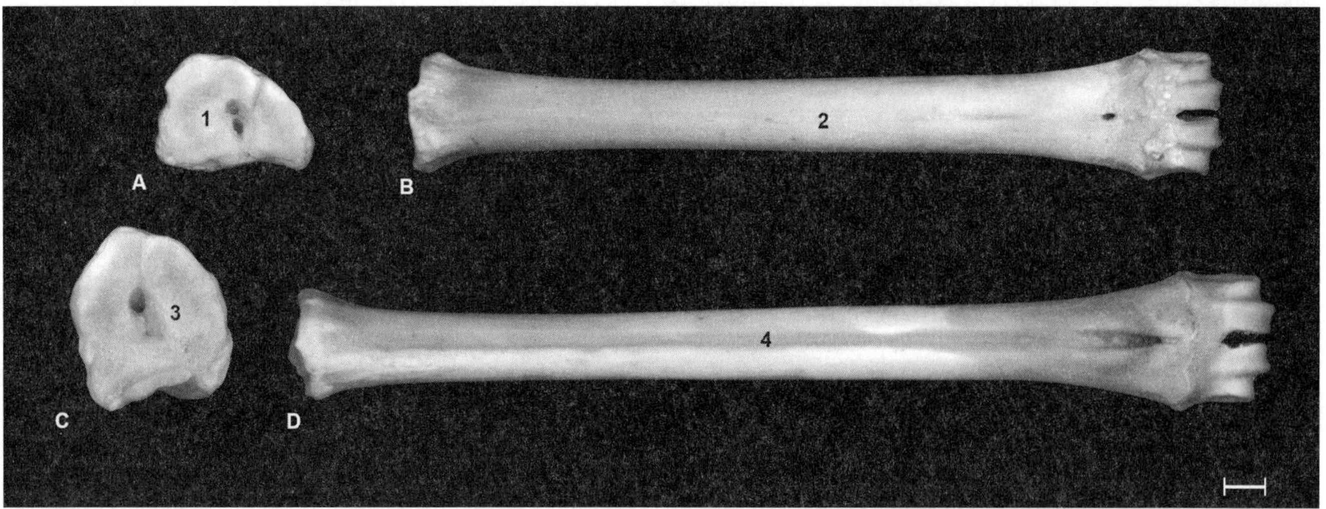

FIGURE 5.65. Mule Deer (*Odocoileus hemionus*) right metacarpal, proximal (A) and anterior (B) views, and right metatarsal, proximal (C) and anterior (D) views. Osteological features: metacarpal with, 1, semi-circular proximal articulating surface; 2, anterior shaft without groove; metatarsal with, 3, more circular proximal articulating surface; 4, anterior shaft with prominent groove.

their bones were also extensively utilized as a raw material for tools. Cases of prehistoric anthropogenic depressions have also been documented for a variety of artiodactyl taxa in many settings. Population densities also varied substantially across the Holocene due to changes in moisture history and climatic seasonality. Some artiodactyls (e.g., American Bison [*Bison bison*]; Bighorn Sheep [*Ovis canadensis*]) experienced body-size diminution during the late Pleistocene and Holocene.

FAMILY Tayassuidae (Peccaries)

Only one species, the neotropical Collared Peccary, or Javelina, occurs in the west, and its range is restricted by cold to southern Arizona and southwest Mexico (Figures 5.66–5.67). These pigs are members of the suborder Suimorpha along with the Old World pigs (Suidae). Active day (in winter) or night (in summer), Collared Peccaries are highly social and travel in herds of up to 50

FIGURE 5.66. Collared Peccary (*Pecari tajacu*), below, Green Jay (*Cyanocorax yncas*), above. Photo by USFWS/Steve Hillebrand

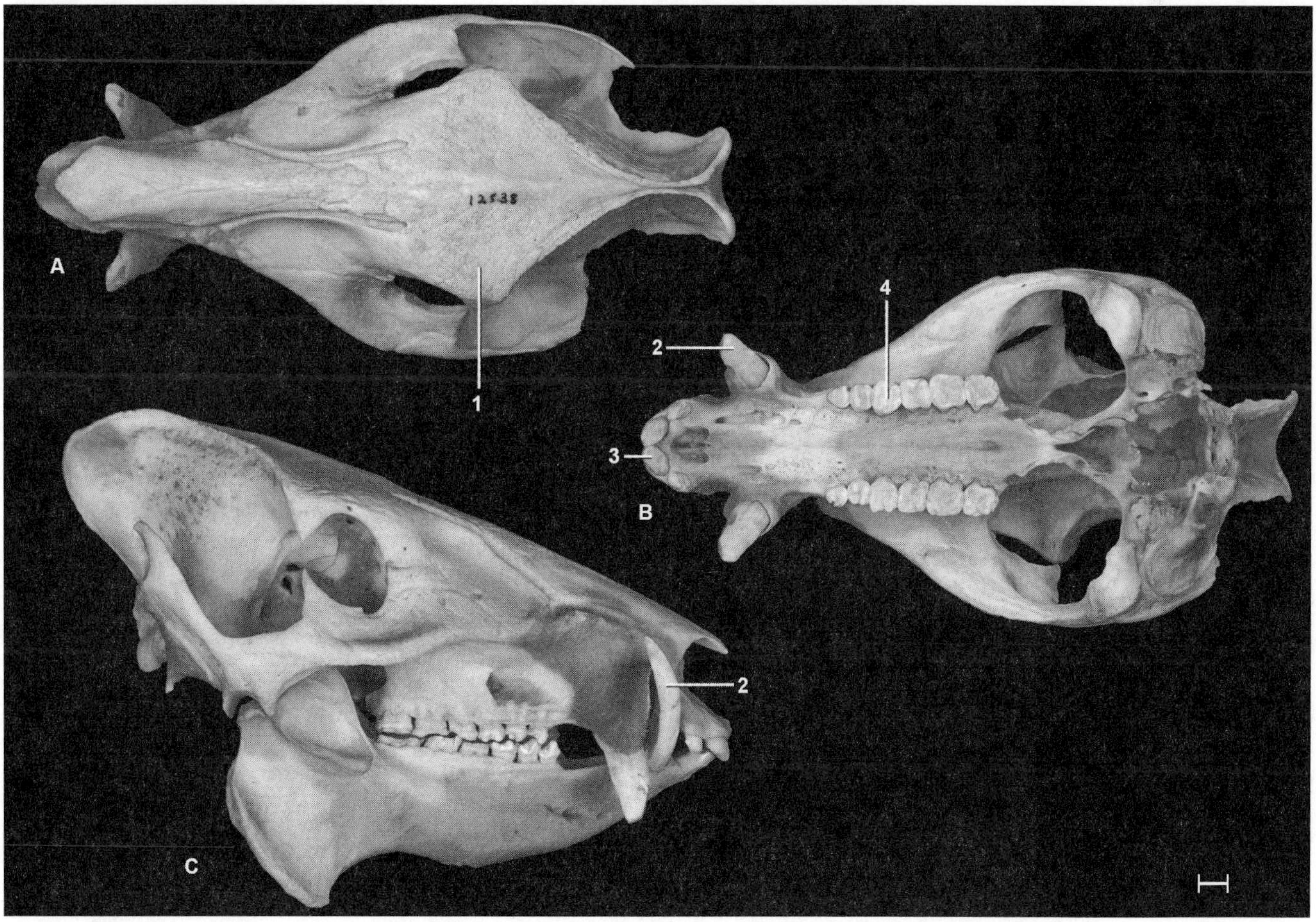

FIGURE 5.67. Cranium of Collared Peccary (*Pecari tajacu*), dorsal (A), ventral (B), and lateral (C) views. Osteological features (Tayassuidae): 1, no appendages (cranial ornamentation) on frontal; 2, well developed, tusk-like canines in mandible, maxilla; 3), upper incisors present; 4) cheekteeth bunodont.

FIGURE 5.68. Mule Deer (*Odocoileus hemionus*).

animals. They forage on cacti, mesquite beans, fruits, and nuts. They have four functional toes on the front feet and three on the back feet. Peccaries have scent glands on the face and rump that emit a pungent musk used to mark territories—herds can thus be smelled before they are seen and have the vernacular name of "skunk pig."

FAMILY Cervidae (Deer and Allies)

Four genera and five species occur in the west: Elk (*Cervus elaphus*), Moose (*Alces americanus*), Mule Deer (*Odocoileus hemionus*; Figures 5.68–5.69; Color Plate 28), White-tailed Deer (*Odocoileus virginianus*), and Caribou (*Rangifer tarandus*). Deer are medium to large ungulates with two functional toes and two vestigial dewclaws that do not bear weight. Most possess dermal glands (suborbital, tarsal, metatarsal, and interdigital)

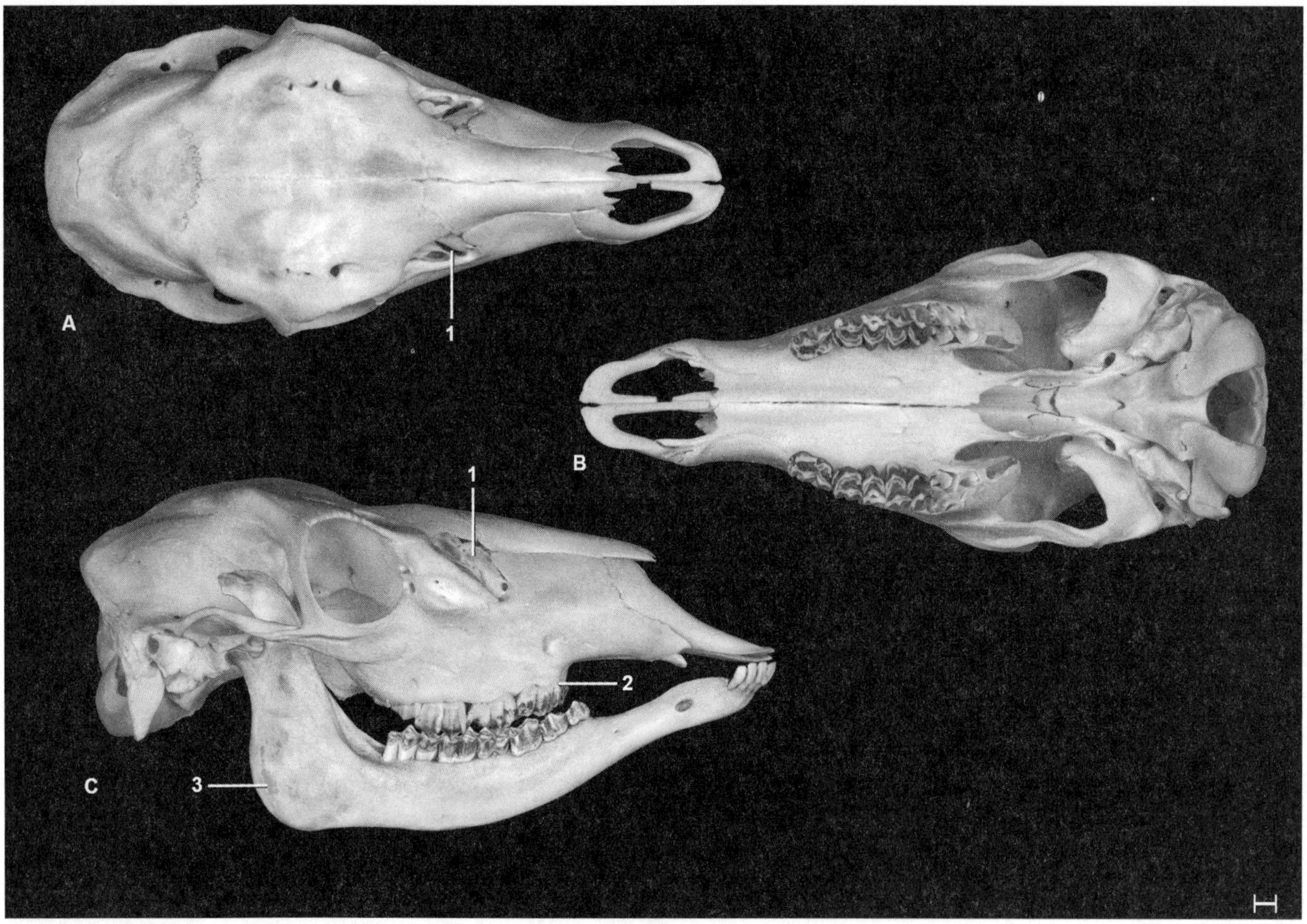

FIGURE 5.69. Cranium of Elk (*Cervus elaphus*), dorsal (A), ventral (B), and lateral (C) views. Osteological features (Cervidae): seasonal antlers in males of all species (both sexes in Caribou [*Rangifer tarandus*]) arise from cranium well posterior to orbit (not shown, female specimen); 1, bone loss between lacrimal and nasal—the two bones do not contact; 2, cheekteeth with distinct separation between enameled crown and root; 3, angular process of mandible with rounded posterior extension.

that produce scent. All species are herbivorous; most are nocturnal or crepuscular. They occupy a wide range of habitat types, from grasslands to deserts to arctic tundra. Most are monestrous (reproducing once per year), breed in the fall, and give birth in the spring; moose cows breed once every other year. Most young are spotted and cryptic.

Males of all species, and both sexes in Caribou, have antlers, bony outgrowths of the frontal bone that are shed and regrown annually. Antlers begin to grow in spring or early summer and are first relatively soft and covered with fine hair or "velvet." As fall approaches, the velvet dries up and is rubbed off, leaving the bare antlers to serve as sexual ornaments that are used in combat with rival males. After the fall breeding season, or rut, the antlers are shed. A young buck's first set usually consists of only a single point or tine. As an animal ages, and if diet is adequate, the antlers increase in size and develop many branches and tines.

FAMILY Antilocapridae (Pronghorns)

Antilocapridae is an endemic North American family with only a single surviving species: Pronghorn (*Antilocapra americana*; Figures 5.70–5.71; Color Plate 29). More closely related to giraffes than bovids, Antilocapridae was a diverse and successful family from the Miocene through the Pleistocene. Pronghorns possess horns that resemble those of bovids, but, uniquely, the hair-derived sheaths are shed each year while the bony core is retained. Also unlike bovids, male horns are branched with a forward-pointing "prong"; female horns are smaller and usually lack prongs and may not regrow shed sheaths. The three genera of Pronghorns (*Capromeryx*, *Tetrameryx*, and *Stockoceros*) that went extinct as recently as the late Pleistocene each had four-pronged horns, two pointing forward and two pointing backward. The smallest of these, the Diminutive Pronghorn (*Capromeryx minor*), stood only .5 m at the shoulder.

FIGURE 5.70. Pronghorn (*Antilocapra americana*).

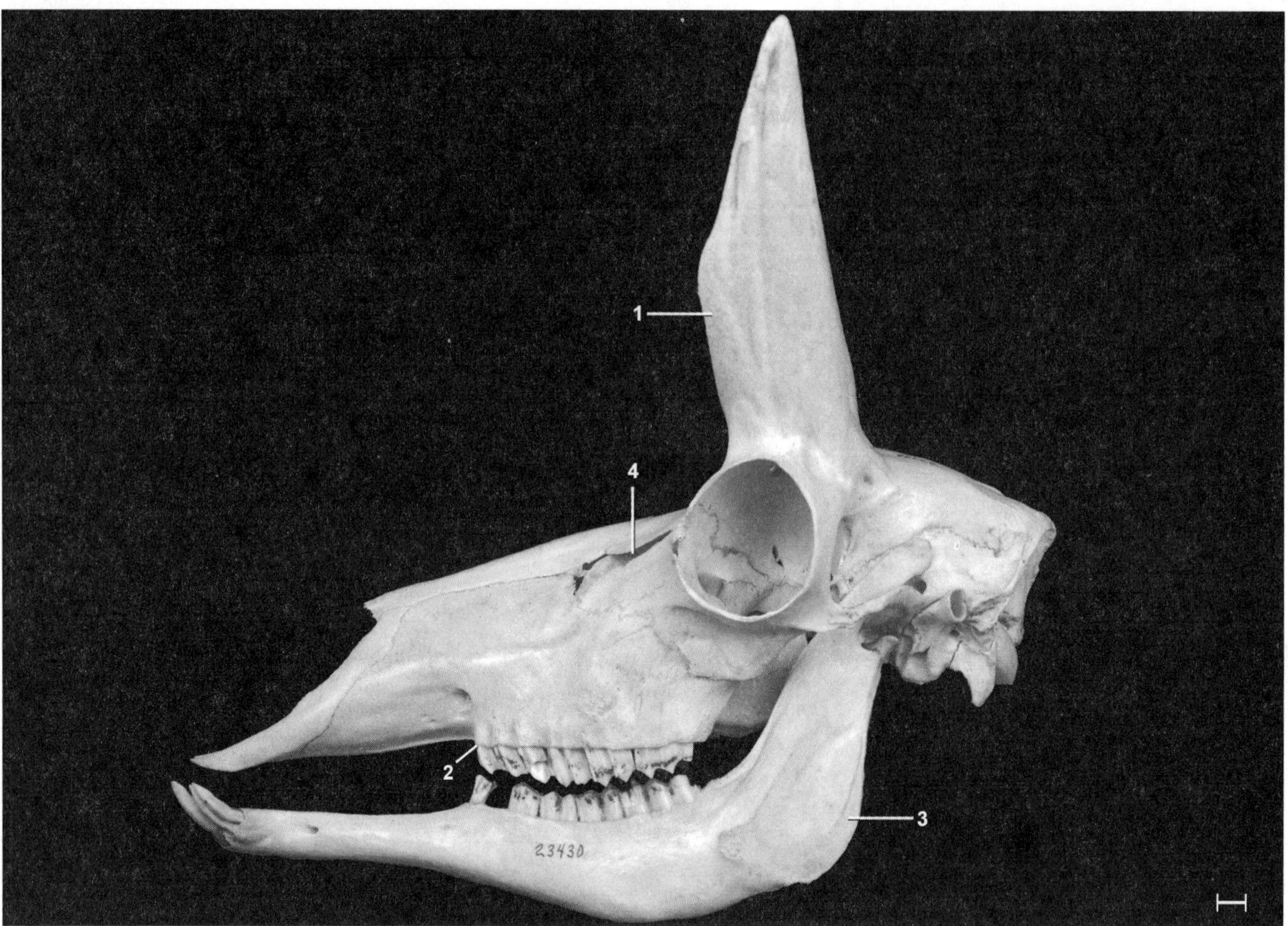

FIGURE 5.71. Cranium, lateral view, of Pronghorn (*Antilocapra americana*). Osteological features: (Antilocapridae): 1, horns (keratinized sheath [not present here] covering bony core) arise from cranium above orbits; in both sexes, sheath is shed annually; 2, cheekteeth lack distinct separation between enameled tooth crown and root; 3, angular process of mandible lacks rounded posterior extension; 4, bone loss between lacrimal and nasal—the two bones do not contact, as in cervids.

Pronghorns are active night or day in open grassland or sagebrush deserts, foraging on a range of grasses, shrubs, forbs, and cacti. They can reach speeds of up to 70 mph, making them the fastest land animal in the Western Hemisphere. Pronghorns form a variety of social groups during the summer, including groups of does and spring-born fawns; yearling and two-year-old male bachelor groups; and older, dominant males attempting to establish territories. In the fall, males with territories will attract harems of up to 20 does. Breeding occurs in the fall; horns are shed about a month later. Migration patterns from summer to winter range vary depending on altitude, range quality, and other conditions. Some currently used migration routes and corridors have been in use for over 7,000 years. Winter groups may include over 100 animals.

Prehistoric peoples of the west made extensive use of Pronghorns. Among other harvesting strategies, they used drives facilitated by constructed rock or wooden fences ("drivelines"), as well as natural topographic features.

FAMILY Bovidae (Sheep, Goats, Bison, and Allies)

Bovidae is a rich, Old World family that includes the diverse antelopes and gazelles. Only four genera are present in North America, but this includes such notable western taxa as the American Bison (*Bison bison*), Bighorn Sheep (*Ovis canadensis*), and Mountain Goat (*Oreamnos americanus*; Figures 5.72–5.74; Color Plate 30). Unlike the cervids, bovids have true horns—permanent keratinous sheaths that cover hollow bony cores rising from the frontal bones. Horns

FIGURE 5.72. Bighorn Sheep (*Ovis canadensis*).

FIGURE 5.73. Mountain Goat (*Oreamnos americanus*).

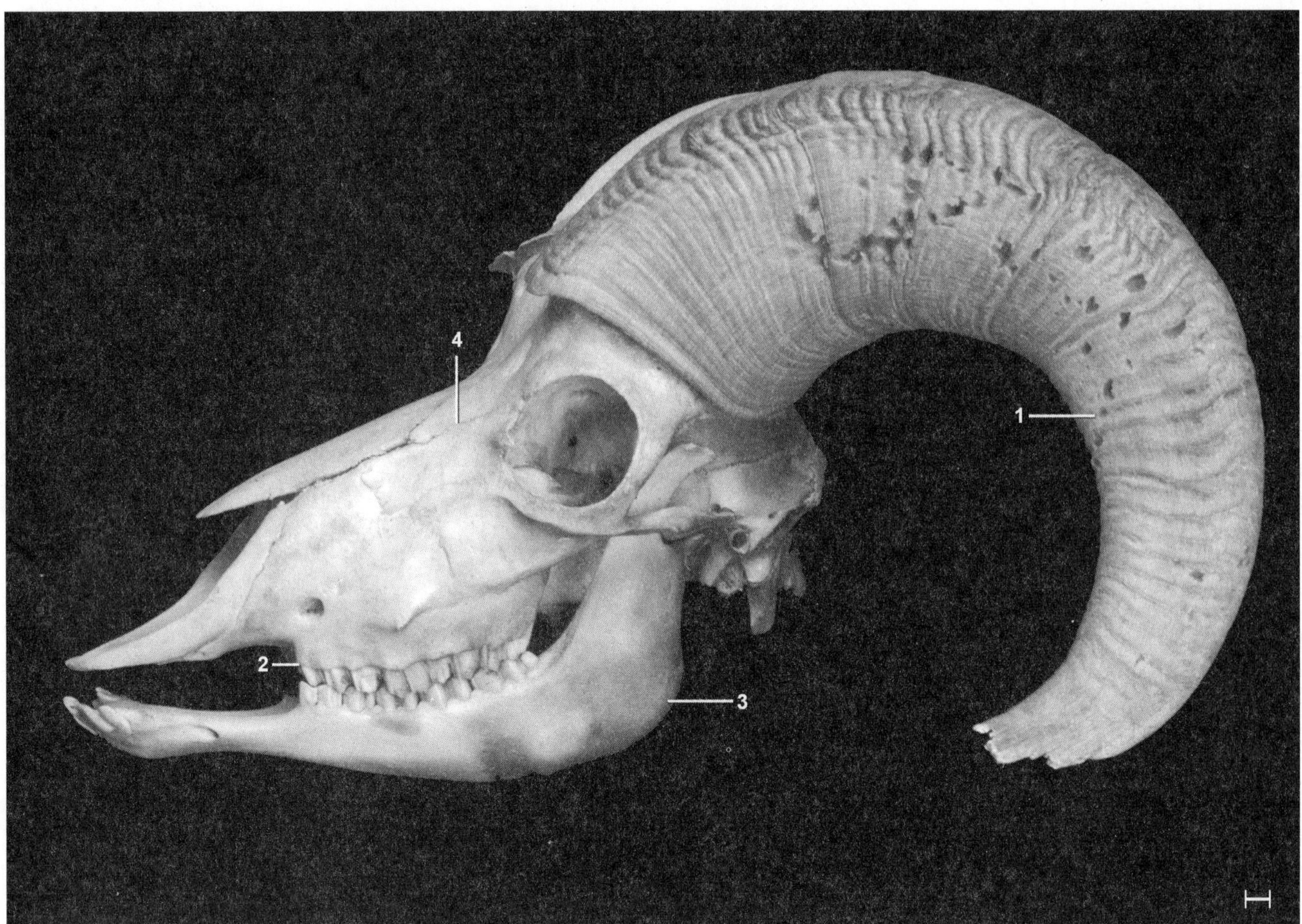

FIGURE 5.74. Cranium, lateral view, of Bighorn Sheep (*Ovis canadensis*). Osteological features (Bovidae): 1, permanent horns (keratinized sheath covering bony core) present in both sexes (larger in males), unlike cervids; 2, cheekteeth lack distinct separation between enameled tooth crown and root; 3, angular process of mandible lacks rounded posterior extension; 4, lacrimal contacts nasal, no bone loss in lacrimal region.

are never branched and occur in both sexes in the taxa above, although they are smaller in females. Bighorn Sheep, widely distributed in the arid west, are limited by water availability and prefer treeless areas with nearby cliffs to escape predators. American Bison, North America's largest Holocene land mammal, graze in herds and reach their highest densities in mixed- and short-grass prairie. The question of whether Mountain Goats were native to Washington's Olympic National Park represents one of the highest profile cases where zooarchaeological data have been used to inform modern landscape management policies.

Notes

Our discussion on the origins, characteristics, and osteology of mammals is derived from Liem et al. (2001), King and Custance (1982), and Hildebrand (1995); the taxonomy follows Wilson and Reeder (2015). Osteological nomenclature follows Gilbert (1975). Cranial features that distinguish different mammals are described in Gilbert (1990) and Elbroch (2006). Data on the natural history, ecology, and behavior of mammals is derived from Jameson and Peeters (2004), Kays and Wilson (2009), Reid (2006), Whitaker (1980), and Wilson and Ruff (1999).

PLATE 1. Lahontan Redside (*Richardsonius egregious*). CA: Lassen Co., Willow Creek (near Eagle Lake), July. This small fish in the minnow family (Cyprinidae) is widespread in lakes and streams throughout the Lahontan drainage system in California and Nevada. *Richardsonius* is after the English naturalist Sir John Richardson (1787–1865) who described the first species. The inspiration for the specific epithet, *egregious*, traditionally meaning "outstanding," is unclear, but we suspect it may derive from the striking breeding colors as evident in this fish.

PLATE 2. Cutthroat Trout (*Oncorhynchus clarkii*). UT: Cache Co., Logan River, April. Named for the prominent reddish-orange slashes on the underside of the jaw, Cutthroat Trout is the primary native trout in cold-water lacustrine and riverine contexts of the interior west. *Oncorhynchus* means hook-nose, a feature that develops in breeding males; *clarkii* is in honor of William Clark, of the Lewis and Clark Expedition (1804–06), who provided the first scientific description of this fish.

PLATE 3. Sockeye Salmon (*Oncorhynchus nerka*). CA: El Dorado Co., tributary of Lake Tahoe, October. Like each of the five species of Pacific salmon, Sockeye Salmon are typically anadromous, growing in the ocean for several years but returning to spawn in the fresh-water riverine settings of their origin. Landlocked populations also exist in many western lakes where spawning takes place in tributary streams—such as these examples in spawning morphology and coloration in a stream feeding Lake Tahoe.

PLATE 4. Bonneville Cisco (*Prosopium gemmifer*). UT: Rich Co., Bear Lake, January. Endemic to Bear Lake on the Utah-Idaho border, Bonneville Cisco are the most abundant fish in this deep, cold, high-elevation lake. The specific epithet, *gemmifer*, refers to being "set with gems," as these two beauties epitomize. Once a year for about a week in mid- to late January, they move from deeper water to spawn along the lake's rocky shorelines where they can be caught with dip nets—as these were.

PLATE 5. CA: Lassen Co., Willow Creek (near Eagle Lake), July. *Zooarchaeology and Field Ecology* students seine Willow Creek to identify, observe, and study local fishes.

PLATE 6. Paiute Sculpin (*Cottus beldingii*). CA: Lassen Co., Willow Creek (near Eagle Lake), July. Most sculpins are marine but several, such as the Paiute Sculpin here, occur in freshwater, typically along the rocky bottoms of cold, well-oxygenated streams or lakes. They are very small fish—this one from Willow Creek measuring about 10 cm in total length is a monster! The number and arrangement of their prominent preopercular spines is helpful in distinguishing many species.

PLATE 7. Western Toad (*Anaxyrus boreas*). CA: Lassen Co., near Eagle Lake, July. Except in higher latitudes and elevations, Western Toads are nocturnal and most easily encountered by walking the shores of lakes and streams with a flashlight on warm summer nights. Although most abundant near water, they wander widely—we found this one in a coniferous forest about a mile from Eagle Lake. As with most toads, they bury themselves in loose sediments or seek shelter in burrows and can thus become intrusive into archaeological sediments.

PLATE 8. Sierran Treefrog (*Pseudacris sierra*). CA: Lassen Co., near Eagle Lake, July. This small frog with a big voice has toe pads, a black or dark brown eye stripe, and highly variable dorsal coloration from reddish to cream but usually in shades of green or brown. Moreover, those colors can change from dark to light within minutes to enhance their crypticity.

PLATE 9. American Bullfrog (*Lithobates catesbeianus*). CA: Lassen Co., Murrer's Meadow, near Eagle Lake, July. Introduced from eastern North America, bullfrogs have made major inroads in aquatic settings all over the west. The deep bellowing bovine-like call of adult males is distinctive. Startled individuals along the shore give a single squawk (adults) or chirp (juveniles) as they bound into the water. Live capture success with butterfly nets is greatly enhanced at night when frogs are dazzled and distracted with flashlights.

PLATE 10. Great Basin Spadefoot (*Spea intermontana*). UT: Emery Co., San Rafael Swell, May. The voice of male spadefoot toads carries for long distances and plays a key role in bringing pairs together in arid environments where suitable breeding pools can be highly ephemeral. Indeed, we followed the calls to find this pair initiating amplexus (see chapter 3, Class Amphibia) in a shallow cattle pond. Note the cat-like vertical pupils.

PLATE 11. Western Tiger Salamander (*Ambystoma mavortium*). UT: Davis Co., North Salt Lake, April. Aquatic larval Tiger Salamanders can metamorphose into terrestrial forms, as in this individual, or adopt a paedomorphic life-history strategy retaining aquatic juvenile characteristics after reaching sexual maturity. The two forms differ osteologically and the paedomorphic morphs are more prevalent in mesic settings and time periods with greater pond permanence.

PLATE 12. Greater Short-horned Lizard (*Phrynosoma hernandesi*). OR: Lake Co., near Paisley (Fivemile Point), June. Often referred to as a "horny toad" or "horned frog," these amazing lizards are neither toad nor frog. The horned lizards are found in a wide range of habitats but the presence of ants, their chief prey, and loose sediments for burying themselves is a requirement for most species. Many species, including the one pictured here, eject blood from the eyes as an anti-predator—especially anti-canid—mechanism.

PLATE 13. Zebra-tailed Lizard (*Callisaurus draconoides*). CA: Riverside Co., Palm Desert, April. The Zebra-tailed Lizard inhabits desert habitats with limited vegetative cover where it uses its lightning speed (up to 9.7 mps, nearly 22 mph!) to evade predators. Zebra-tails exhibit an unusual behavior of lifting, curling, and wagging their tails as predators approach. This behavior may invite predator strikes to the tails that break off freely (caudal autotomy) and can be regenerated, or serve to deter pursuit—"I see you, don't even try it."

PLATE 14. Long-nosed Leopard Lizard (*Gambelia wislizenii*). OR: Lake Co., near Paisley (Fivemile Point), July. This large lizard inhabits arid and semiarid plains with bunch grass, creosote (*Larrea tridentata*), sagebrush (*Artemisia*), and other scattered low plants. The breeding adult here, indicated by the bright orange flecks, let the camera get close in a Black Greasewood (*Sarcobatus vermiculatus*) dominated community. Leopard lizards routinely bite hard when handled—this one did—but usually do not draw blood.

PLATE 15. Gophersnake (*Pituophis catenifer*). CA: Lassen Co., near Eagle Lake, July. Gophersnakes mimic rattlesnakes morphologically and behaviorally to deter predators—the two taxa not only have very similar scale colors and markings, but when threatened, Gophersnakes flatten their heads to form a more triangular rattlesnake look; they also rapidly vibrate their tails to mimic the rattle. They are good climbers. We observed this one move high in the tree to raid nestlings from a bird's nest.

PLATE 16. Great Basin Rattlesnake (*Crotalus oreganus lutosus*). CA: Lassen Co., near Eagle Lake, July. Great Basin Rattlesnakes exhibit enormous color variation that often matches the shades of local rock or sediment substrates. This one nestled under rocks close to the shore of Eagle Lake has a beautiful chocolate pattern. Rattlesnakes are venomous and have highly developed venom-injection systems that involve large, paired, hollow, and moveable fangs. The amount of venom injected in a bite can vary widely. Fortunately, rattlesnakes are not aggressive and shun contact with humans.

PLATE 17. Yellow-bellied Marmot (*Marmota flaviventris*). UT: Salt Lake Co., Little Cottonwood Canyon, July. A marmot lies flat on a rock to soak up the sun. Inhabiting meadows near talus slopes or rock outcrops, this chunky sciurid is the most widespread marmot in western North America. Their large size (up to 5 kg) and the thick layer of fat they put on in fall prior to hibernation made them an attractive and widely utilized prey item to regional prehistoric peoples.

PLATE 18. Golden-mantled Ground Squirrel (*Callospermophilus lateralis*). CA: Lassen Co., near Eagle Lake, July. With a brilliant golden hood and dark lateral stripes, this squirrel is often mistaken for a chipmunk, but it is nearly twice as big and the stripes do not continue onto the head. It is found in coniferous and mixed forest, alpine meadows, talus slopes, and chaparral habitats. It is often seen perched on a log, stump, or rock, as with this one inhabiting a talus slope next to a Ponderosa Pine (*Pinus ponderosa*) forest. Like most ground squirrels, they hibernate, so their presence in archaeofaunas can provide information on site seasonality.

PLATE 19. California Ground Squirrel (*Otospermophilus beecheyi*). CA: Lassen Co., near Horse Lake, July. Mostly ground dwelling, California Ground Squirrels will stand sentry on shrubs, stumps, or fence posts, as in this adult female and pup pair. When alarmed, they emit a single, loud, shrieking whistle or a staccato churr or trill of about five notes. Not overly social, California Ground Squirrels form loose colonies and may share burrow systems, but each squirrel has its own entrance hole.

PLATE 20. Utah Prairie Dog (*Cynomys parvidens*). UT: Iron Co., Summit, March. Prairie dogs are highly social and form large colonies or "towns" featuring extensive underground burrow systems in prairies, grasslands, or open meadow habitats. This individual sounds a loud "bark" to warn others of our approach. The four North American species have all declined dramatically due to habitat destruction and systematic persecution for their exaggerated reputation as an agricultural pest. Photo by Michael Broughton

PLATE 21. North American Porcupine (*Erethizon dorsatum*). UT: Davis Co., North Canyon, May. Betraying her poor vision, this porcupine ambled slowly toward us across a snow-covered meadow in a mixed coniferous-aspen forest and was not at all alarmed when she finally recognized our presence. Such nonchalant attitude reflects a highly successful antipredator adaptation of a body protected by sharp, microscopically barbed quills.

PLATE 22. Desert Woodrat (*Neotoma lepida*). CA: Lassen Co., Eagle Lake Field Station, July. This youngster was just released from a live trap and paused briefly against a student's foot (in background). Handsome animals, woodrats are famous for constructing large, aboveground, stick-pile nests at the base of trees, shrubs, or cacti, or in caves, rocky outcrops, or crevices.

PLATE 23. American Pika (*Ochotona princeps*). UT: Salt Lake Co., Little Cottonwood Canyon, August. If they do not reveal themselves by their loud, whistled barks or nasal bleats, pikas can be located by scanning high-elevation talus slopes for thick piles of drying meadow vegetation. Eventually, they move these "haystacks" into their burrows deep beneath the talus and will live off them all winter.

PLATE 24. Desert Cottontail (*Sylvilagus audubonii*). CA: Riverside Co., Palm Desert, April. This rabbit pauses motionless among thick shrubs after a quick zigzag burst, exemplifying the typical predator-escape strategy of cottontails. Relative to hares (*Lepus* spp.), cottontails are more indicative of habitats with thicker vegetative cover. As in all lagomorphs, cottontails have a distinctive double pair of incisors in the premaxilla.

PLATE 25. Black-tailed Jackrabbit (*Lepus californicus*). CA: Lassen Co., near Eagle Lake, July. The giant ears (pinnae) of the jackrabbit not only collect and funnel sound to enhance predator detection but are highly vascularized and play a key role in thermoregulation (dissipating heat) in hot environments. In many settings of the arid west, lagomorphs were the most economically important animal resource to native peoples.

PLATE 26. Myotis bats (*Myotis* spp.). CA: Lassen Co., Eagle Lake Field Station, May. At dusk, myotis bats pour out of a mountain cabin to begin their evening hunt for flying insects, which they consume at rates of over 1,000 bugs per hour. Bats roost and form breeding colonies in caves, and through natural deaths can be incorporated into archaeological sediments in these settings.

PLATE 27. Raccoon (*Procyon lotor*). CA: Lassen Co., Eagle Lake Field Station, July. This familiar carnivore with a black mask and ringed tail licks her chops after raiding the field station garbage bins. They do not wash their food but they are semiaquatic foragers and use their dexterous front paws to catch and manipulate prey in shallow water.

PLATE 28. Mule Deer (*Odocoileus hemionus*). UT: Davis Co., Wasatch Range, November. Among the most familiar animals of the west, this Mule Deer doe struts off to woodland cover after foraging in an open area with grasses and low shrubs. Note the dark tarsal gland on her inner right hindlimb, used for scent communication, and multiple ticks on her head, neck, and torso.

PLATE 29. Pronghorn (*Antilocapra americana*). UT: Box Elder Co., Lakeside Range, May. This group of adult females pauses briefly before bolting across grassland habitat in the Bonneville Basin. Pronghorn are migratory in many areas and currently use migration routes in Wyoming that have existed for at least the past 7,000 years, based on analyses of pronghorn archaeofaunas along the route. This research has led, in part, to multimillion-dollar investments in highway overpasses and fencing removal to protect the pronghorn using these ancestral routes today.

PLATE 30. Bighorn Sheep (*Ovis canadensis*). CA: Riverside Co., San Jacinto Mountains, December. This group of sheep finds pockets of grass in the steep, treeless, rocky terrain of the San Jacinto Mountains. During the fall rut, rams rear up and butt heads at high speeds in combat for access to ewes. While Mule Deer have been the dominant artiodactyl historically in the Great Basin, to judge from regional paleofaunas, Bighorn Sheep were proportionately far more widespread and abundant in prehistoric times.

PLATE 31. Canada Goose (*Branta canadensis*). CA: Sacramento Co., American River, near Fair Oaks, November. Unlike most North American ducks, geese form long-term monogamous pairs, lack sexual dimorphism in plumage, and the males provide extensive parental care by protecting their incubating mate and nest. Waterfowl (Anatidae) were the most important avian subsistence resources for indigenous peoples in coastal and wetland settings of the west. However, making species-level identifications with duck and goose bones is notoriously challenging.

PLATE 32. Common Goldeneye (*Bucephala clangula*). UT: Box Elder Co., Bear River National Wildlife Refuge. January. A striking duck with brilliant yellow eyes set in a triangular-shaped, iridescent green head, goldeneyes breed primarily in Canada and Alaska on open lakes next to woodlands where nest cavities are available. They winter in coastal areas and on inland lakes and rivers in the continental U.S. These are diving ducks—this male is up between dives pursuing aquatic invertebrates, vegetation, and small fish.

PLATE 33. Dusky Grouse (*Dendragapus obscurus*). UT: Davis Co., Sessions Mountains, October. With a partially raised crest, this grouse scurries up a dead limb of a White Fir (*Abies concolor*) moving higher into the canopy. Dusky Grouse spend most of the winter months in firs or pines where they forage principally on conifer needles. They move to the ground in spring to nest and forage through the summer and fall on leaves, insects, and berries.

PLATE 34. Western Grebe (*Aechmophorus occidentalis*). CA: Lassen Co., Eagle Lake. July. Deep in courtship, this pair performs the "weed dance." With aquatic vegetation collected during short dives, the pair rises out of the water breast to breast in nearly vertical position, gently connecting their mutual offerings. Other notable displays include "rushing" or the "rush dive" where two or more grebes run rapidly across the water surface side by side before diving in headfirst. Unlike the weed dance, rushing is often performed by competing males.

PLATE 35. Double-crested Cormorant (*Phalacrocorax auritus*). UT: Davis Co., Farmington Pond, April. This bird pulls up a Rainbow Trout (*Oncorhynchus mykiss*) and swallows it whole, showcasing the species' piscivorous habits. Double-crested Cormorants have bright orange facial skin and striking aquamarine eyes. Like other cormorants, double-cresteds form large breeding colonies on small islands or in trees that are inaccessible to mammalian predators. Zooarchaeological research with Pacific Coast avifaunas in several regions has documented that intensive prehistoric human harvesting on such colonies led to local extirpations.

PLATE 36. Double-crested Cormorant (*Phalacrocorax auritus*). CA: Lassen Co., Eagle Lake. July. This squadron begins to assume the V-shaped formation that is used not only by cormorants but many other birds. The V-formation reduces the energetic costs of flying by allowing birds to catch the rising air or "upwash" produced by those ahead. Both individual spacing and the timing of wing beats are synchronized to optimize these savings. Birds take turns in the lead position.

PLATE 37. American White Pelican (*Pelecanus erythrorhynchos*). CA: Lassen Co., Eagle Lake. July. A group of pelicans loaf on a hot afternoon on a protected islet on Eagle Lake. As we approach, one bird lifts off on wings that span nine feet. Earlier we observed this group fishing cooperatively in a calm bay. Swimming side by side in a long row, the pelicans herded small, schooling fish into the shallows when, in a big splash, they jabbed their giant, pouched bills into the water and scooped out their catch.

PLATE 38. Osprey (*Pandion haliaetus*). CA: Lassen Co., near Eagle Lake, July. While one adult perches near their huge stick nest high atop a Ponderosa Pine (*P. ponderosa*), the other returns with a fish carried headfirst and aligned with the bird's body to reduce drag. Osprey is the only North American raptor with a diet specialized on live fish. Equipped with a reversible outer toe, sharp talons, and toes covered with spike-covered pads, Osprey catch fish by plunging feet-first into rivers, lakes, lagoons, and marshes.

PLATE 39. Sandhill Crane (*Grus canadensis*). UT: Wasatch Co., near Strawberry Reservoir. April. Cranes are perennially monogamous, and pairs perform exuberant courtship dances involving leaping, stick throwing, wing flapping, and calling. Their long tracheas coil into the sternum enhancing the depth of their distinctive bugling or rattled-honking calls. This pair meanders side by side across snow patches in a sagebrush steppe, occasionally probing for seeds, tubers, insects, and small vertebrates. Photo by Michael Broughton

PLATE 40. Long-billed Dowitcher (*Limnodromus scolopaceus*). UT: Davis Co., Farmington Bay, September. Members of the large sandpiper family (Scolopacidae), Long-billed Dowitchers are commonly seen in flocks in or near freshwater during fall and spring migrations. Breeding in Alaska and eastern Siberia, this flock, containing mostly juveniles (grayish) but a few adults (reddish), is in migration to wintering areas in southern California, Mexico, or the southeastern United States.

PLATE 41. Greater Roadrunner (*Geococcyx caifornianus*). CA: Riverside Co., Palm Desert, April. Perched on a rock with a fully erect crest, this icon of the desert Southwest gives a yawn-like gesture after sounding a sharp bark. Roadrunners' many intriguing vocalizations include cooing, whining, growling, and the nonvocal, bill clacking. Animal prey make up about 90% of their diet—larger prey restrained by their long bills are repeatedly body slammed, fracturing and disarticulating the skeleton to elongate the body prior to swallowing.

PLATE 42. Northern Pygmy Owl (*Glaucidium gnoma*) carrying vole (*Microtus* sp.). UT: Davis Co., North Canyon, November. We first noticed this owl causing a commotion on the ground—stumbling about and flapping its wings. From there, it lifted off with some difficulty and perched in the lower branches of a maple. In the tree, the reason for its actions became apparent: it had just snatched a large vole that likely approached the weight of the owl itself. Indeed, pygmy owls are ferocious hunters and have been known to take prey up to three times their body weight.

PLATE 43. Burrowing Owl (*Athene cunicularia*). UT: Box Elder Co., near Lakeside Range, April. Bobbing its head to gauge our distance, an adult Burrowing Owl stands guard beside its deep burrow, probably originally dug by a badger or ground squirrel. A mate (sexes are similar) stood close by at first but flew off and landed a short distance away, sounding an agitated alarm chatter. Burrowing owl diets are comprised of a mix of arthropods and small rodents (mice, voles), the remains of which we identified by breaking apart the numerous pellets scattered around the burrow opening.

PLATE 44. Barn Owl (*Tyto alba*). UT: Davis Co., Farmington Bay, December. At daybreak on a blue, wintry morning, a ghostly Barn Owl glances upward and hovers briefly with landing gear fully extended. Within a second, it drops to the snow, covering a large vole in its talons. On this morning, the catch was aided by the bird's excellent low-light vision, but Barn Owls are well known for their ability to locate prey in total darkness by sound alone.

PLATE 45. Costa's Hummingbird (*Calypte costae*). CA: Riverside Co., Palm Desert, April. The dark crown and gorget (throat patch) of this young male Costa's Hummingbird transforms into a radiant violet with the turn of his head and a shift in lighting. Many other birds exhibit these "structural colors" formed by light interacting with a feather's three-dimensional form.

PLATE 46. Downy Woodpecker (*Picoides pubescens*). UT: Davis Co., North Canyon, December. Gentle wood tapping breaks the silence on this crisp December morning and betrays the presence of the West's smallest woodpecker. Propped up by his stiff tail feathers (only males have the red cap), he delicately extracts a spider encased in its web from beneath the bark.

PLATE 47. White-headed Woodpecker (*Picoides albolarvatus*). CA: Lassen Co., Eagle Lake Field Station, July. This striking woodpecker hitches up a pine trunk: propped up by his stiff tail feathers, he moves both feet at once and jerks quickly upward. He stops occasionally to peel back bark and probe for insects with his chisel-like bill. White-headed woodpeckers are year-round residents of montane coniferous and mixed forests, especially those dominated, as here, by Ponderosa (*Pinus ponderosa*) and Jeffrey Pine (*P. jeffreyi*).

PLATE 48. Western Tanager (*Piranga ludoviciana*). UT: Davis Co., North Salt Lake, June. A welcome sign of the western summer, this striking male tanager shakes off raindrops from a late afternoon thunderstorm. Tanagers are denizens of open coniferous and mixed forests. The perching birds (order Passeriformes), to which tanagers belong, represent the largest order of birds in the world, comprising over half of all known species.

PLATE 49. Red-winged (*Agelaius phoeniceus*), Yellow-headed (*Xanthocephalus xanthocephalus*), and Brewer's Blackbirds (*Euphagus cyanocephalus*). UT: Davis Co., Farmington Bay, September. Rising off a snag, red epaulets (wing patches) and bright yellow heads of the male birds are conspicuous in this mixed flock of blackbirds. These flash colors are displayed even more prominently during territorial singing. Most marsh-nesting blackbirds adopt polygynous mating systems, where a single male holds a territory that attracts several nesting females.

CHAPTER 6

Birds

I. General Osteology of Birds

Characteristics

Birds are winged, bipedal, endothermic, egg-laying vertebrates that represent a surviving clade of theropod dinosaurs. Over 10,000 species are recognized worldwide, a number that far exceeds any other tetrapod class. They have adapted to a wide range of habitats from the peaks of the tallest mountains, to scorching deserts, to oceanic islands. Extant North American birds range in size from the Calliope Hummingbird (*Selasphorus*), which weighs less than 2.7 gm, to the California Condor (*Gymnogyps californianus*), which can exceed 10.5 kg. They possess beaks but lack teeth, have very high metabolic rates, four-chambered hearts, and strong but lightweight skeletons. A number of birds make and use tools, most notably New Caledonian Crows (*Corvus moneduloides*), and transmit learned behavior across generations. The Corvidae—ravens, crows, jays—and parrots (Psittacidae) are widely thought to be among the most intelligent vertebrates. Birds are also well known for long-distance, seasonal migrations and communication using complex vocal songs and calls. Given the structural specialization required by flight, birds are also morphologically the most homogenous group of tetrapods.

Origins

As noted, birds should technically be placed within the class Reptilia. Birds belong to a group of bipedal dinosaurs, the theropods, which first appeared during the late Triassic about 230 million years ago. Theropods are represented by a range of forms that includes birds, small, fleet-footed dinosaurs, and giant carnivorous ones such as *Tyrannosaurus rex*. But birds are distinguished from the other theropods by the capability of flight, several unique skeletal features, and unique adaptations for endothermy. Shared features between birds and other theropods include reduced digits, a furculum (fused clavicles), air-filled—pneumatized—long bones, and the presence of feathers. Indeed, recent fossil evidence has been read to suggest the presence of feathers in earlier nontheropod dinosaurs, raising the possibility that many, perhaps most, dinosaurs were feathered for insulation and display. Feathers and wings suggesting varying degrees of flight capability remain confined to the theropods.

Two traditional hypotheses have been proposed to explain the evolution of avian flight. The arboreal theory suggests that the ancestors of birds lived in trees and that feathered wings to enhance gliding ultimately replaced skin membranes. Alternatively, the "ground-up," or cursorial theory, postulates that wings first evolved in terrestrial theropods and served as structures that increased the lift, agility, and balance in early birds as they leapt to pursue mobile prey. In any case, by the late Jurassic period (~160 mya) the early ancestors of birds had clearly emerged, including one of the better known "transitional forms" in all of evolution, *Archaeopteryx*. This creature is widely recognized as near the common ancestor for all birds. *Archaeopteryx*, known from seven individuals recovered from marine deposits in southern Germany, retained such reptilian features as a long tail, teeth, and fingers with claws.

Yet it also had well-developed furculae and coracoids and feathers on the forelimbs and tail that indicate it was capable of flapping, powered flight. Still, because the sternum is poorly developed, it has long been assumed that *Archaeopteryx* was not a strong flier.

Apparently derived from *Archaeopteryx*-like ancestors, birds diversified during the Cretaceous. For example, the 135 million year old *Sinornis* was a sparrow-sized bird with a well-ossified, keeled sternum and caudal vertebrae modified to form a pygostyle—two features that characterize modern birds. *Sinornis* also had partially fused fingers in the wing and more fusion of the vertebrae, pelvic bones, and foot bones than *Archaeopteryx*. Other well-known Cretaceous birds include *Hesperornis*, a flightless, loon-like, diving seabird, and *Ichthyornis*, a tern-like form that flew above the same shallow seas. Both groups had well-developed, nonserrated, thecodont teeth with constricted bases and expanded roots set into distinct alveoli. The modern orders of birds began to diverge during the Tertiary period about 60 mya. During the Quaternary (last 2.6 million years), major glaciations and habitat changes fragmented populations, causing multiple extinctions and the evolution of new species. Most modern species evolved during the early to middle Pleistocene.

Osteology

Many features of the avian skeleton reflect links to their reptilian ancestry, while many others represent adaptations associated with their unique form of locomotion (Figures 6.1–6.13). To facilitate flight, various fusions and reinforcements create a skeleton that is light but powerful. Cross sections of bird bones reveal a distinctive, lightweight, hollowed design, with slim, bony struts forming trusses to provide added strength with minimal additional weight.

The modern bird cranium lacks teeth and is noticeably delicate. Complete fusion of the separate cranial elements occurs early in development (Figures 6.1–6.2). Although birds were derived from diapsid stock, the bar separating the two temporal fenestrae has been lost, retaining only a single opening that merges directly into the orbit. Birds have highly developed vision. Their expansive orbits support a large eye mass, and total eye volume is routinely larger than the cranial capacity in birds. Rings of sclerotic bones located within the orbit provide support for the large eyes. Bird beaks are highly variable across species and reflect the myriad feeding adaptations of different species. The premaxilla forms the bony core of the beak, while a keratinous sheath, the rhamphotheca, covers the exterior. Birds have very kinetic or flexible skulls, including a highly mobile frontonasal junction, which enhances their manipulatory capabilities. Reptilian features evident in the bird cranium include a single inner (or middle) ear bone (the stapes or columella), a single occipital condyle, a lower jaw comprised of a series of small bones, and an articular-quadrate connection between the lower jaw and the skull.

The vertebral column in birds is divided into cervical, thoracic, synsacral, and caudal regions (Figures 6.3–6.5). Most prominent are the cervical vertebrae that vary in number from 13–25. In many species, the necks are elongated and together with the head form a dexterous appendage, critical in feeding. In birds, the centra have a unique saddle shape—in the cervical vertebrae they are heterocoelous (hetero = different)—where the anterior surface is saddle shaped horizontally and articulates reciprocally with a vertical saddle shaped surface on the posterior of the adjacent vertebra. Typically the first four thoracic vertebrae fuse together in early adulthood and articulate tightly with the synsacrum (Figures 6.9–6.10). The synsacrum represents the fusion of the remaining thoracic vertebrae, all of the lumbar and sacral vertebrae, and a few of the anteriormost caudal vertebrae. Enclosed laterally by the os coxa, the synsacrum forms a ridged platform along the back that facilitates lift during flight. Posterior to the synsacrum are several free caudal vertebrae that are capped by a short, blade-like element, the pygostyle, which forms a foundation for the tail feathers. Fully ossified dorsal and ventral ribs are broad and flat, usually number seven per side, and provide a strong connection between the vertebrae and the sternum. Uncinate processes are small, bony flaps, projecting postero-dorsally off the dorsal ribs that provide an overlap between adjacent ribs and attachment sites for muscles associated with inspiration and expiration (Figure 6.4).

Clearly, the most obvious flight-related structural modifications of the avian skeleton involve the forelimb (Figures 6.5–6.8). To provide a strong and rigid foundation for the wings, different elements are variously elongated, fused to other elements, or lost altogether. The proximal forelimb elements (humerus, radius, ulna) are similar in shape and relative orientation to those of other tetrapods, but major changes are evident in the distal forelimb. Forming an elongated wrist to which the primary flight feathers attach, the carpals and metacarpals fuse to form a unique, blade-like element, the carpometacarpus. In addition, the digits are reduced in total number (only I, II, and III are present), size, and in the number of phalanges they contain. Digit I, the pollex, retains a single phalange and supports the alula, or bastard wing, which moves independently from the rest of the wing tip. Digit II contains two elongated phalanges that extend the wing tip. Only two free carpals are retained, the cuneiform and scapholunar. The pectoral girdle is similarly modified to provide attachment sites for the fully remodeled flight-muscle architecture. Most notable is the furculum, which represents the fusion of clavicles, and the enlarged coracoids that securely attach the wing to the sternum. The latter typically carries a prominent, ventrally projecting keel (the carina), which provides a broad attachment site for the primary flight muscles—the supracoracoideus and pectoralis. The proportionate size of the keel in general reflects the flight power of different avian taxa. It is enormous, for instance, in the most powerful fliers—the hummingbirds—and absent in flightless birds (e.g., Ratites, including Ostrich, Emu, and Rhea). The saber-like scapula is situated on the dorsal part of the rib cage, articulates with the coracoid and the furculum, and acts to strap the wing to the body.

Since birds are bipedal, the hindlimbs must carry and balance the full body weight when they are on land (Figures 6.9–6.13). To stabilize the body, the femur projects anteriorly to position the legs directly under the center of gravity. Further, to produce a consolidated strut, substantial rearrangement of the lower limb elements has occurred in birds. Specifically, the tibia fuses with several tarsals to form a unique element, the tibiotarsus. A splinter-like fibula attaches tightly to the lateral side of the tibiotarsus and may fuse to that bone in older birds. The remaining tarsals fuse with the metatarsals to form the tarsometatarsus from which four digits radiate distally to support relatively large feet. The orientation of the pedal digits varies among birds, but the number of phalanges on each one follows a typical pattern: two on digit I, three on digit II, four on digit III, and five on digit IV. The os coxa is distinctive in birds and is characterized by (1) complete fusion with the synsacrum; (2), a broad, dorsally oriented ilium that extends both posterior and anterior to the acetabulum; and (3), a thin, elongate pubis that lacks a symphysis and resides posterior and ventral to the ischium (Figures 6.9–6.10).

Finally, since birds have long necks, they also have long, prominent trachea that are unique in several ways compared to other vertebrates. First, the tubular walls of the trachea are supported by a series of closely connected, cartilaginous rings that ossify in most birds. Second, the posterior end of the trachea is modified in some taxa, especially in male waterfowl. In male ducks, the posterior trachea just anterior to the syrinx enlarges and ossifies into a sounding chamber, referred to as a tracheal or syringeal bulla. This structure amplifies and modifies sounds produced in the primary organ of sound production, the syrinx. The morphology of tracheal bullae has also been used in reconstructing phylogenetic relationships among ducks.

Remarks

Given both high species diversity and a high level of osteological similarity among taxa, birds are perhaps the most time-consuming and challenging group of vertebrates to work with. The time invested, however, yields substantial rewards as species-level bird identifications provide a wealth of inferential potential. In the realm of paleoenvironmental reconstruction, bird remains are especially informative, as many taxa are sensitive indicators of specific habitats. And since the class is represented by both terrestrial and aquatic taxa, they can shed light on the nature of prehistoric conditions in both contexts. As many taxa are characterized by seasonal migrations or have relatively narrow breeding periods, they are also among the more indicative groups in analyses of the seasonality of site occupation. Such studies can also be pursued through analyses of medullary bone—bone formation that occurs seasonally within the hollow medullary (or endosteal) cavities, uniquely, in female bird long bones. Medullary bone serves as a labile reservoir for the supply of eggshell calcium and indicates the season of egg formation in female birds.

Birds also assumed a significant ceremonial importance for native peoples. Given that a variety of regalia utilized bird feathers, the presence of bird taxa in archaeological faunas may be due to such factors rather than strictly gastric ones. In addition, bird taxa in locations well outside their historical range have also been used to establish prehistoric trade networks. The presence of macaws (*Ara*) in archaeological sites from New Mexico and Arizona, for instance, has been used to deduce trade contact between prehistoric peoples occupying those regions and ones from Mexico.

The skeletal element abundance profiles show an interesting pattern in the avifaunas from many western archaeological sites—wing elements far outnumber other portions of the skeleton. The disproportionate abundance of wing bones has been attributed to a wide range of causes from differential preservation related to bone density to the differential transport of bird wings back to sites.

Finally, aside from the Domestic Dog, which likely arrived with the first Americans, the Wild Turkey (*Meleagris gallopavo*) represents the only animal domesticated by aboriginal peoples of North America. The high diversity of birds worldwide is certainly reflected locally, as 17 orders occur in the west.

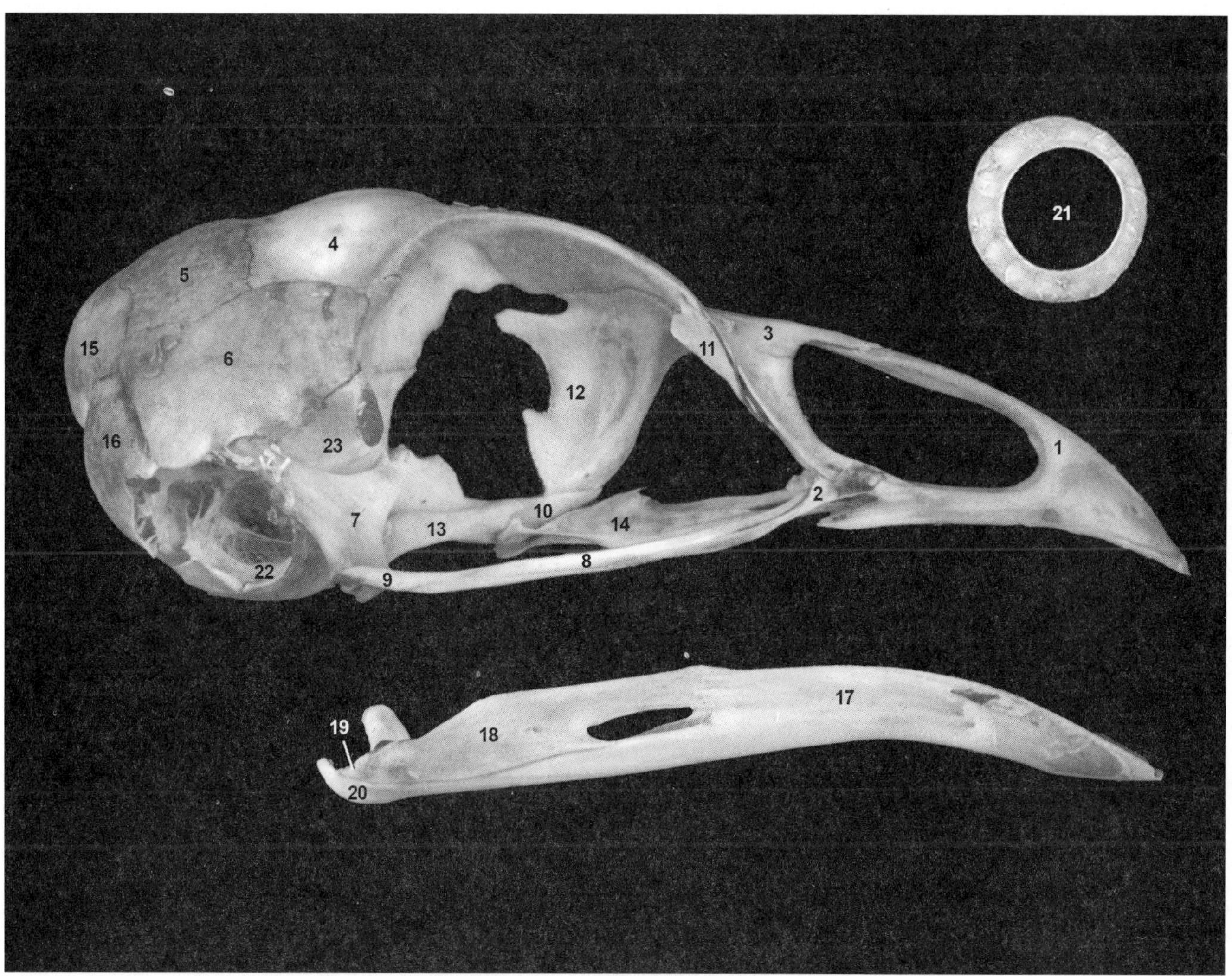

FIGURE 6.1. Cranium of Greater Sage-Grouse (*Centrocercus urophasianus*), lateral view.

1. Premaxilla
2. Maxilla
3. Nasal
4. Frontal
5. Parietal
6. Squamosal
7. Quadrate
8. Jugal
9. Quadratojugal
10. Vomer
11. Lacrimal
12. Ethmoid
13. Presphenoid
14. Palatine
15. Supraoccipital
16. Exoccipital
17. Dentary
18. Surangular
19. Articular
20. Angular
21. Sclerotic bones (ring)
22. Otic
23. Orbitosphenoid

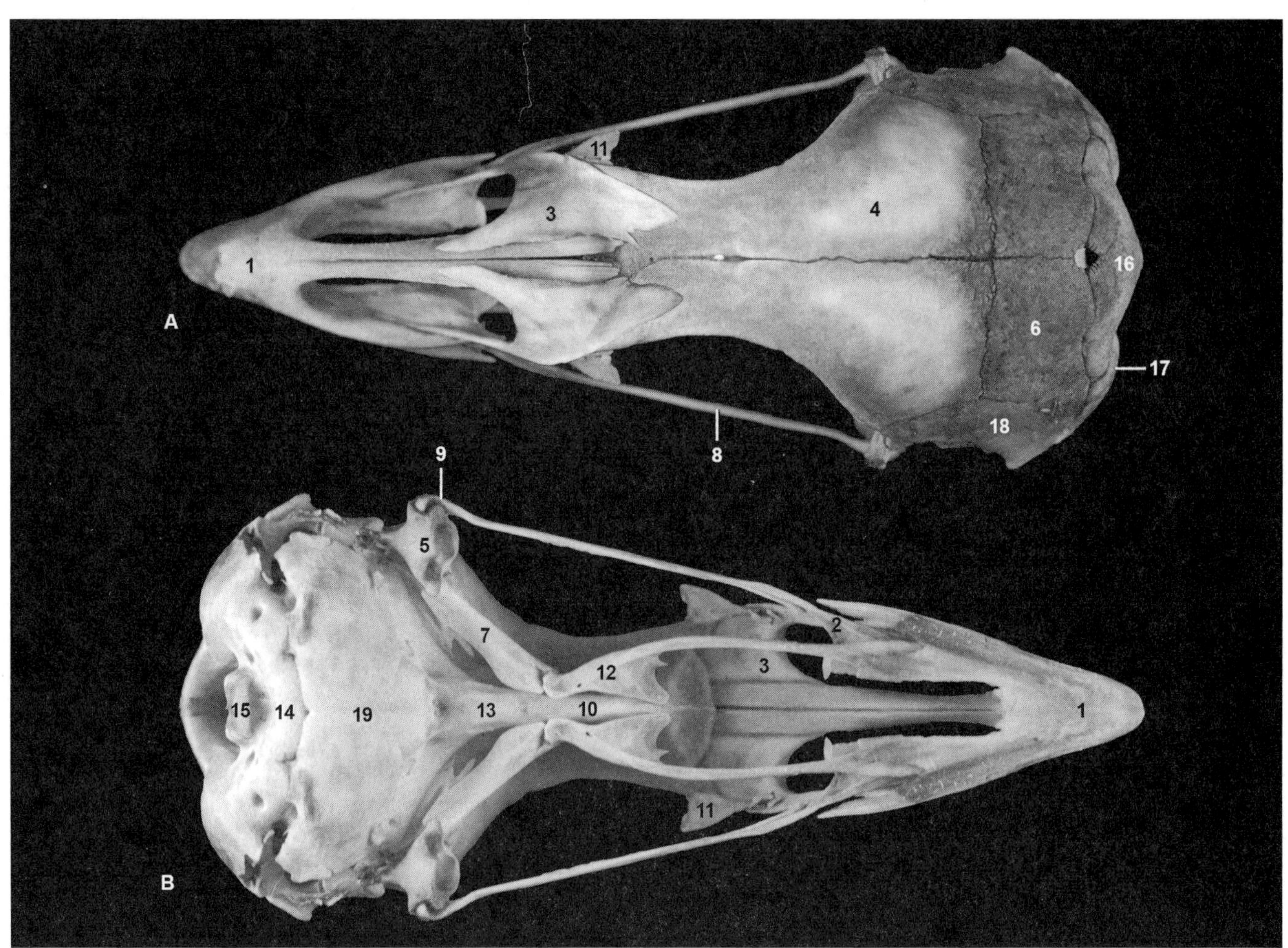

FIGURE 6.2. Cranium of Greater Sage-Grouse, dorsal (A) and ventral (B) views.

1. Premaxilla
2. Maxilla
3. Nasal
4. Frontal
5. Quadrate
6. Parietal
7. Pterygoid
8. Jugal
9. Quadratojugal
10. Vomer
11. Lacrimal
12. Palatine
13. Presphenoid
14. Basioccipital
15. Occipital condyle
16. Supraoccipital
17. Exoccipital
18. Squamosal
19. Basisphenoid

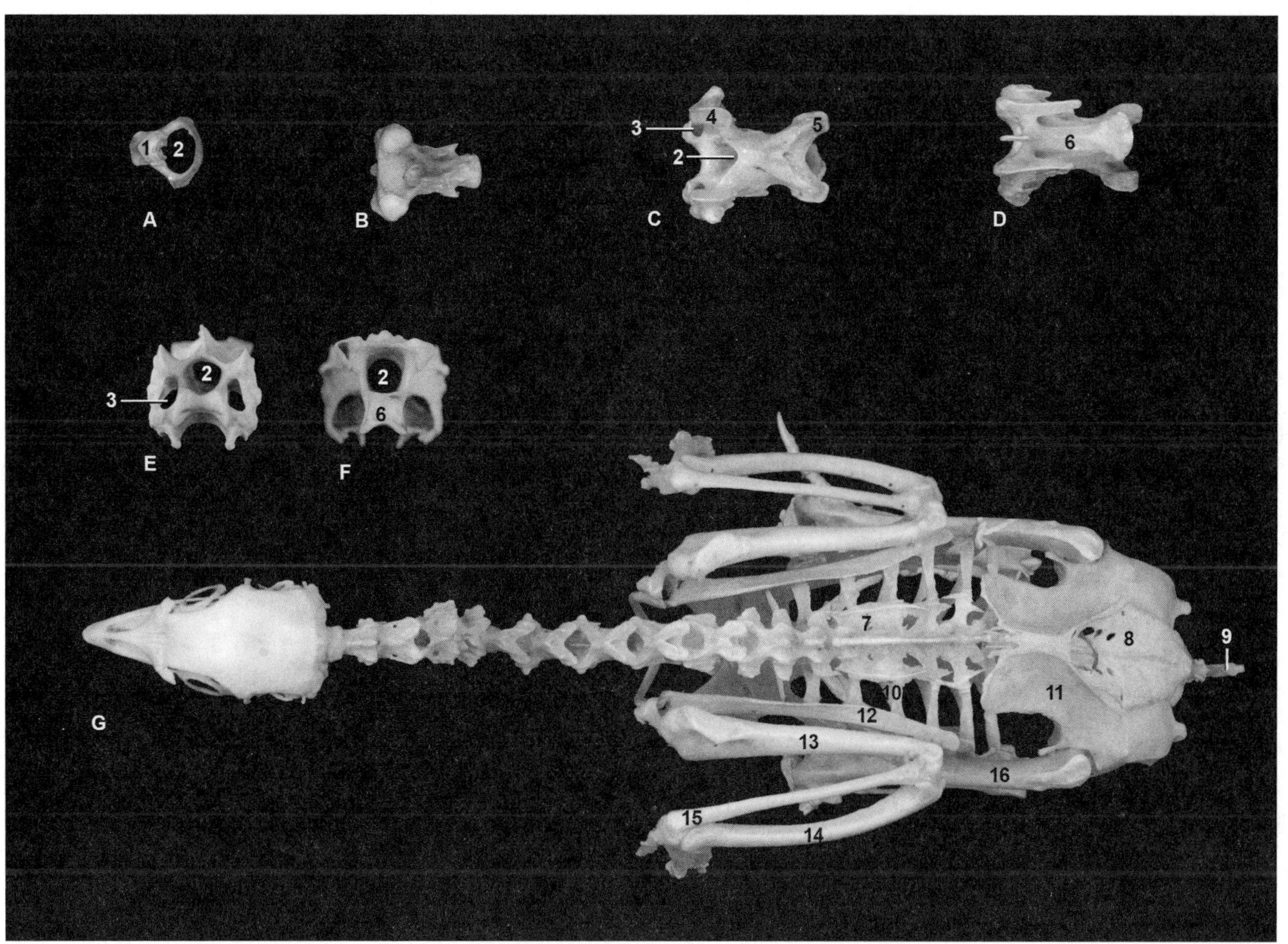

FIGURE 6.3. Cervical vertebrae (A–F) and skeleton, dorsal view (G), of Domestic Chicken (*Gallus gallus*). Atlas (A), axis (B), and dorsal (C), ventral (D), anterior (E), and posterior (F) views of cervical vertebrae.

1. Prezygapophysis
2. Vertebral foramen
3. Transverse foramen
4. Prezygapophysis
5. Postzygapophysis
6. Centrum
7. Thoracic vertebrae
8. Sacral vertebrae
9. Pygostyle
10. Dorsal rib
11. Ilium
12. Scapula
13. Humerus
14. Ulna
15. Radius
16. Femur

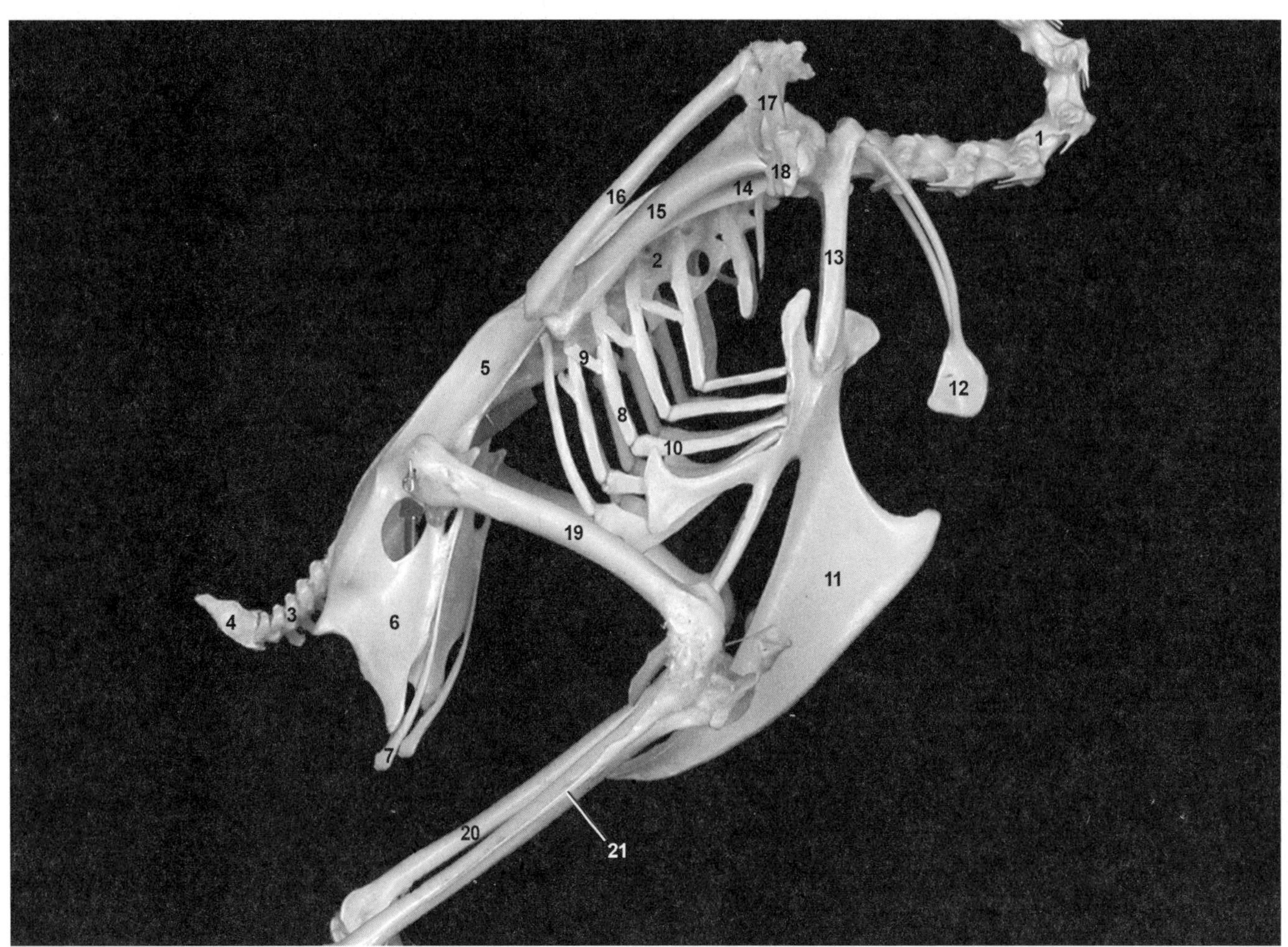

FIGURE 6.4. Skeleton of Domestic Chicken, lateral view.

1. Cervical vertebrae
2. Thoracic vertebrae
3. Caudal vertebrae
4. Pygostyle
5. Ilium
6. Ischium
7. Pubis
8. Dorsal rib
9. Uncinate process
10. Ventral rib
11. Sternum
12. Furculum
13. Coracoid
14. Scapula
15. Humerus
16. Ulna
17. Carpometacarpus
18. Phalanges
19. Femur
20. Tibiotarsus
21. Fibula

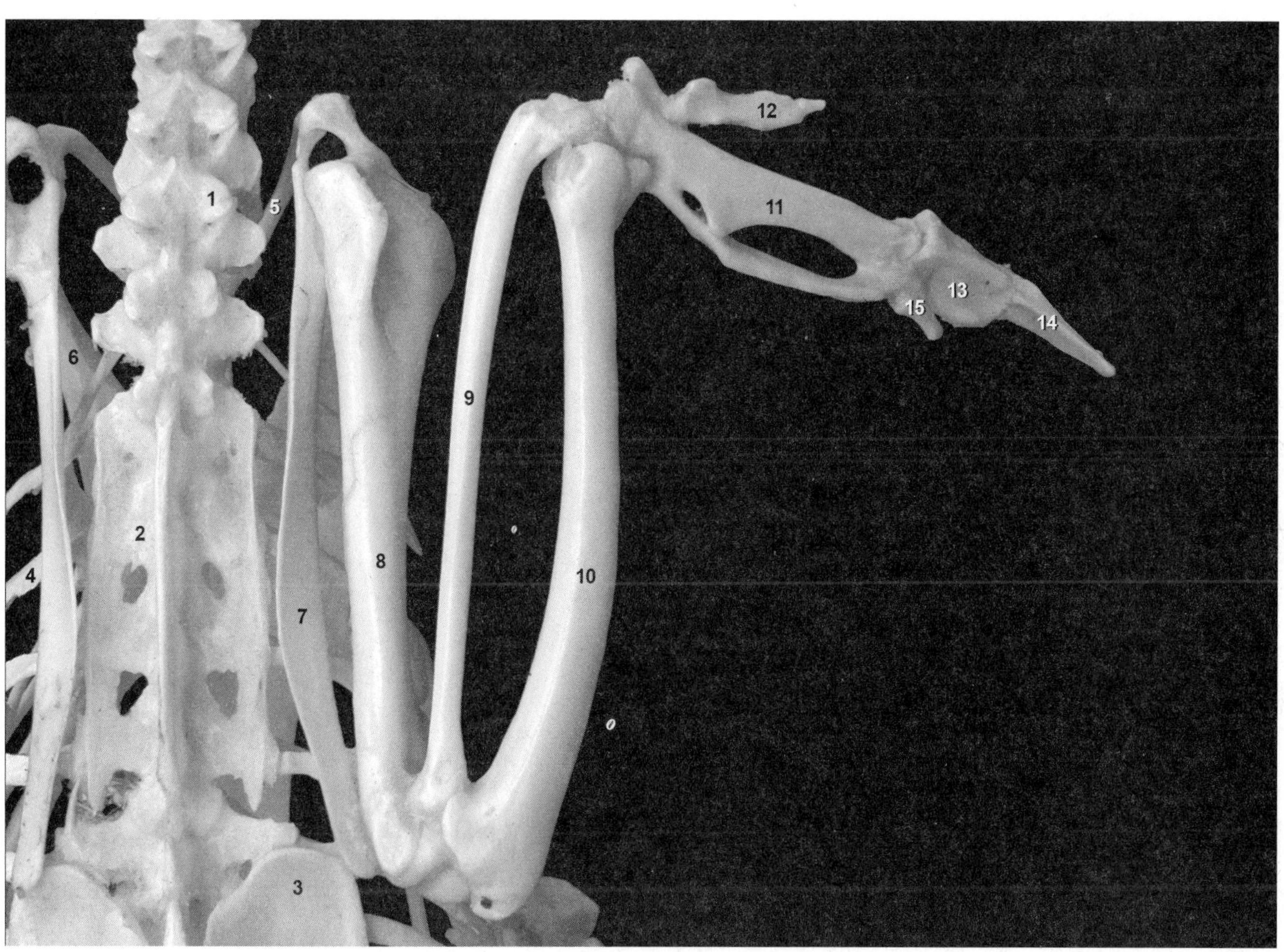

FIGURE 6.5. Vertebral column and forelimb elements of Domestic Chicken, dorsal view.

1. Cervical vertebrae
2. Thoracic vertebrae
3. Ilium
4. Dorsal rib
5. Furculum
6. Coracoid
7. Scapula
8. Humerus
9. Radius
10. Ulna
11. Carpometacarpus
12. Pollex; Digit I, Phalanx 1
13. Digit II, Phalanx 1
14. Digit II, Phalanx 2
15. Digit III, Phalanx 1

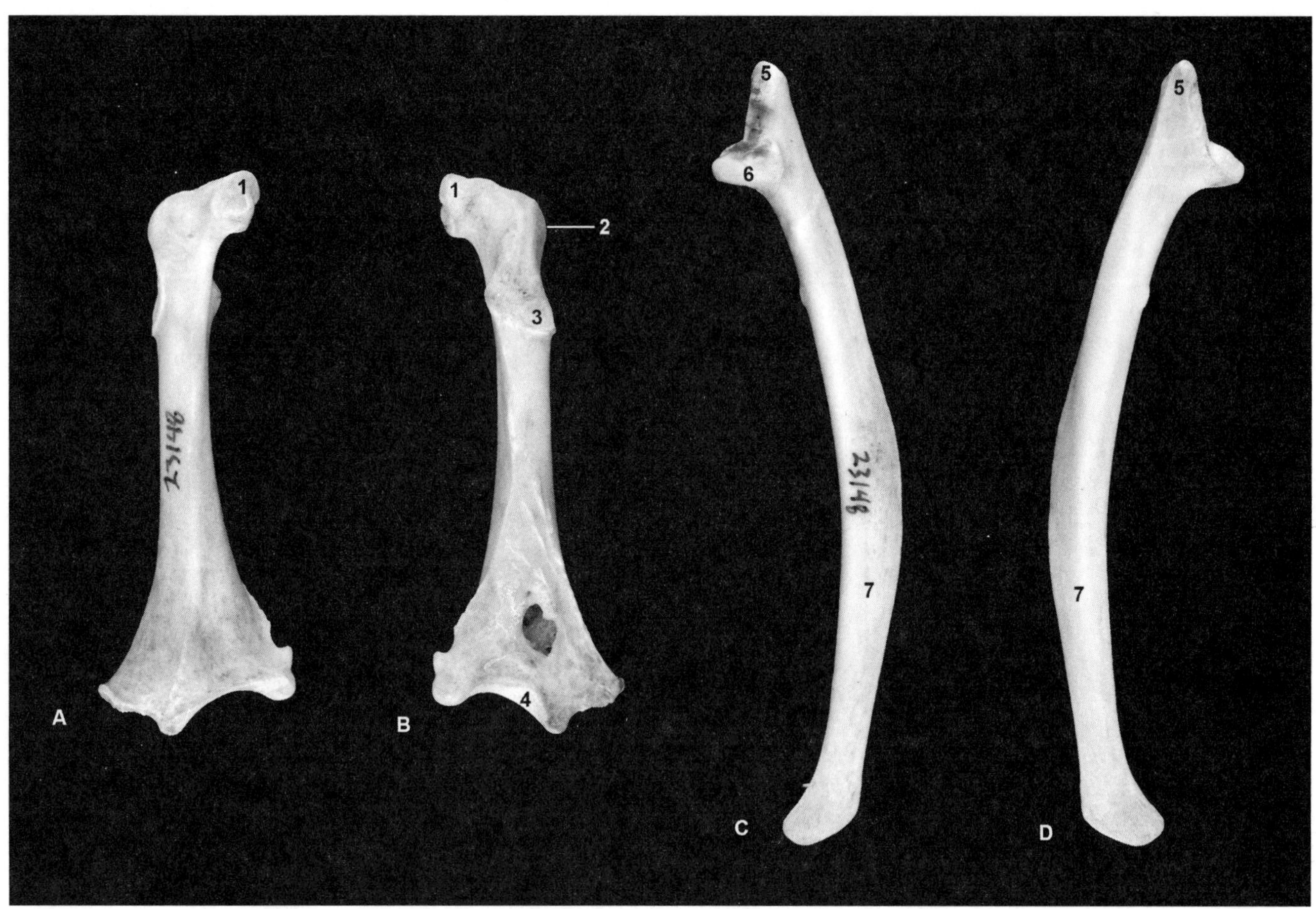

FIGURE 6.6. Pectoral girdle elements of Dusky Grouse (*Dendragapus obscurus*): r. coracoid, anterior (A) and posterior (B) views, bottom is proximal; l. scapula, dorsal (C) and ventral (D) views; bottom is distal.

1. Head
2. Glenoid facet
3. Scapular facet
4. Sternal facet
5. Acromion
6. Glenoid facet
7. Blade

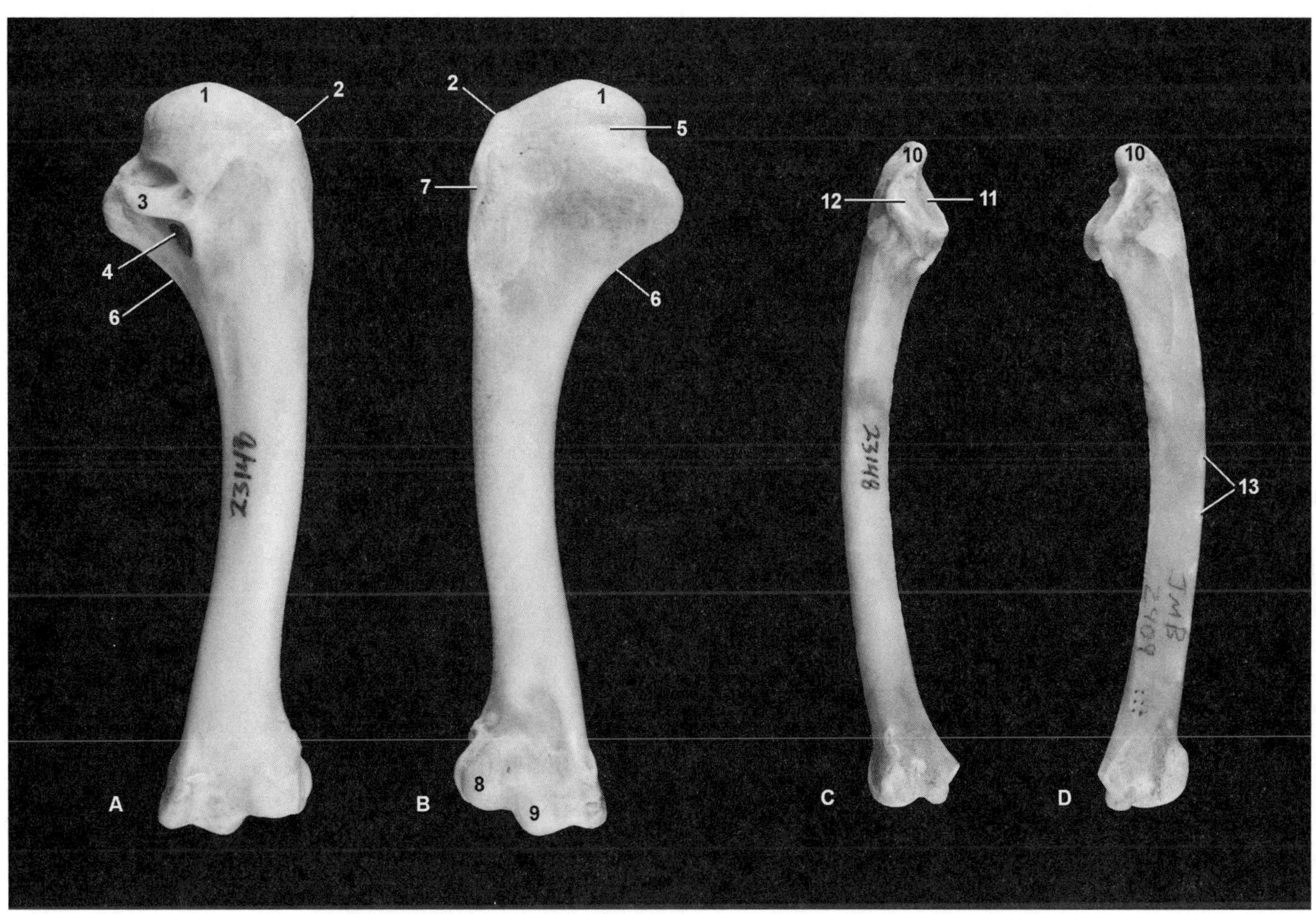

FIGURE 6.7. Forelimb elements of Dusky Grouse: r. humerus, posterior (A) and anterior (B) views; r. ulna, dorsal (C) and ventral (D) views; top is proximal.

1. Head
2. External tuberosity
3. Internal tuberosity
4. Pneumatic foramen
5. Ligamental furrow
6. Bicipital crest
7. Deltoid crest
8. External condyle
9. Internal condyle
10. Olecranon process
11. Internal cotyla
12. External cotyla
13. Papillae

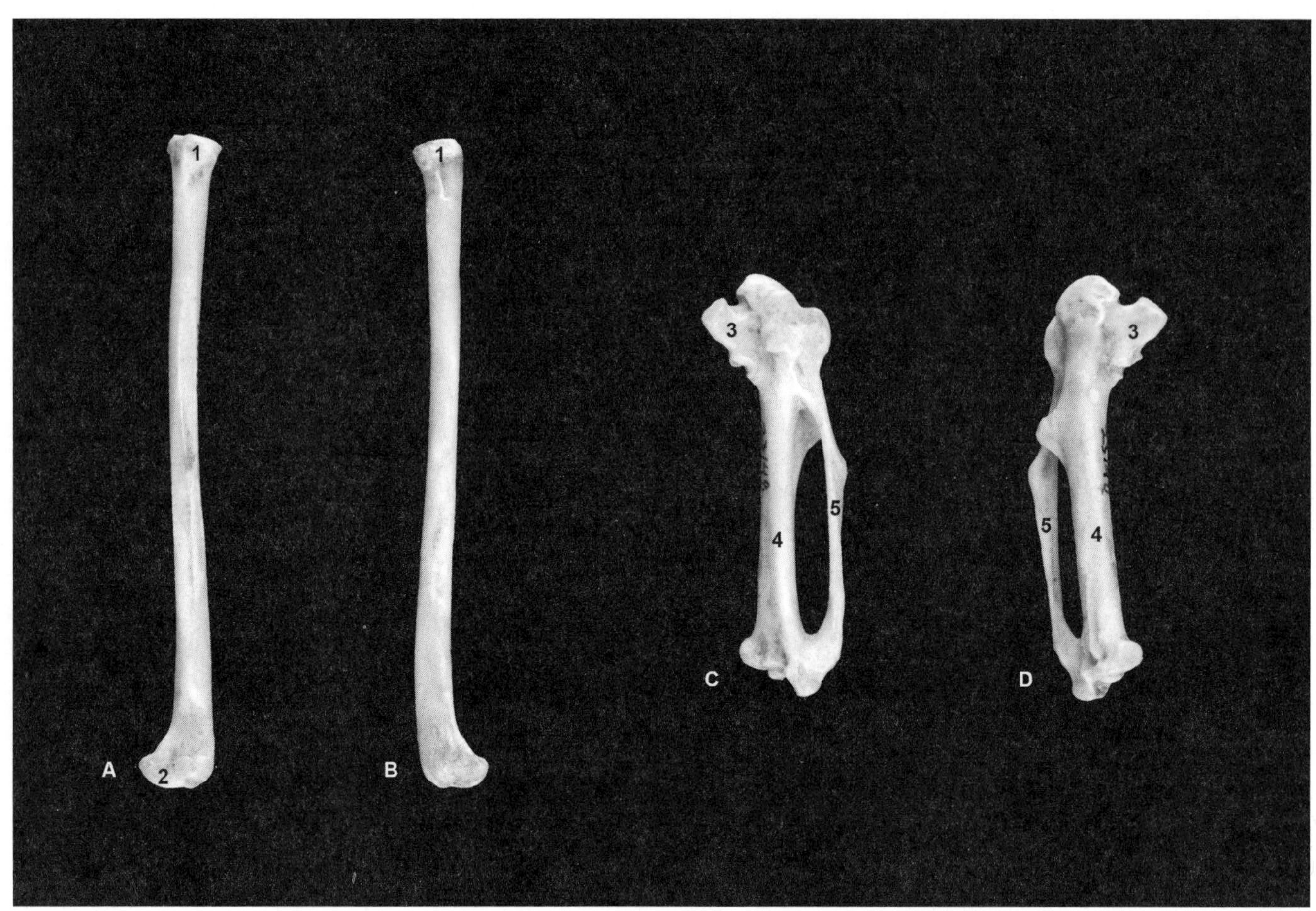

FIGURE 6.8. Forelimb elements of Dusky Grouse: r. radius, dorsal (A) and ventral (B) views; r. carpometacarpus, ventral (C) and dorsal (D) views; top is proximal.

1. Head
2. Scapho-lunar facet
3. Metacarpal I
4. Metacarpal II
5. Metacarpal III

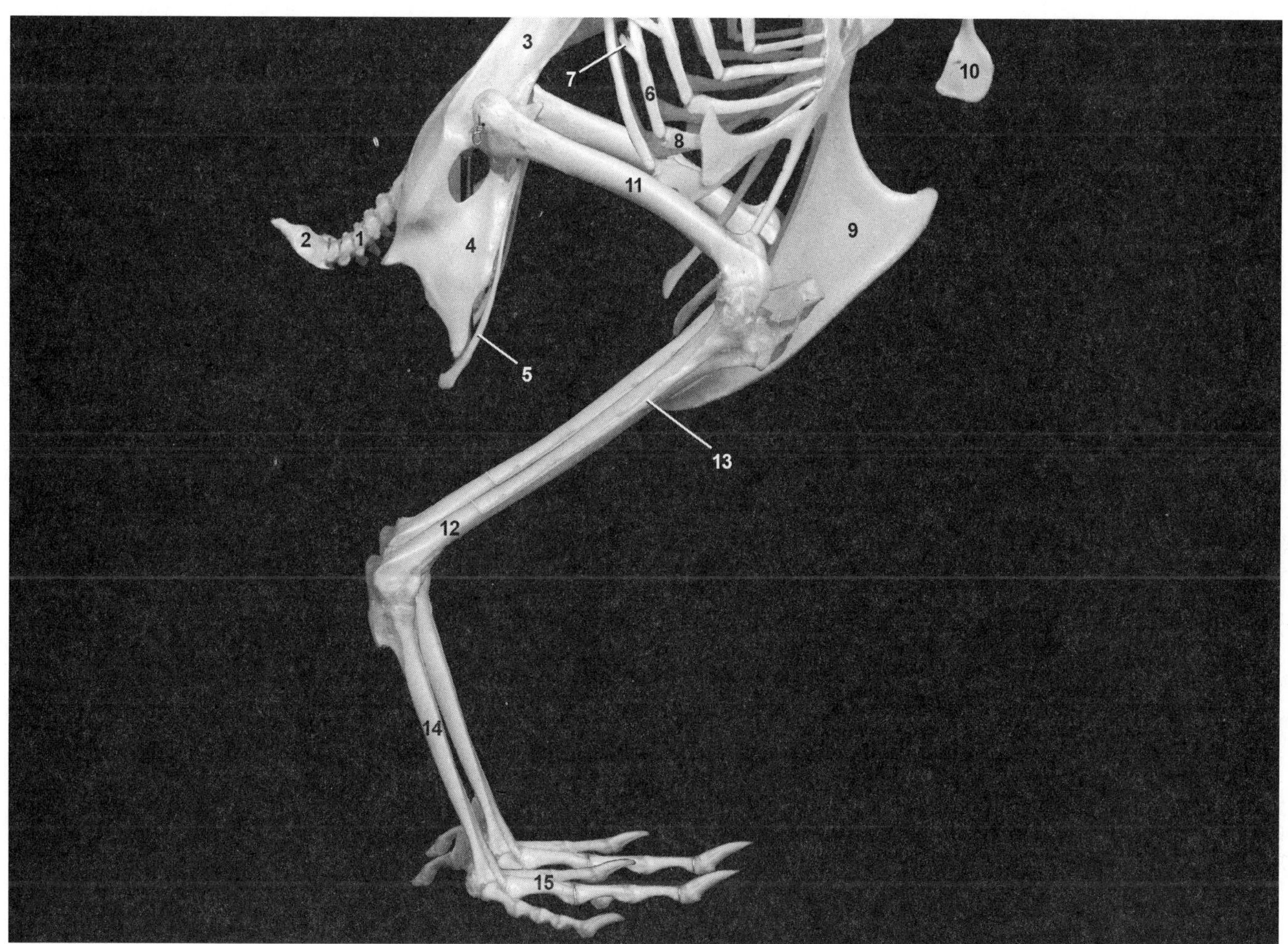

FIGURE 6.9. Pelvic girdle and hindlimb of Domestic Chicken, lateral view.

1. Caudal vertebrae
2. Pygostyle
3. Ilium
4. Ischium
5. Pubis
6. Dorsal rib
7. Uncinate process
8. Ventral rib
9. Sternum
10. Furculum
11. Femur
12. Tibiotarsus
13. Fibula
14. Tarsometatarsus
15. Phalanges

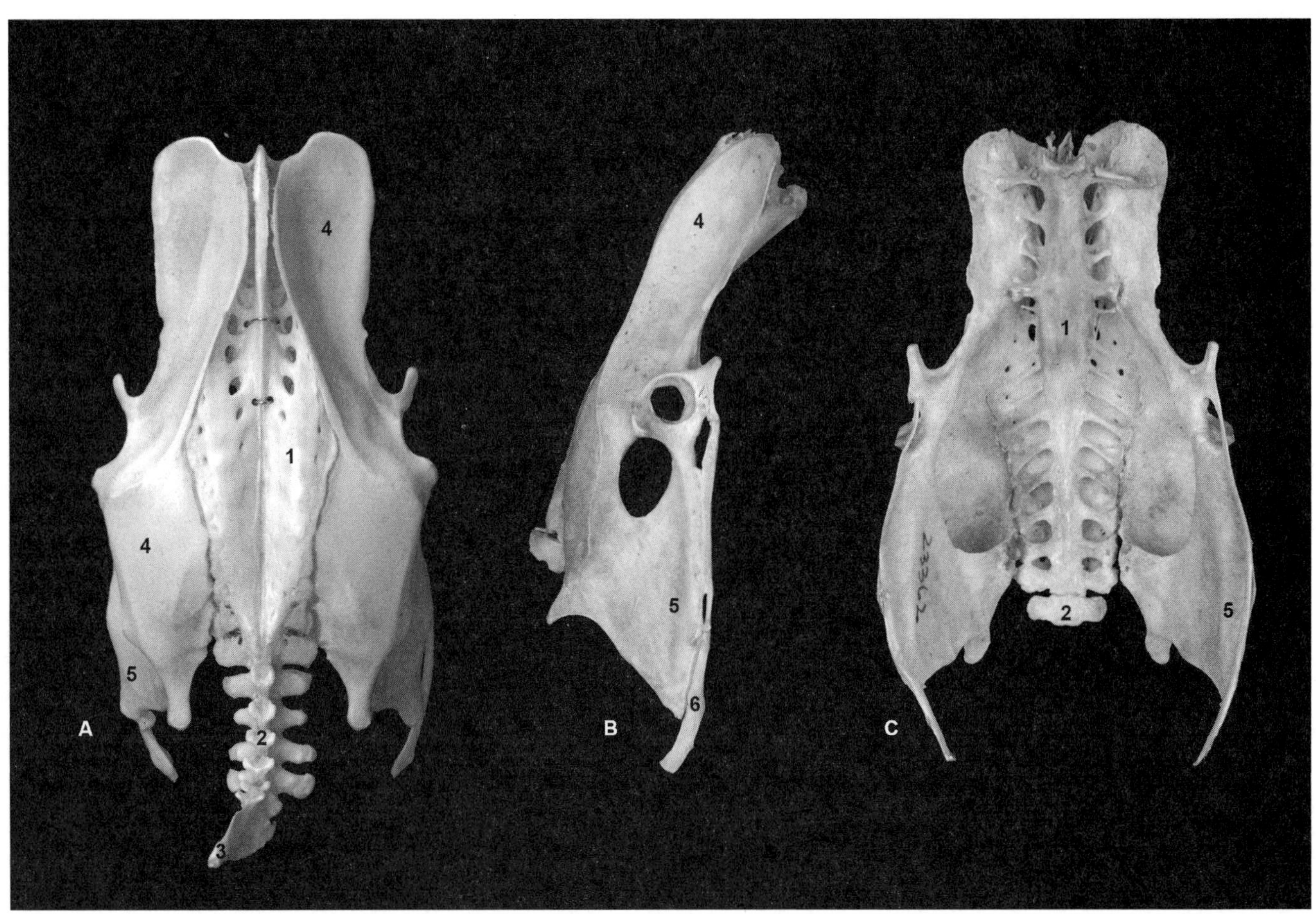

FIGURE 6.10. Pelvic girdle, synsacrum, and os coxa of Domestic Chicken, dorsal (A), lateral (B), and ventral (C), views; top is anterior.

1. Synsacrum
2. Caudal vertebrae
3. Pygostyle
4. Ilium
5. Ischium
6. Pubis

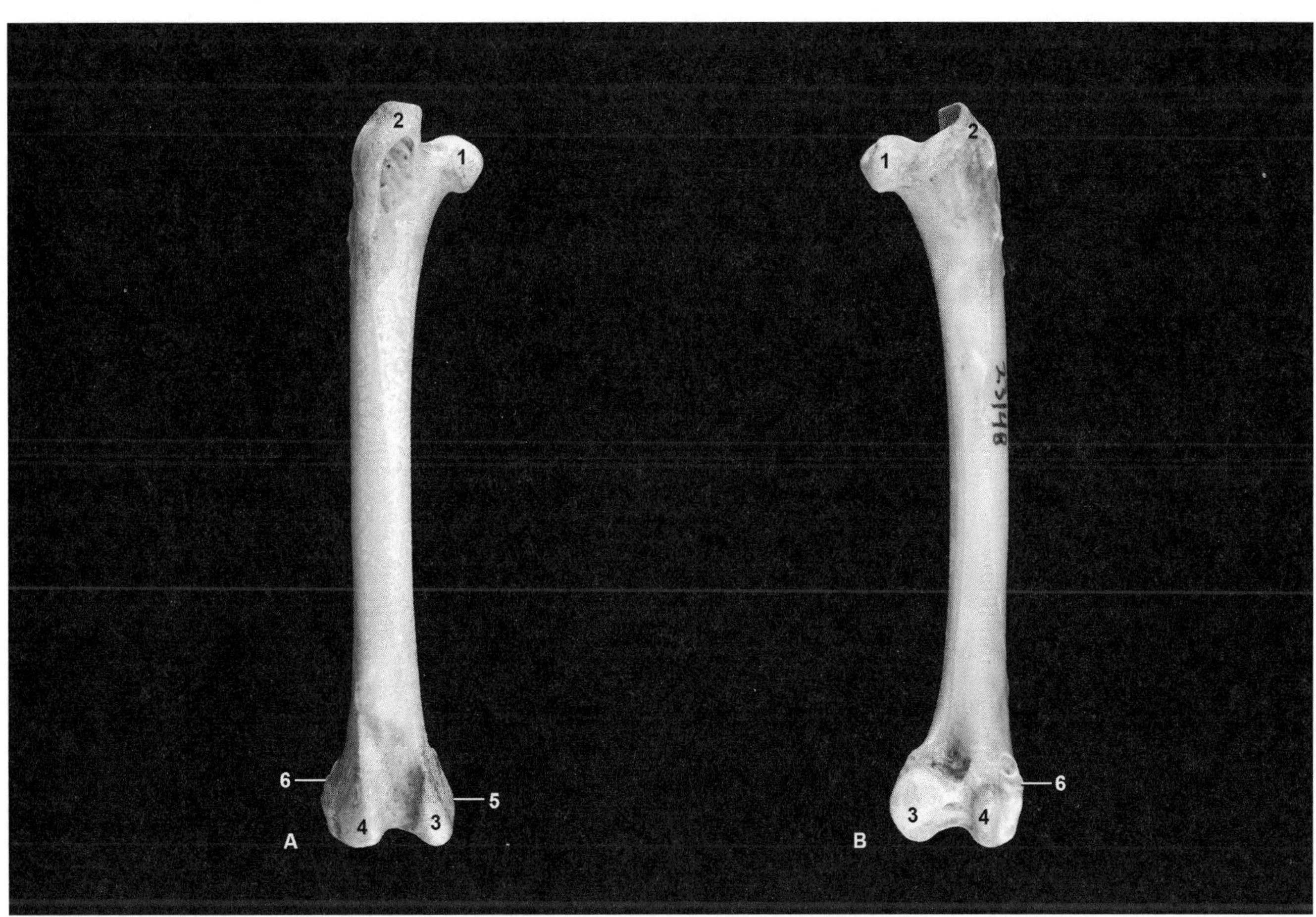

FIGURE 6.11. Dusky Grouse r. femur, anterior (A) and posterior (B) views; top is proximal.

1. Head
2. Trochanter
3. Medial condyle
4. Lateral condyle
5. Medial epicondyle
6. Lateral epicondyle

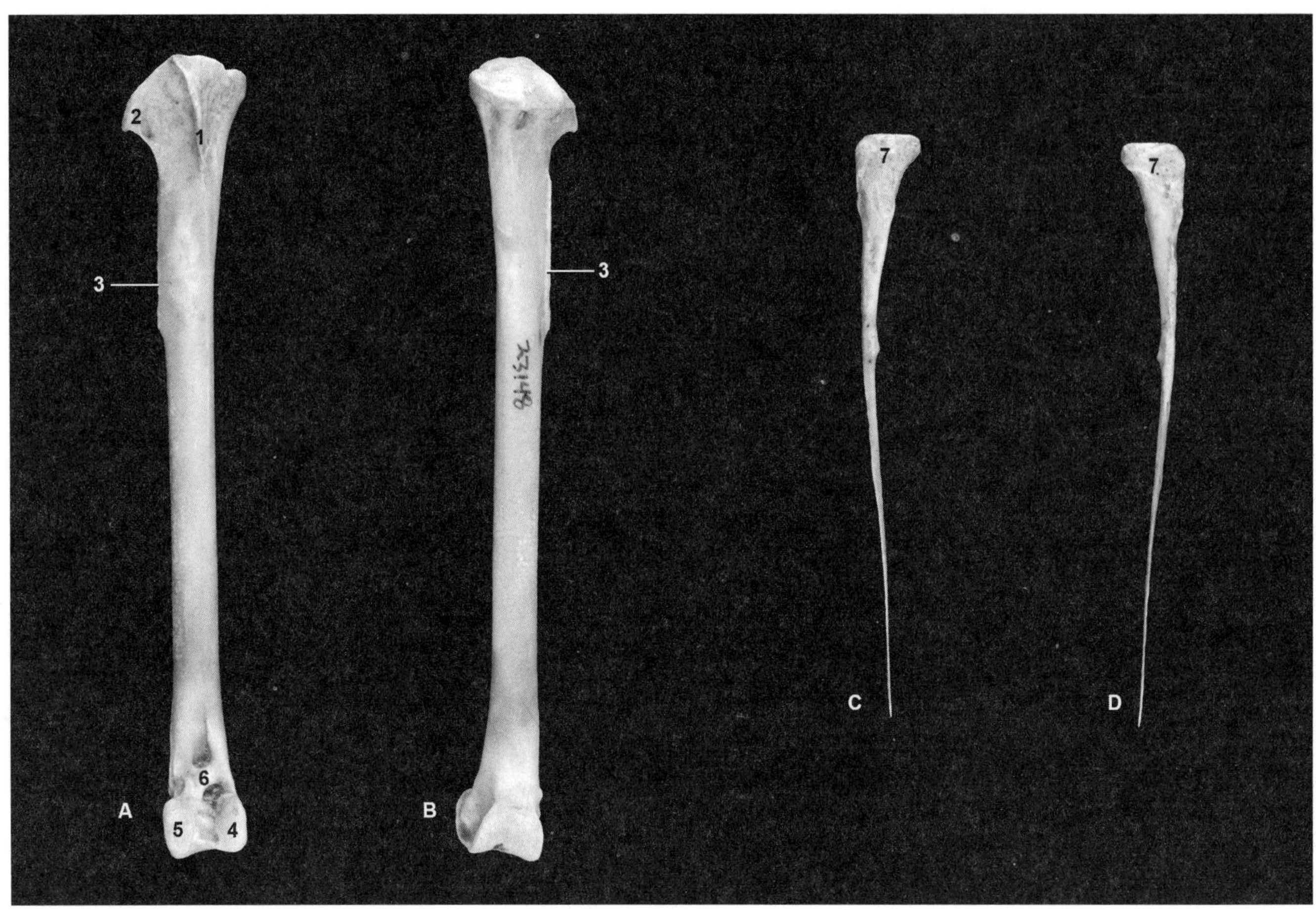

FIGURE 6.12. Hindlimb elements of Dusky Grouse: r. tibiotarsus, anterior (A) and posterior (B) views; l. fibula lateral (C) and medial (D) views; top is proximal.

1. Medial cnemial crest
2. Lateral cnemial crest
3. Fibular crest
4. Medial condyle
5. Lateral condyle
6. Supratendinal bridge
7. Head

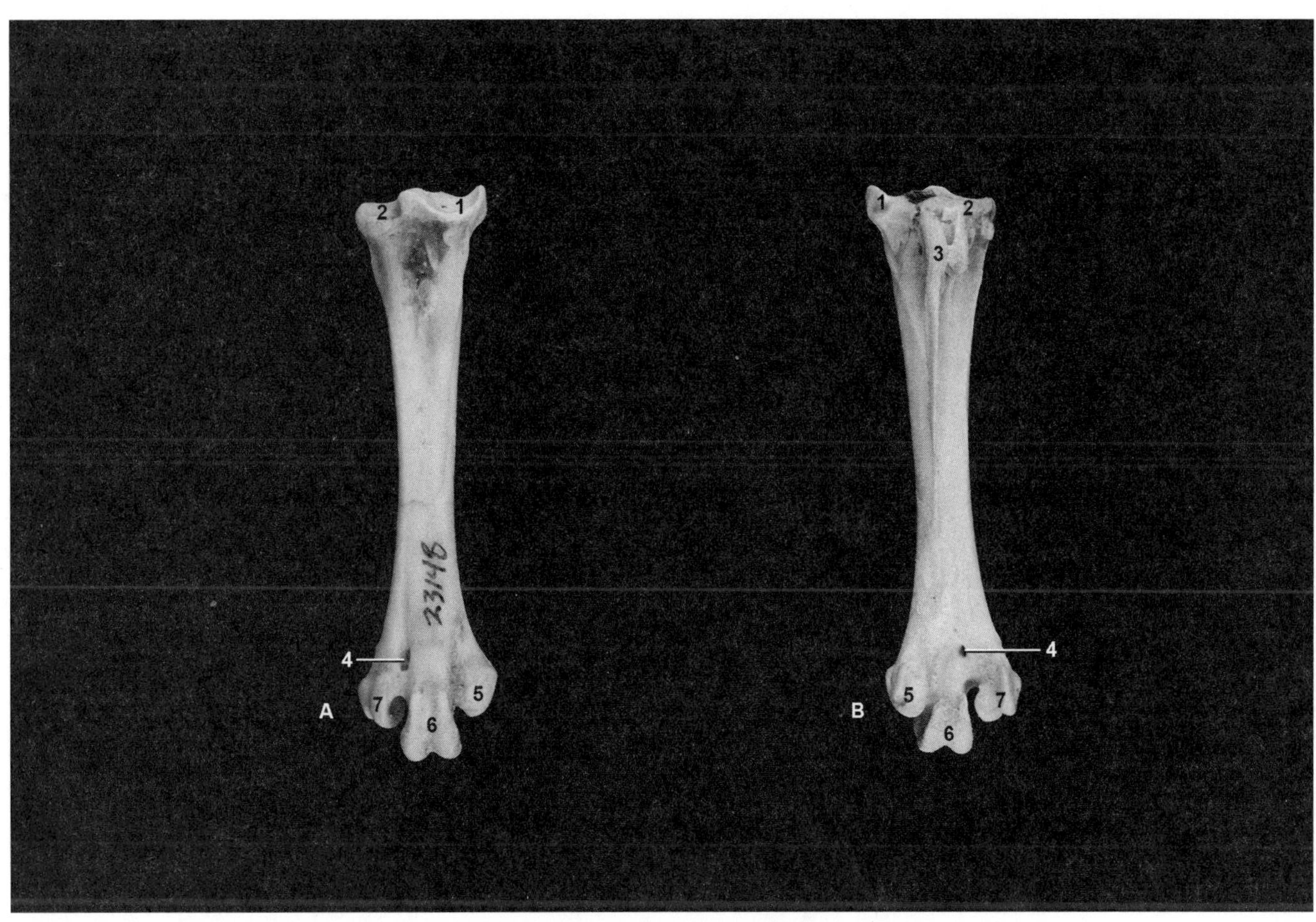

FIGURE 6.13. Dusky Grouse r. tarsometatarsus, anterior (A) and posterior (B) views; top is proximal.

1. Medial cotyla
2. Lateral cotyla
3. Hypotarsus
4. Distal foramen
5. Trochlea, Digit II
6. Trochlea, Digit III
7. Trochlea, Digit IV

II. Taxonomy and Osteological Variation of Western Birds

CLASS Aves (Birds)
ORDER Anseriformes (Screamers, Swans, Ducks, and Geese)
FAMILY Anatidae (Swans, Ducks, and Geese)

The waterfowl order Anseriformes (L., anser = goose) is represented by 17 genera in the west. They have large heads with conspicuous and horizontally flattened bills, long necks, heavy bodies, short tails, long wings, and webbed feet used for paddling (Figures 6.14–6.16; Color Plates 31–32). Dramatic sexually dichromatic plumage occurs in Northern Hemisphere ducks: males are brightly colored, while females display cryptic patterns of browns and grays. The sexes look alike in geese and swans. Anatids occur in a variety of aquatic habitats, and diets are variable, species specific, and may include mollusks and other aquatic invertebrates, fish, aquatic vegetation, plant seeds, and roots. The young are precocial and hatch with a covering of down. Most taxa breed in the north of the continent and winter in more southern latitudes.

Due to their large size and abundance, ducks and geese were clearly the most economically important order of birds for aboriginal peoples occupying western aquatic settings. Anatids can be used to reconstruct the nature of past aquatic environments and moisture history, since species differ in the water depth required for effective foraging (e.g., "diving" versus "dabbling" ducks).

FIGURE 6.14. Canada Goose (*Branta canadensis*).

FIGURE 6.15. Mallard (*Anas platyrhynchos*).

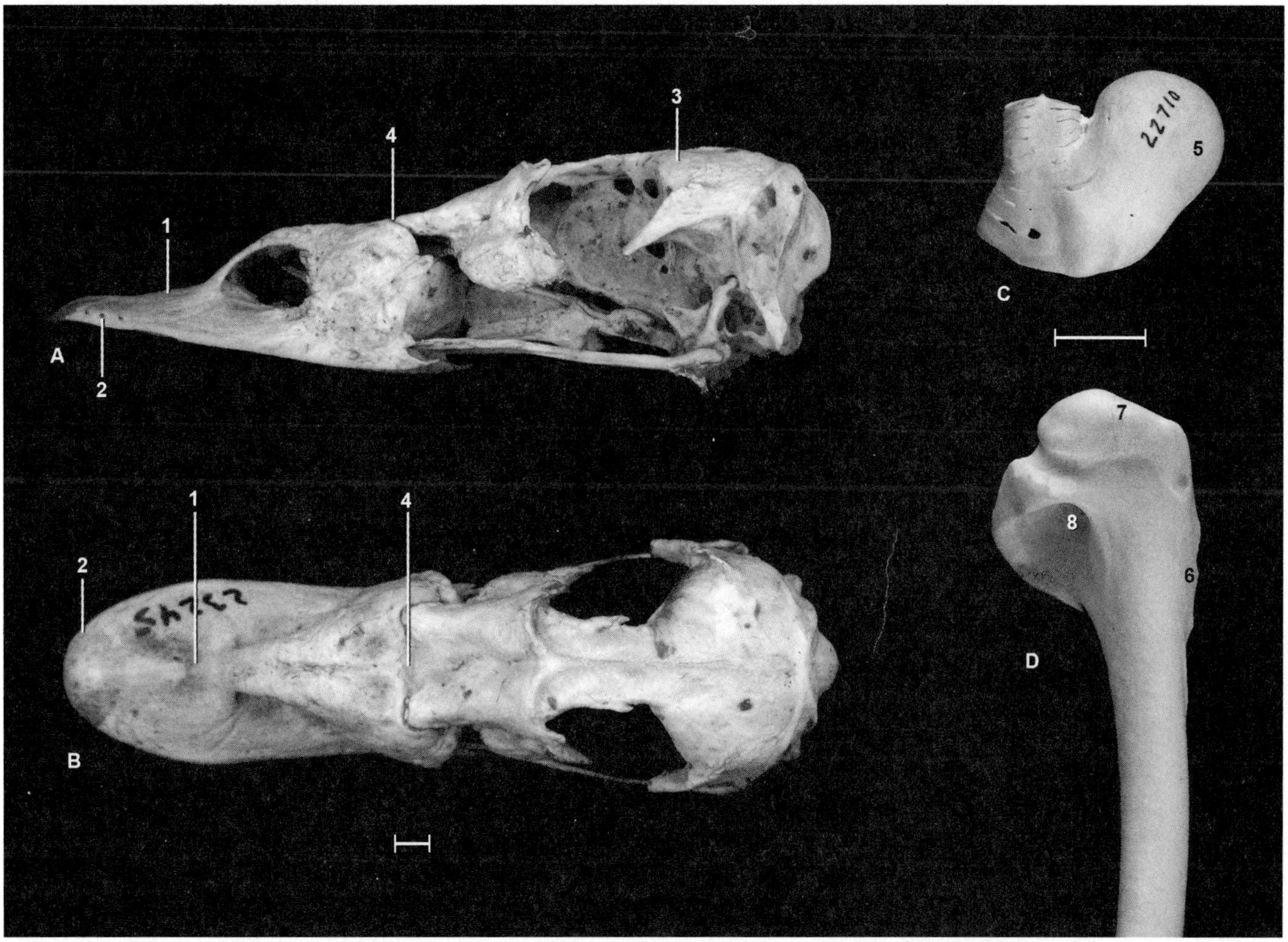

FIGURE 6.16. White-winged Scoter (*Melanitta fusca*) cranium, lateral (A) and dorsal (B) views; syringeal bulla (C), and proximal r. humerus (D), posterior view. Osteological features (Anatidae): 1, premaxilla dorso-ventrally flattened and spatulate; 2, premaxilla, dentary pitted at anterior end; 3, cranium long, narrow; 4, bill weakly connected to posterior cranium; 5, asymmetrical syringeal bulla well developed in most male ducks; proximal end of r. humerus with, 6, flaring bicipital crest, 7, broad head, and, 8, deeply excavated pneumatic foramen.

ORDER Galliformes (Grouse, Ptarmigan, Turkey, and Quail)

Two galliform (L., gallus = rooster) families, Phasianidae (grouse, turkey, ptarmigan) and Odontophoridae (quail), and 12 genera occur in the west (Figures 6.17–6.19; Color Plate 33). Galliformes are land birds well adapted for terrestrial life, with legs designed for walking and running and three long, forward-pointing toes used for scratching the ground as they forage. All have stout, decurved, sharply pointed, chicken-like bills. Their wings are generally short and rounded, giving them the capacity for explosive takeoffs and rapid bursts of flight; most, however, return to the ground soon after being flushed. Preferred habitats vary greatly among the represented species, from coniferous forests (Dusky Grouse, *Dendragapus obscurus*; Spruce Grouse, *Falcipennis canadensis*) to low-elevation, arid deserts (Gambel's Quail, *Callipepla gambelii*; Scaled Quail, *Callipepla squamata*), to arctic tundra (Willow Ptarmigan, *Lagopus lagopus*). Diets are also variable and mostly include vegetation such as leaves and seeds but can also include invertebrates.

FIGURE 6.17. Wild Turkey (*Meleagris gallopavo*).

Fossil evidence suggests that the Wild Turkey (*Meleagris gallopavo*; Figure 6.17) was widespread across the west during the Pleistocene, but that their range contracted to the eastern United States, Southwest, and Mexico as that period ended. Recent DNA and archaeological research suggests that at least two separate cases of turkey domestication occurred in prehistoric North America, one involving the subspecies from south-central Mexico (*M. g. gallopavo*) and a second involving either the Rio Grande (*M. g. intermedia*) or Eastern United States (*M. g. silvestris*) subspecies; the latter two represent the source for the domestic turkeys used by Precontact peoples of the Southwest. Domestic turkeys in both southern Mexico and the Southwest appear at roughly the same time between 1,800 and 1,500 years ago and represent the only cases of animal domestication occurring in Precontact North America. Other notable galliform taxa include the Greater Sage-Grouse (*Centrocercus urophasianus*)—males of this species can exceed 2.9 kg and display on traditional grounds in open sagebrush known as leks.

Being of substantial size, grouse were among the most highly sought terrestrial avian prey types for native peoples. Still highly prized as game birds today, introductions of many Old World taxa have established substantial populations in the west (e.g., Chukar, *Alectoris chukar*; Gray Partridge, *Perdix perdix*; Ring-necked Pheasant, *Phasianus colchicus*).

ORDER Gaviiformes (Loons)

FAMILY Gaviidae (Loons)

Gaviiformes (L., gavia = a bird) are large, bulky, piscivorous, waterbirds; dagger-like bills, short necks, long wings, and legs set far back on the body give them a distinctive shape (Figures 6.20–6.21). Only one species, the Common Loon (*Gavia immer*; Figure 6.20), occurs in inland settings in the west. Several marine taxa are, however, present in coastal settings. Common Loons breed principally in Canada on clear, freshwater lakes with rocky shorelines surrounded by forest. In inland settings south of the Canadian border, they are most abundant on lakes during spring

FIGURE 6.18. Ruffed Grouse (*Bonasa umbellus*).

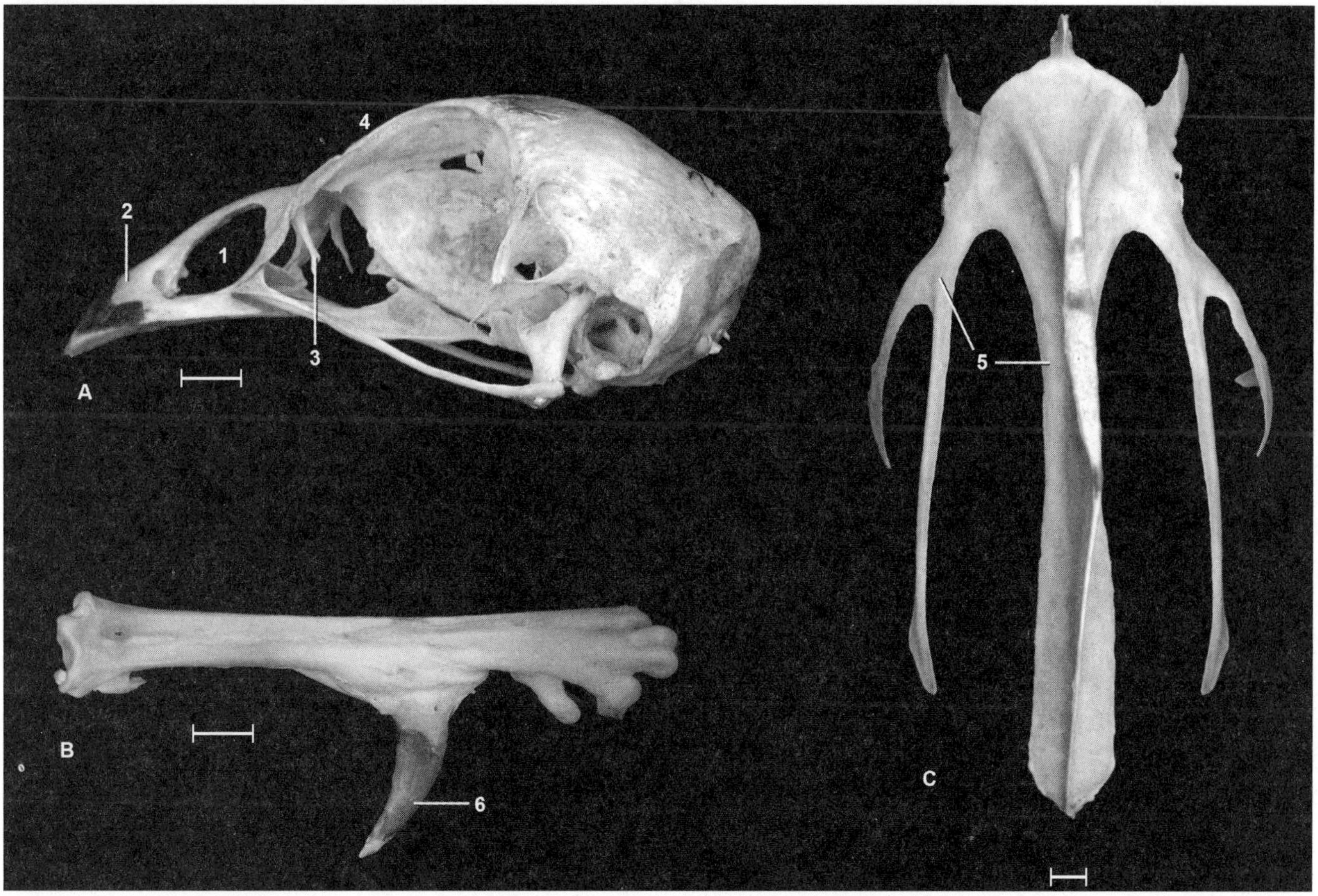

FIGURE 6.19. Domestic Chicken (*Gallus gallus*) cranium, lateral view (A); l. tarsometatarsus, anterior view (B), and sternum, ventral view (C). Osteological features (Galliformes): 1, external nares (bony nostril openings) large, perforate, occupy most of the bill; 2, bill heavy, stout, with downward curve; 3, small lacrimal; 4, braincase rounded, high domed; 5, sternum with narrow sternal plate, two-pronged lateral processes; 6, spurred tarsometatarsus (in males) in some taxa.

FIGURE 6.20. Common Loon (*Gavia immer*). Photo by USFWS/Steve Maslowski

and fall migrations. A symbol of the wild north, they are perhaps best known for their evocative eerie wails. During migration, Common Loons will occasionally mistake wet roads or parking lots for water and make crash landings. Here they become stranded since they have long takeoffs and require a considerable amount of open water to generate lift. Loons can also become stranded on small ponds.

ORDER Podicipediformes (Grebes)
FAMILY Podicipedidae (Grebes)

Grebes are sleek, diving birds that resemble loons but are smaller and more diverse; most species have pointed bills, long necks, short wings, almost no tails, and lobed toes extending from legs that arise far back on the body (Figures 6.22–6.23; Color Plate 34). The order and family names trans-

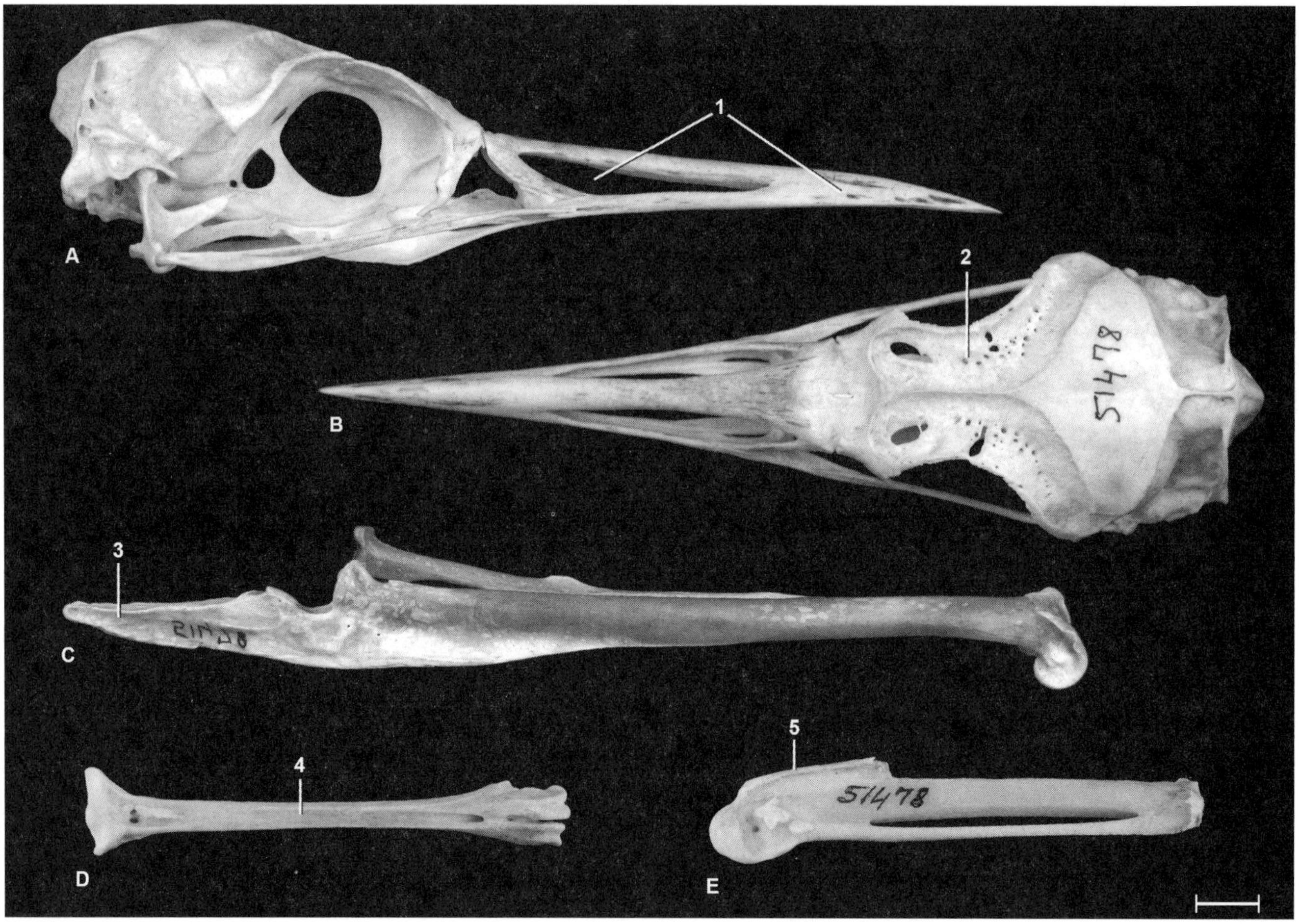

FIGURE 6.21. Pacific Loon (*Gavia pacifica*): cranium, lateral (A) and dorsal (B) views; r. tibiotarsus, medial view (C); r. tarsometatarsus, anterior view (D); and l. carpometacarpus, ventral (E) view. Osteological features (Gaviidae): 1, bill large, pointed, with large, perforated, external nares; 2, frontal with prominent salt-gland fossa; 3, proximal tibiotarsus with huge cnemial process; 4, tarsometatarsus shaft medial-laterally constricted; 5, metacarpal I of carpometacarpus extends (as thin shelf) distally ⅓ length of bone; 6, femur (not pictured), as in Podicipediformes (Figure 6.23), short, squat with bulbous proximal, distal ends.

late as "rump feet" (L., podicis = rump or vent; pes = foot) due to the unique posterior orientation of the legs and feet. Indeed, their leg morphology makes for efficient underwater propulsion, but they can barely walk on land. Small species feed on aquatic invertebrates; larger species are mostly piscivorous. Grebes ingest feathers to form a plug between the gizzard and intestine; they eject used feather plugs. Six species in three genera (*Podiceps, Aechmophorus, Podilymbus*) occur in western North America. The Eared Grebe (*Podiceps nigricollis*) is the most abundant grebe in the world and is also the best-represented species in the west. Eared Grebes breed in shallow wetlands and occur in highest densities during the fall on Mono Lake, California, and Great Salt Lake, Utah. In these migration staging areas they can double their weight, gorging on aquatic invertebrates

FIGURE 6.22. Western Grebe (*Aechmophorus occidentalis*).

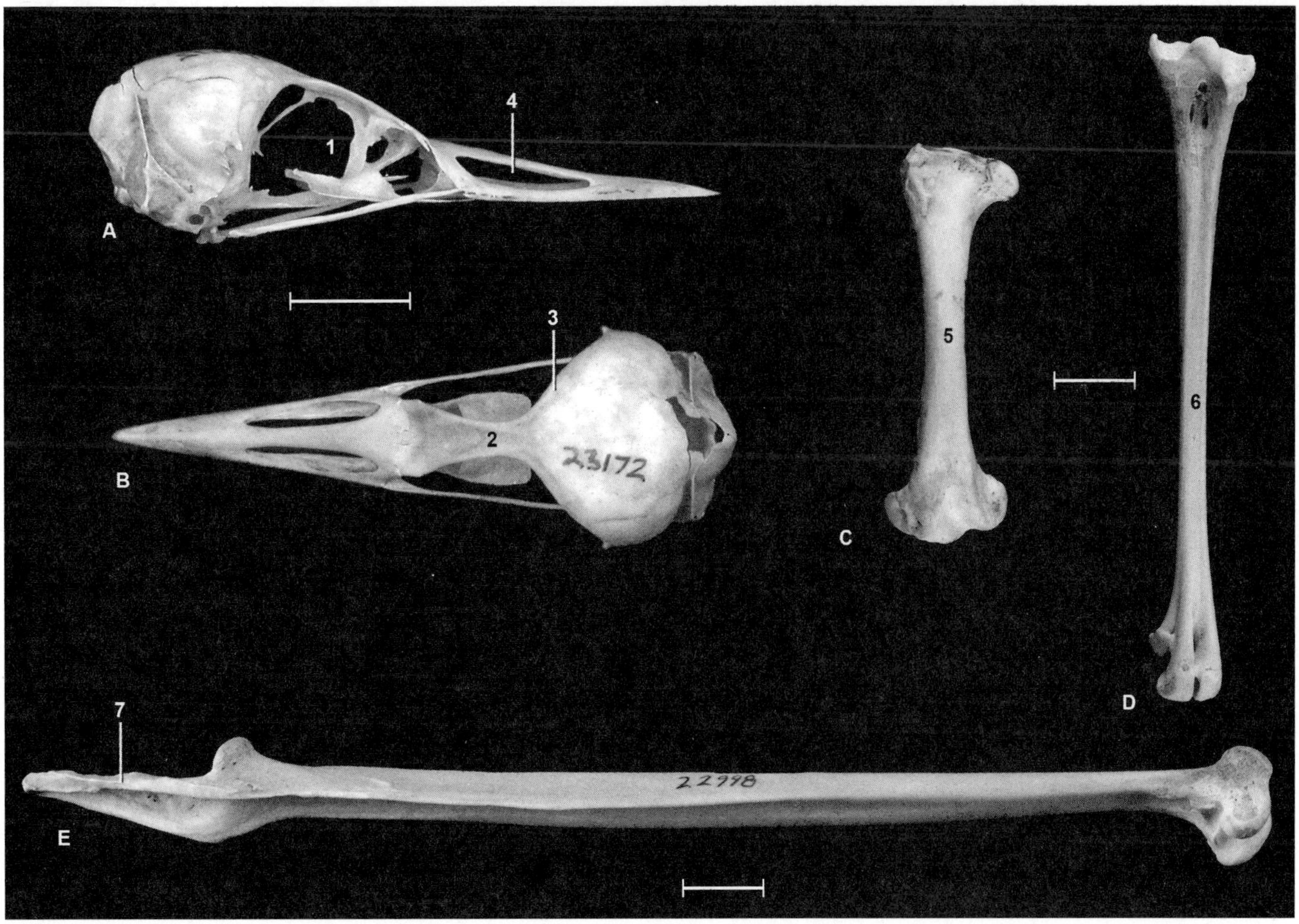

FIGURE 6.23. Eared Grebe (*Podiceps nigricollis*): cranium, lateral (A) and dorsal (B) views; and Western Grebe (*Aechmophorus occidentalis*) r. femur, anterior view (C), l. tarsometatarsus, anterior view (D), and r. tibiotarsus, medial view (E). Osteological features (Podicipedidae): 1, lacks perpendicular plate of ethmoid; 2, narrow frontal (medial-laterally); 3, lacks salt-gland fossa on frontal; 4, external nares perforate; leg elements similar to Gaviiformes but smaller: 5, femur short, squat with bulbous proximal, distal ends; 6, medial-laterally constricted shaft of tarsometatarsus; 7, proximal tibiotarsus with pronounced cnemial process.

FIGURE 6.24. Double-crested Cormorant (*Phalacrocorax auritus*).

(especially Brine Shrimp [*Artemia franciscana*] and Brine Flies [*Ephydra hians*, *E. cinerea*]) in preparation for a nonstop flight to wintering areas in the southwestern United States and Mexico.

ORDER Suliformes (Cormorants, Shags, Boobies, and Allies)

FAMILY Phalacrocoracidae (Cormorants and Shags)

In the west, Suliformes (Old Norse, sula = gannet) are represented by a single family, Phalacrocoracidae (cormorants) and a single species, Double-crested Cormorant (*Phalacrocorax auritus*; Figures 6.24–6.25; Color Plates 35–36). Double-crested Cormorants are large, shiny-black waterbirds with long necks and long, hooked bills that they use

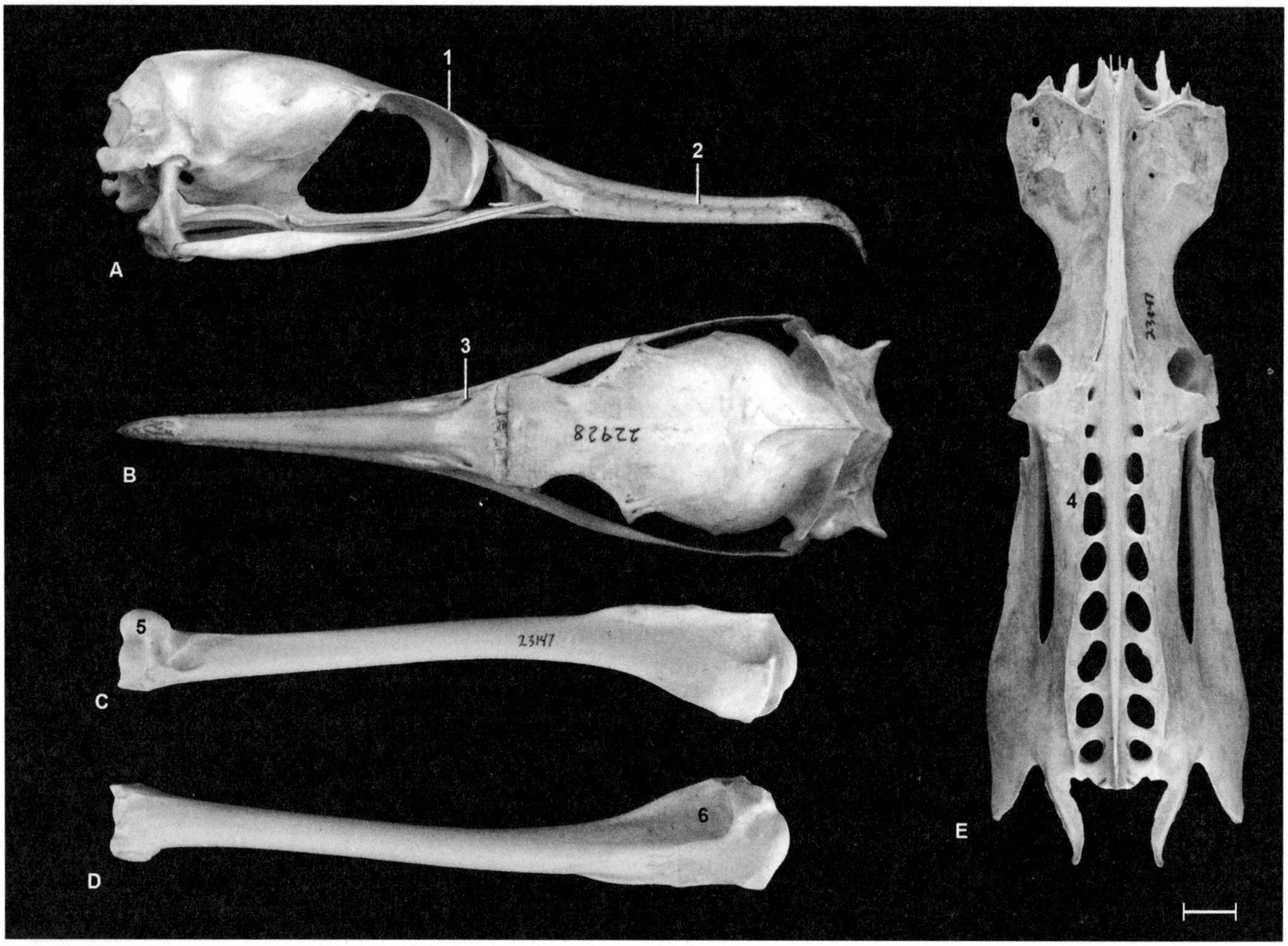

FIGURE 6.25. Double-crested Cormorant (*Phalacrocorax auritus*) cranium, lateral (A) and dorsal (B) views; r. humerus, anterior (C) and posterior (D) views; and os coxa, dorsal view (E). Osteological features (Phalacrocoracidae): 1, cranium long, low-domed, flat; 2, premaxilla long with hooked tip; 3, tiny external nares; 4, os coxa highly fenestrated dorsally, elongated and narrow, with anterior and posterior ends near equal maximum breadths; 5, distal humerus with narrow breadth, external (dorsal) condyle prominently hooked; 6, posterior surface of proximal humerus with deeply excavated ligamental furrow.

to grasp fish prey while swimming underwater. Double-crested Cormorants are found in a wide range of aquatic habitats from open coastline to inland lakes; they breed in large colonies in areas inaccessible to mammalian predators (tall trees, islands). Nesting material is primarily sticks but may also include various debris and junk (rope, deflated balloons, dead birds, bones). They cast pellets of undigested fish bone and exoskeletons of invertebrates. Several of the more compelling cases of prehistoric human impacts on continental avian faunas have involved cormorants.

ORDER Pelecaniformes (Pelicans, Herons, Egrets, Ibises, and Allies)

Order Pelecaniformes (L., pelecanus = water bird) is represented by three families in western North America: Pelecanidae (pelicans), Ardeidae (herons, egrets, bitterns), and Threskiornithidae (ibises). American White Pelican (*Pelecanus erythrorhynchos*; Figure 6.26; Color Plate 37), the only western inland pelican species, is a giant waterbird with wingspan up to 2.75 m; it is readily identified by its enormous bill and large pouch used for catching fish. Pelicans breed on lakes in the mountain west and winter along the coasts. Typically gregarious, they often forage in small teams on the surface of lakes where they "drive" fish into small concentrations; all at once, they scoop them up by simultaneously dipping their heads and pouches under water. They are not known to cast pellets.

FIGURE 6.26. American White Pelican (*Pelecanus erythrorhynchos*).

Represented by 10 species in the region, Ardeids are medium to large wading birds with a distinctive upright stance and long, skinny legs, necks and bills (Figures 6.27–6.29). Strictly carnivorous, most feed on fish and other aquatic animals. Rodents are also on the menu for some of the larger species. All species possess powder-down breast and rump patches that provide a dust used for preening. Ardeids cast pellets of bone and mammal fur. These birds have long S-shaped necks; uniquely, the esophagus and trachea are situated posterior to the vertebral column for a substantial span of the neck. The Great Blue Heron (*Ardea herodias*; Figure 6.29) is the largest, most widespread, and familiar of the herons. It can be found along calm freshwater and seacoasts and usually nests in trees near water.

FIGURE 6.27. Snowy Egret (*Egreta thula*).

FIGURE 6.28. Green Heron (*Butorides virescens*).

The only represented threskiornithid, White-faced Ibis (*Plegadis chihi*), is a medium-sized aquatic wader with a long, decurved (curved downward) bill (Figure 6.29). Highly gregarious, these birds are often seen in large flocks.

ORDER Accipitriformes (Hawks, Eagles, and New World Vultures)

Three accipitriform (L., accipiter = hawk) families occur in the region: Cathartidae (New World vultures; Figure 6.30), Pandionidae (Osprey, *Pandion haliaetus*; Color Plate 38), and Accipitridae (hawks and eagles; Figures 6.31–6.32); 13 genera are represented. The New World vultures (Cathartidae) are large to massive, black, naked-headed

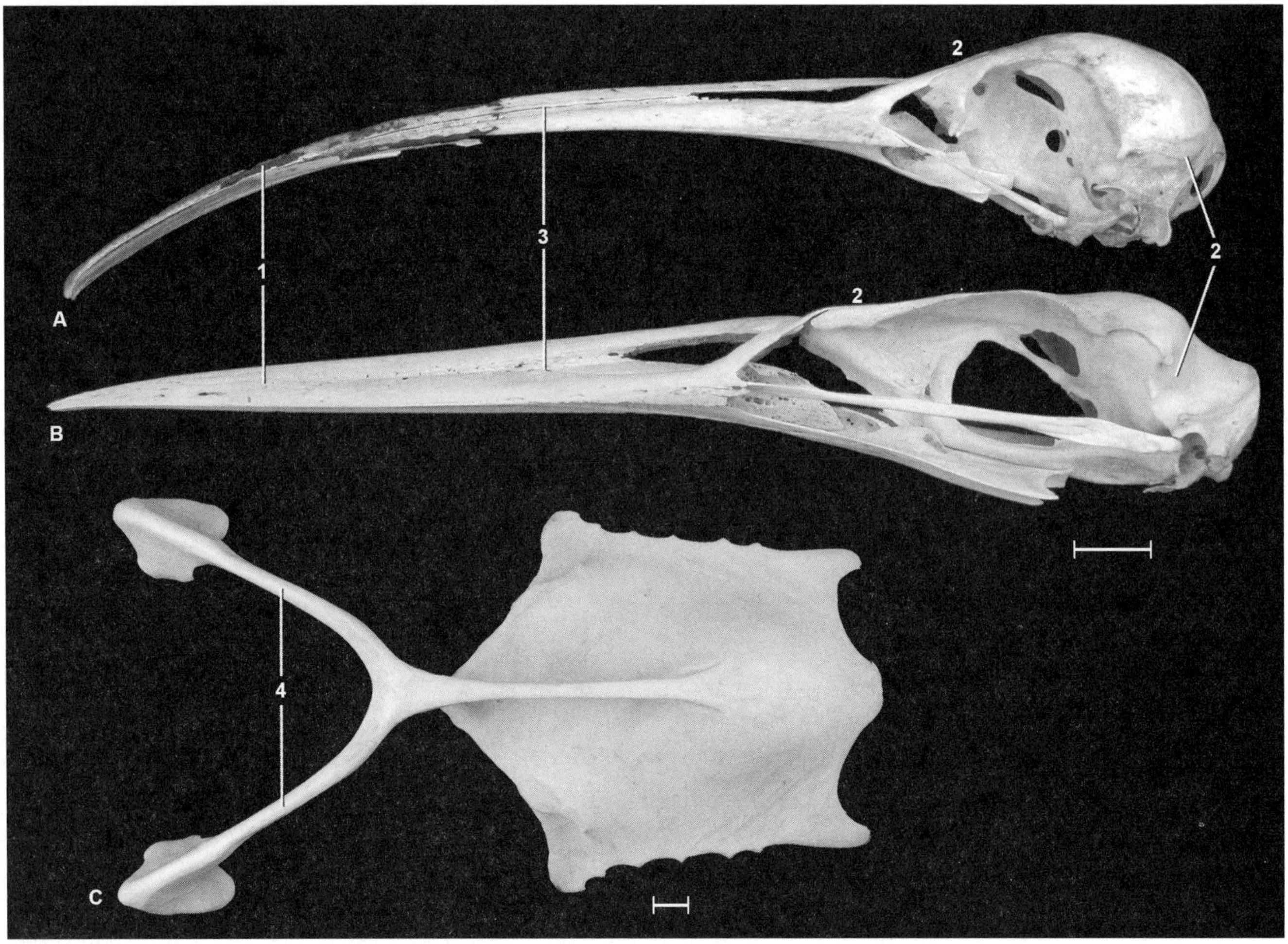

FIGURE 6.29. White-faced Ibis (*Plegadis chihi*), cranium, lateral view (A); Great Blue Heron (*Ardea herodias*), cranium, lateral view (B); American White Pelican (*Pelecanus erythrorhynchos*), sternum and furculum, ventral view (C). Osteological features (Pelecaniformes): 1, bill long and either decurved with rounded tip (in *Plegadis*, A), straight and pointed (in Ardeidae, B), or flattened with hooked tip (in *Pelecanus*, not shown); 2, cranium flat, low domed, narrow; 3, long, narrow groove extends from external nares distal to bill tip—less pronounced in *Pelecanus*; 4, U-shaped furculum fused to sternum (carina) in adult *Pelecanus*.

scavengers. In the morning, perched birds often sit with their wings outstretched, apparently to warm the body or dry feathers. Only the Turkey Vulture (*Cathartes aura*) is widespread in the region.

During the Pleistocene, the California Condor (*Gymnogyps californianus*) once ranged over much of North America. That range has contracted dramatically over the millennia, especially so in the last century due to poisoning from ingesting lead bullets or shot, among other factors. The largest bird in North America, the last surviving wild condor was captured in the late 1980s, rendering the species extinct in the wild. Since that time, however, a captive breeding program has generated a sizable stock of birds, many having been released at various sites in California, Arizona, Utah, and Baja California, Mexico. As of this writing (October 2014), approximately 400 condors exist, with about half of those living in the wild. Wild birds are reproducing but have not produced young that have survived to breeding age (~6–8 years).

Pandionidae is represented by a single, widespread species, Osprey. Appearing almost gull-like in flight, this bird is piscivorous and catches fish by hovering and then plunging feet-first into water.

Accipitrids are small to large diurnal raptors with short, hooked beaks for ripping flesh and powerful legs with sharp talons for grasping and killing mostly vertebrate prey. Most species cast pellets of bone and fur. The keen eyesight of these birds—four to eight times better than humans—enables them to spot prey at great distances. For reasons unclear, hawks and eagles exhibit reversed body size sexual dimorphism: females are larger than males. Most species are territorial and monogamous, often mating for life. Many western raptors migrate from their summer breeding ranges to more southerly winter ranges. Swainson's Hawk (*Buteo swainsoni*), the western region's impressive accipitrid migrant, breeds across much of western North America but winters in Argentina and Brazil. Raptors migrate by day apparently to take advantage of more favorable wind conditions. Perhaps as a result, most other birds migrate by night.

FIGURE 6.30. Turkey Vulture (*Cathartes aura*).

FIGURE 6.31. Red-tailed Hawk (*Buteo jamaicensis*).

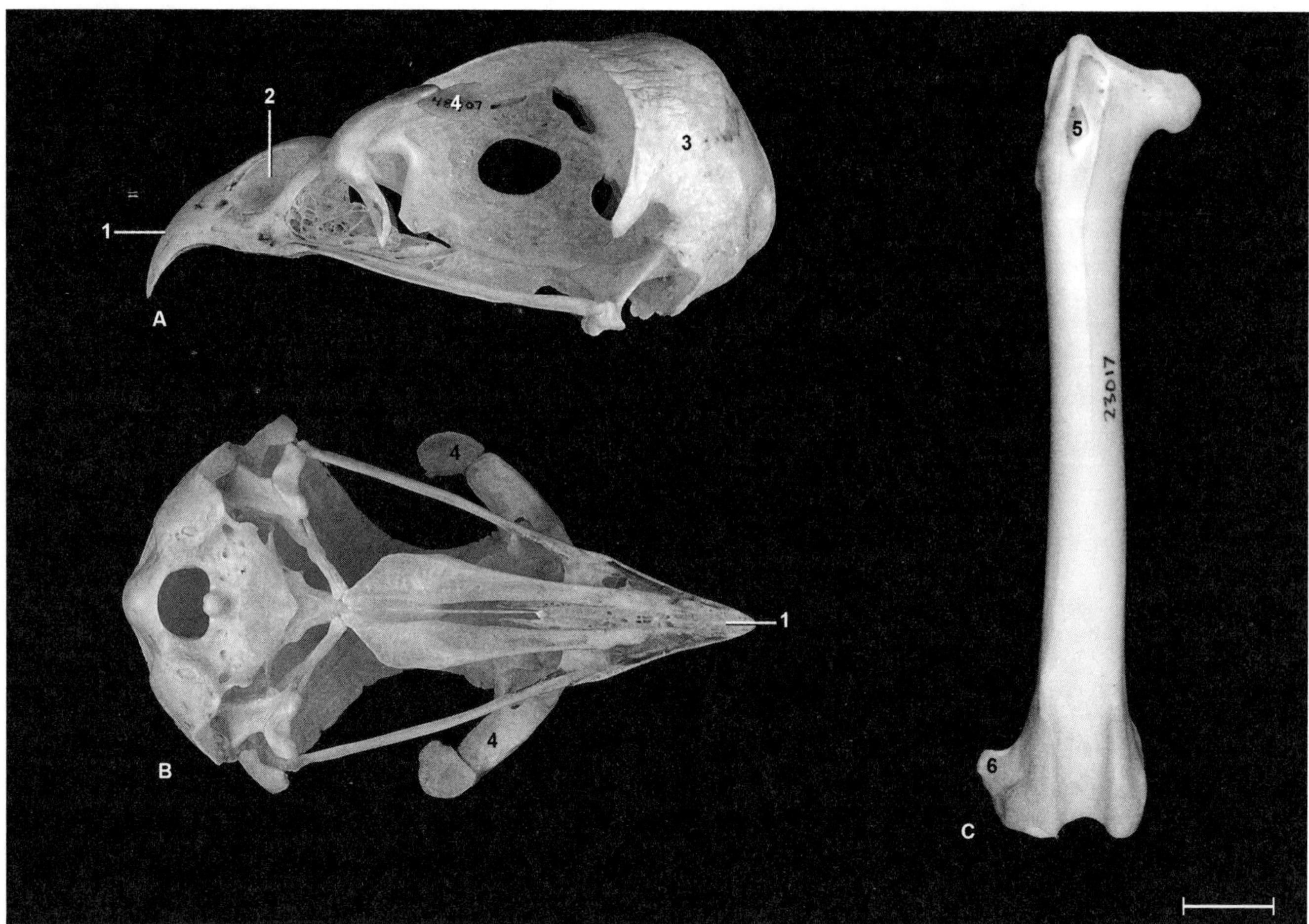

FIGURE 6.32. Red-shouldered Hawk (*Buteo lineatus*), cranium, lateral (A) and ventral (B) views; Red-tailed Hawk (*Buteo jamaicensis*), r. femur, anterior view (C). Osteological features (Accipitriformes): 1, premaxilla short, hooked, sharply pointed at tip—ventral surface lacks deep groove, as in Strigiformes (Figure 6.44), and antero-lateral margins lack "teeth" and notches, as in Falconiformes (Figure 6.54); 2, external nares large, oval—mostly imperforate in accipitrids, perforate in cathartids; 3, cranium short, round, maximum length longer relative to width compared to Strigiformes (Figure 6.44); 4, in accipitrids, lacrimal and supraocular well developed, extend postero-laterally into wing-like processes—lacrimals form blunt postorbital processes in cathartids (not shown); femur with, 5, prominent pneumatic foramen (medial to trochanter) and, 6, lateral epicondyle.

ORDER Gruiformes (Cranes, Coots, and Rails)

Two gruiform (L., grus = crane) families, Rallidae (coots and rails) and Gruidae (cranes), and six genera occur in the west. Rallids are small- to medium-sized, short-tailed, stubby-winged, wetland birds with laterally compressed bodies; they are typically solitary and shy. All are opportunistic omnivores and feed mostly on invertebrates and plant material. The most familiar, American Coot (*Fulica americana*; Figure 6.33), is larger and more gregarious than the other species and is superficially similar in morphology and habit to ducks. However, its pure black body and white, chicken-like beak clearly distinguish this swimming rail from the anatids.

Gruids are large, graceful wading birds with long necks and legs; wingspans range from 1.8–2.1 m. With variable diets, they inhabit a variety of freshwater wetlands and uplands, but are typically restricted to open habitats. Cranes are well known for complex "dance" displays. The primary western species, Sandhill Crane (*Grus canadensis*; Figures 6.34–6.35; Color Plate 39), is a tall, gray bird of open grasslands, meadows, and wetlands. It congregates in huge numbers in annual migrations to and from its northern breeding ranges. When birds reach cruising altitudes, which have been recorded up to 3,657 m, they fly in a V-formation and continually produce a distinctive, deep, rolling trumpet and rattling call.

FIGURE 6.33. American Coot (*Fulica americana*).

FIGURE 6.34. Sandhill Crane (*Grus canadensis*). Photo by Michael Broughton

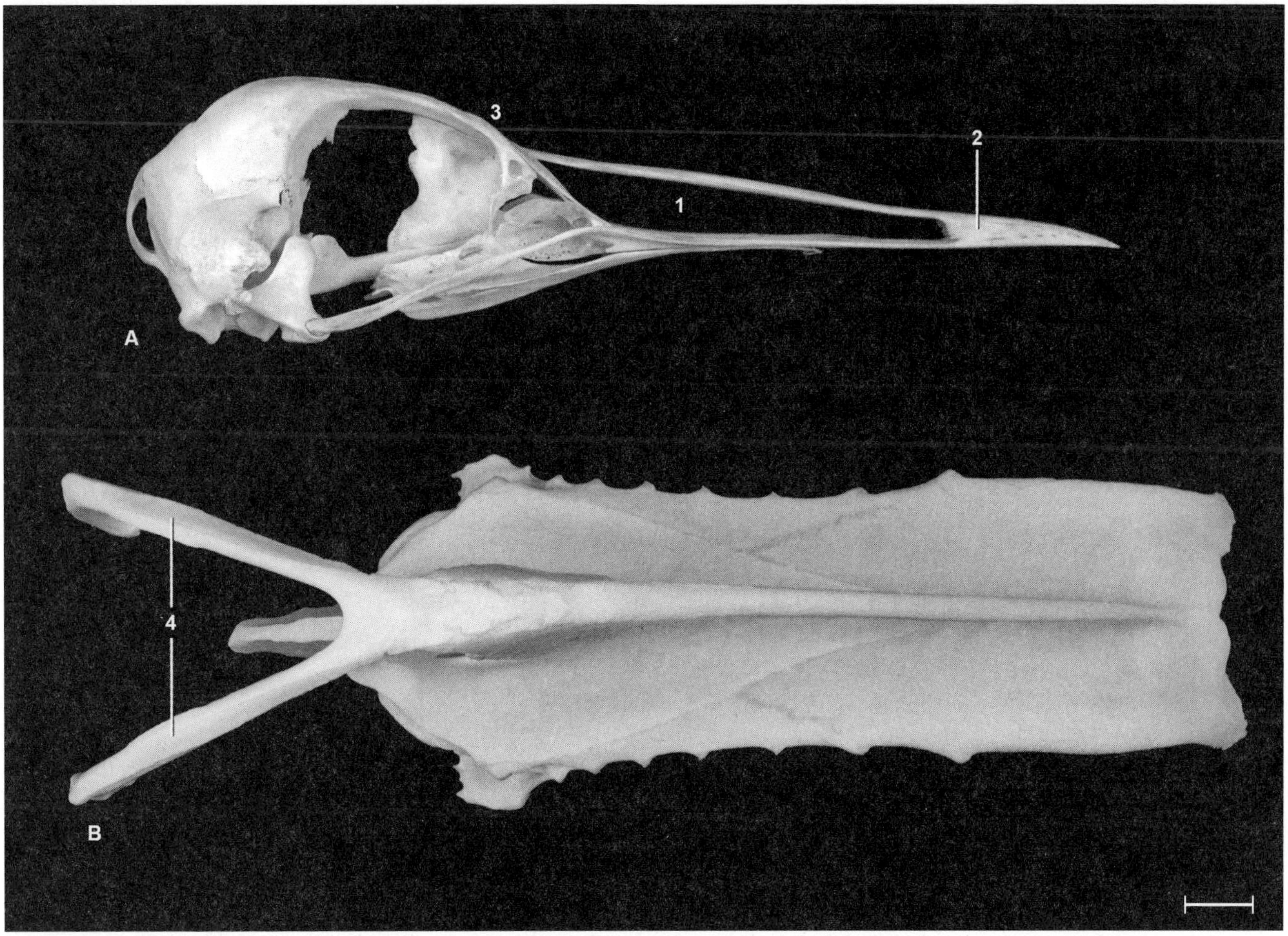

FIGURE 6.35. Sandhill Crane (*Grus canadensis*) cranium, lateral view (A), and furculum-sternum, ventral view (B). Osteological features (Gruiformes): 1, perforate external nares large, long, slit-like; 2, bill long, straight, pointed—no groove extends from nares as in Pelecaniformes; 3, frontal rises steeply from premaxilla; 4, V-shaped furculum fused to sternum (carina) in mature *Grus*.

ORDER Charadriiformes (Shorebirds, Gulls, Alcids, and Allies)

Charadriiformes (L., charadrius = yellowish bird) is a diverse order represented in the west by 23 genera and four families: Charadriidae (plovers), Recurvirostridae (avocets, stilts), Scolopacidae (sandpipers), and Laridae (gulls, terns); most members are closely associated with water. Charadriids are small to medium-sized shorebirds with big eyes, rounded heads, short, thick bills, and upright posture; they are found in a variety of open habitats, but typically near water. Killdeer (*Charadrius vociferus*) is the most familiar and widespread member of this family and is well known for feigned-injury distraction displays that are used to lure would-be predators away from the nest. Recurvirostrids are tall, very skinny shorebirds with long, thin bills, straight in the Black-necked Stilt (*Himantopus mexicanus*) and recurved (curved up) in the American Avocet (*Recurvirostra americana*). They are common in shallow freshwater or saline wetlands, lakeshores, and coastal estuaries. Scolopacids (Figure 6.36; Color Plate 40) are small to large shorebirds with long toes and long, tapered wings; they have a short, elevated hind toe that distinguishes them from the charadriids. Many have bills equipped with nerve receptors at the tips that allow them to detect and capture prey deeply submerged in mud through smell and touch. The Laridae (gulls and terns; Figures 6.37–6.38) are predominantly white bodied as adults and are generally found near water. Omnivorous and generally well adapted to people, gulls frequent agricultural fields, landfills,

FIGURE 6.36. Long-billed Curlew (*Numenius americanus*).

FIGURE 6.37. Ring-billed Gull (*Larus delawarensis*).

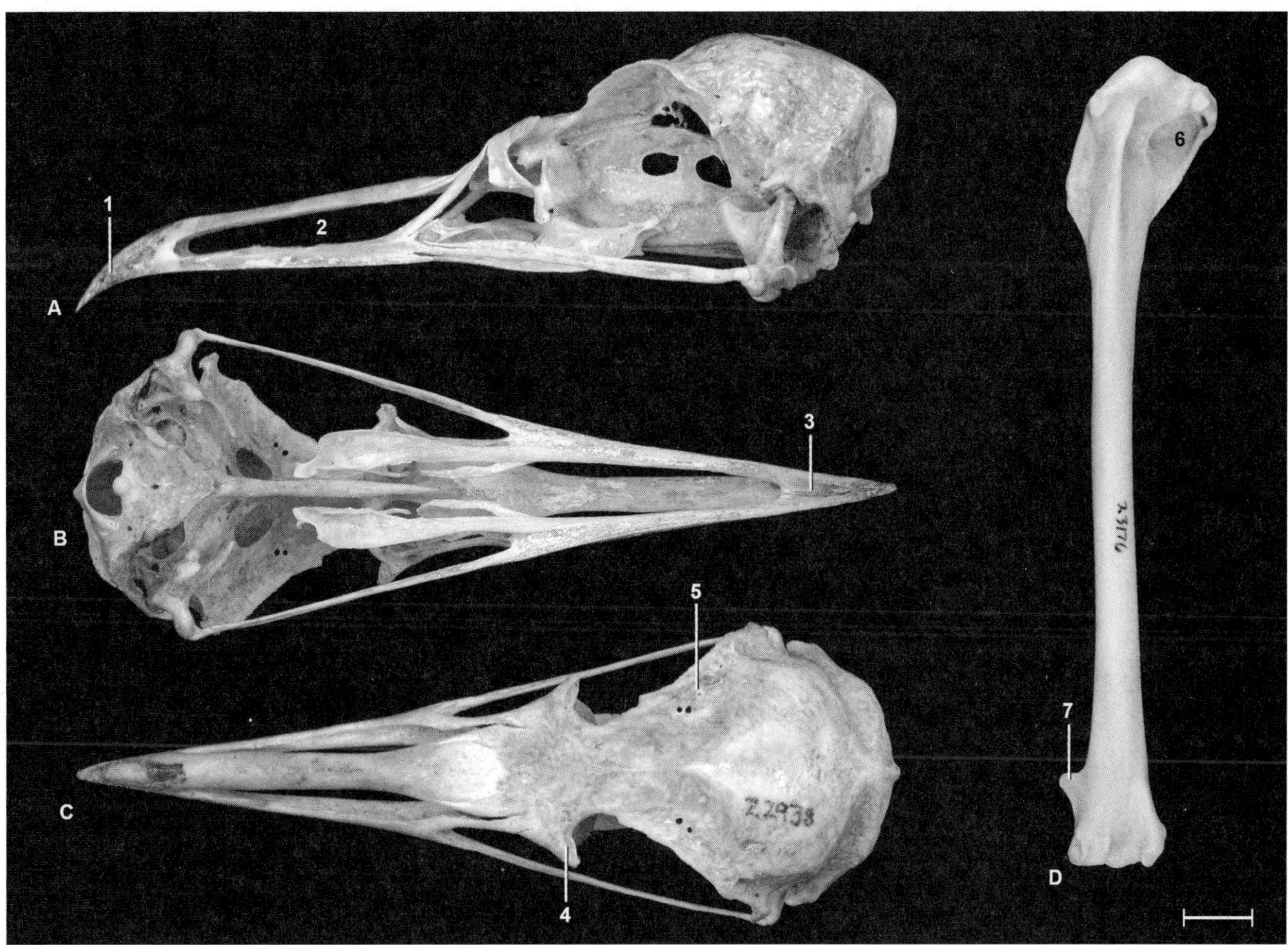

FIGURE 6.38. California Gull (*Larus californicus*) cranium, lateral (A), ventral (B), and dorsal (C) views; and l. humerus, posterior view (D). Osteological features (Charadriiformes): 1, premaxilla long, slender; 2, external nares, perforate, long, typically slit-like; 3, antero-ventral surface of premaxilla with deep groove; 4, fusion of lacrimal to frontal forms prominent lateral projections; 5, salt-gland fossa well developed in some taxa (e.g., gulls, terns, and alcids); humerus with, 6, pneumatic fossa deeply excavated and, 7, prominent dorsal supracondylar process.

and urban ponds. The term "seagull" is a vernacular name for gulls.

ORDER Columbiformes (Pigeons and Doves)
FAMILY Columbidae (Pigeons and Doves)

The order Columbiformes (L., columba = pigeon, dove) is represented by a single family and six genera in the west. They are small- to medium-sized birds with short legs and necks, small heads, and plumage typically dominated by tones of gray and brown. Their short bills are soft at the base and hard at the tip, often with a naked, swollen "cere" at the base of the upper bill. Columbids are commonly seen foraging on the ground for seeds, grains, or fruits with mincing steps and bobbing heads. They mostly occur in open habitats: fields, croplands, grasslands, and deserts, although the Band-tailed Pigeon (*Patagioenas fasciata*) inhabits woodlands and forests. Pigeons and doves are the only North American birds that suction water into the esophagus so they do not have to raise their bills to drink. Also unique, they produce "crop milk," a milk-like substance that is secreted by the walls of the crop (upper stomach) and fed to young. The Mourning Dove (*Zenaida macroura*; Figures 6.39–6.40) is the most familiar species. The exotic Eurasian Collared-Dove (*Streptopelia decaocto*) is rapidly expanding its range since a population was established in Florida in the 1980s. The ubiquitous "city pigeon," formally Rock Pigeon (*Columba livia*), is also introduced from the Old World.

FIGURE 6.39. Mourning Dove (*Zenaida macroura*).

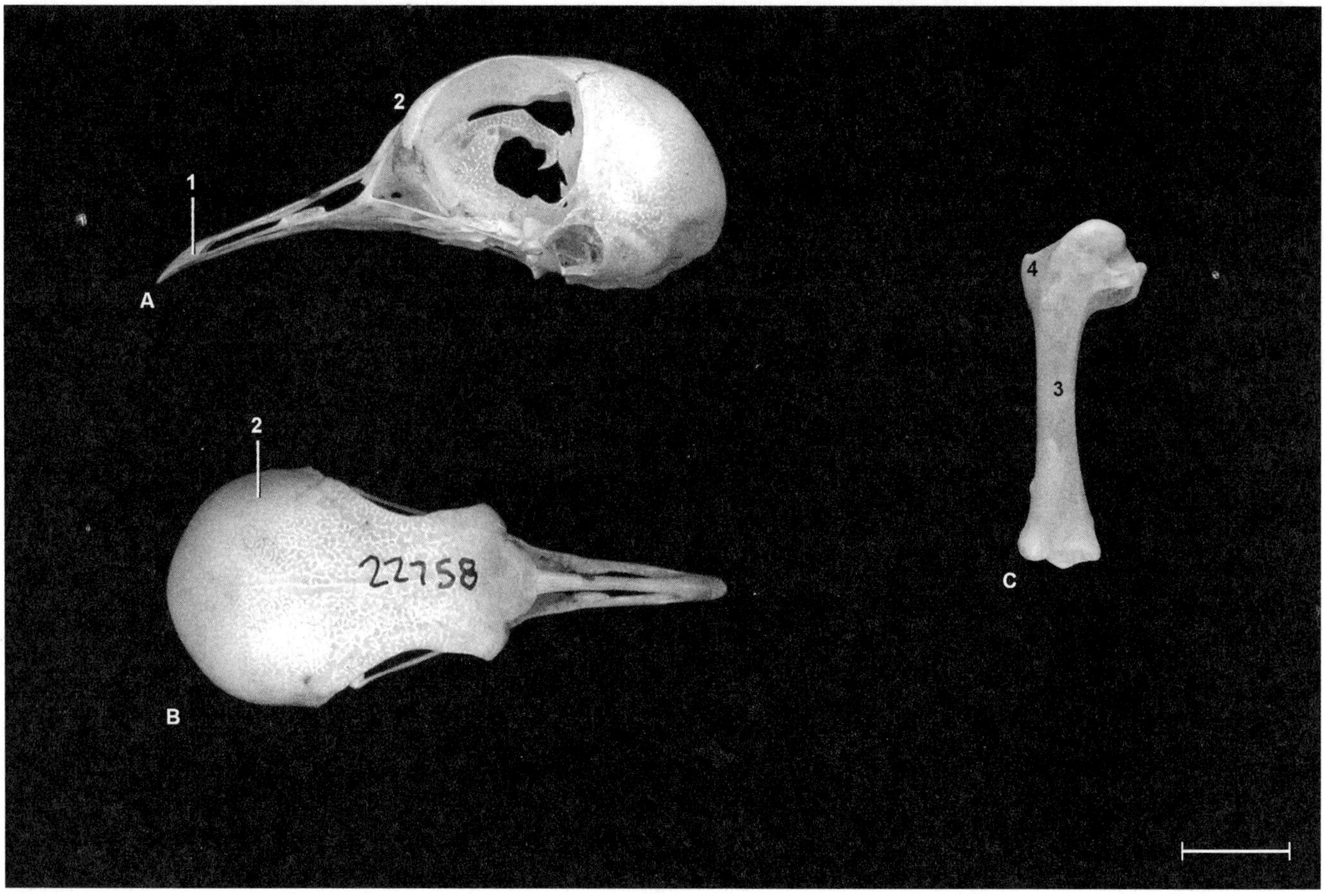

FIGURE 6.40. Mourning Dove (*Zenaida macroura*) cranium, lateral (A) and dorsal (B) views; and l. humerus, posterior view (C). Osteological features (Columbidae): 1, bill weakly developed; 2, cranium narrow, high vaulted, dorsal aspect violin shaped; 3, humerus broad for length; 4, deltoid crest of humerus prominent, extending to point dorsally.

ORDER Cuculiformes (Cuckoos, Anis, and Roadrunners)
FAMILY Cuculidae (Cuckoos, Anis, and Roadrunners)

The order Culculiformes (L., cuculus = cuckoo) is represented by one family, Cuculidae, and three genera in the west. The cuculids are a diverse group of slim, medium-sized birds that include the cuckoos (*Coccyzus*), anis (*Crotophaga*), and roadrunners (*Geococcyx*). All species have zygodactyl feet (two toes pointed forward, two pointed back), long, decurved bills, and tails that are at least as long as the body. Of the three distinctive genera that are present in western North America, only the Greater Roadrunner (*Geococcyx californianus*; Figures 6.41–6.42; Color Plate 41) occurs

FIGURE 6.41. Greater Roadrunner (*Geococcyx californianus*).

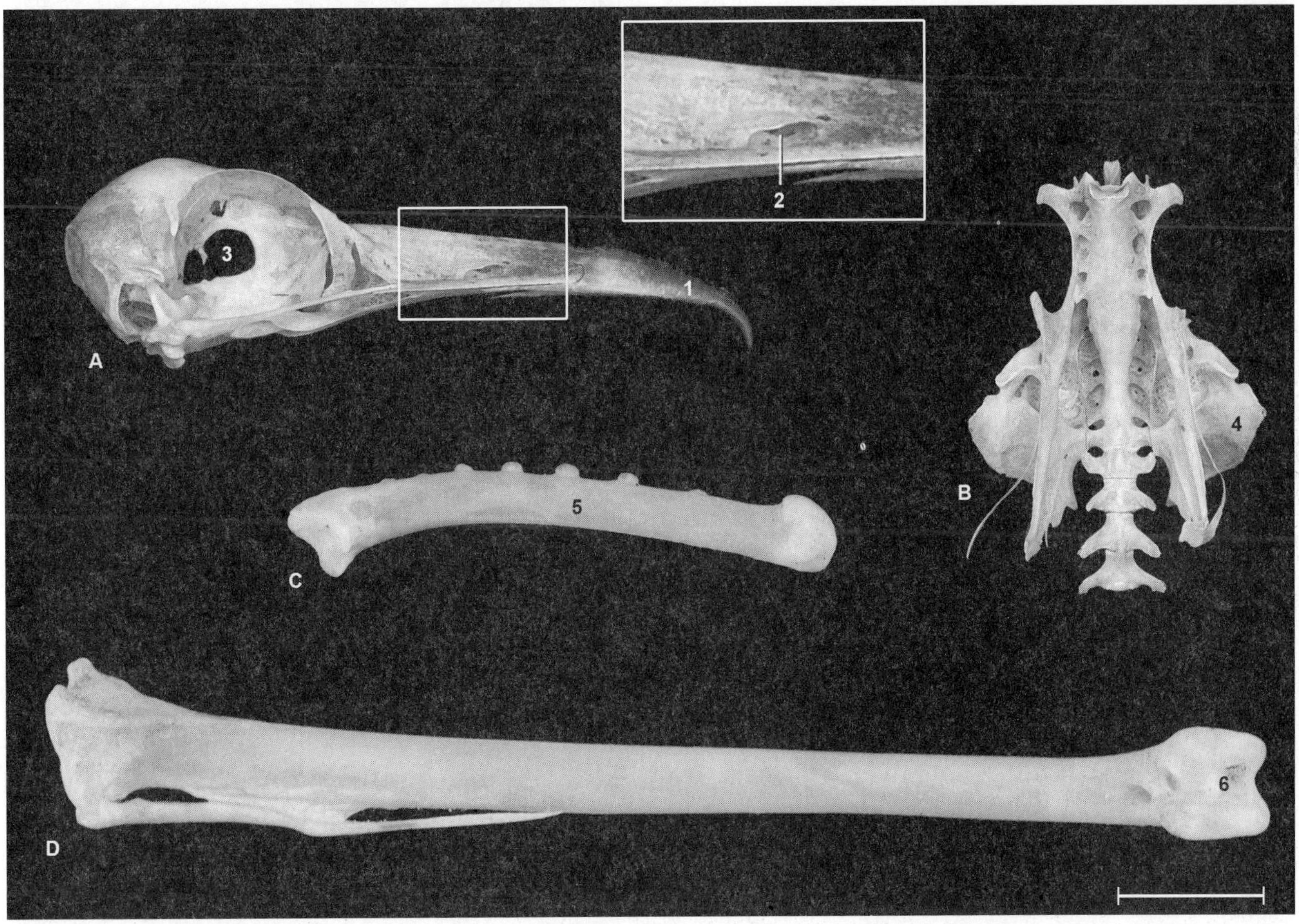

FIGURE 6.42. Greater Roadrunner (*Geococcyx californianus*) cranium, lateral view (A); os coxa and synsacrum, ventral view (B); r. ulna, ventral view (C); and r. tibiotarsus and fibula, anterior view (D). Osteological features (*Geococcyx*): 1, premaxilla long, decurved with pointed tip; 2, external nares imperforate, small, ventrally oriented; 3, interorbital fenestra is present—absent in superficially similar woodpeckers (Figure 6.52); 4, pelvis with lateral extensions of posterior ilia curving ventrally to form scroll-like structures; 5, ulna short, bowed, with prominent papillae and olecranon process; 6, distal tibiotarus with deep sulcus between circular, closely connected medial and lateral condyles.

over a substantial range in the region. An icon of the desert Southwest, this bird inhabits dry, open habitat from grasslands to rocky deserts. Diurnal, it feeds on a variety of small animals, especially lizards and snakes that it pursues with running speeds up to 18 mph; two birds may tag-team to kill snakes. Roadrunners use salt glands located in front of their eyes to excrete excess salt from the blood. These glands are common in marine birds that ingest seawater. Roadrunners are able to survive without drinking if their prey has sufficient water content but will drink when water is available.

FIGURE 6.43. Great Horned Owl (*Bubo virginianus*).

ORDER Strigiformes (Owls)

The west is graced by the presence of two strigiform families (L., strigis = owl)—Tytonidae and Strigidae—and 10 genera (Figures 6.43–6.44; Color Plates 42–44). Owls are mostly nocturnal predators with short and strong hooked bills, forward-facing eyes surrounded by a disk of feathers and zygodactyl feet with sharp talons; they also have round heads with flat faces and moveable ear-like or horn-like feather tufts that function in communication or camouflage. Owls are found in a variety of habitats from deep forests, to arctic tundra, to open deserts. With facial disks that funnel and amplify sound to asymmetrically shaped and placed, external, moveable earflaps, hearing is exceptional. This asymmetry, which in some species (e.g., Saw-whet Owl [*Aegolius acadicus*, Figure 6.44]; Boreal Owl [*Aegolius funereus*], Great Gray Owl [*Strix nebulosa*]) extends to the skull shape, allows owls to triangulate sounds from small vertebrates and facilitates prey capture in total darkness. Owls cast pellets of fur and bone and are widely known for depositing substantial vertebrate assemblages in western caves where they often nest and roost.

ORDER Caprimulgiformes (Goatsuckers)
FAMILY Caprimulgidae (Goatsuckers)

Three genera of Caprimulgiformes (L., capra = female goat, mulgere = to milk) occur in western North America; they are medium-sized, dark, and cryptically patterned crepuscular and nocturnal birds, superficially similar to small owls. They forage aerially for small insects. The small bill looks insignificant when closed, but it opens to reveal a huge mouth surrounded by long, stiff, rictal bristles—presumably adaptations to their aerial foraging habits. These birds have a distinctive red "eyeshine," produced when light reflects off the tapetum, a structure inside the eyes of some vertebrates (not humans) that increases light absorption. They are referred to as goatsuckers, amusingly, after an erroneous superstition (from the Roman naturalist C. Plenius Secundus ("Pliny the Elder") that these birds (European Nightjars [*Caprimulgus europaeus*]) would fly into barns

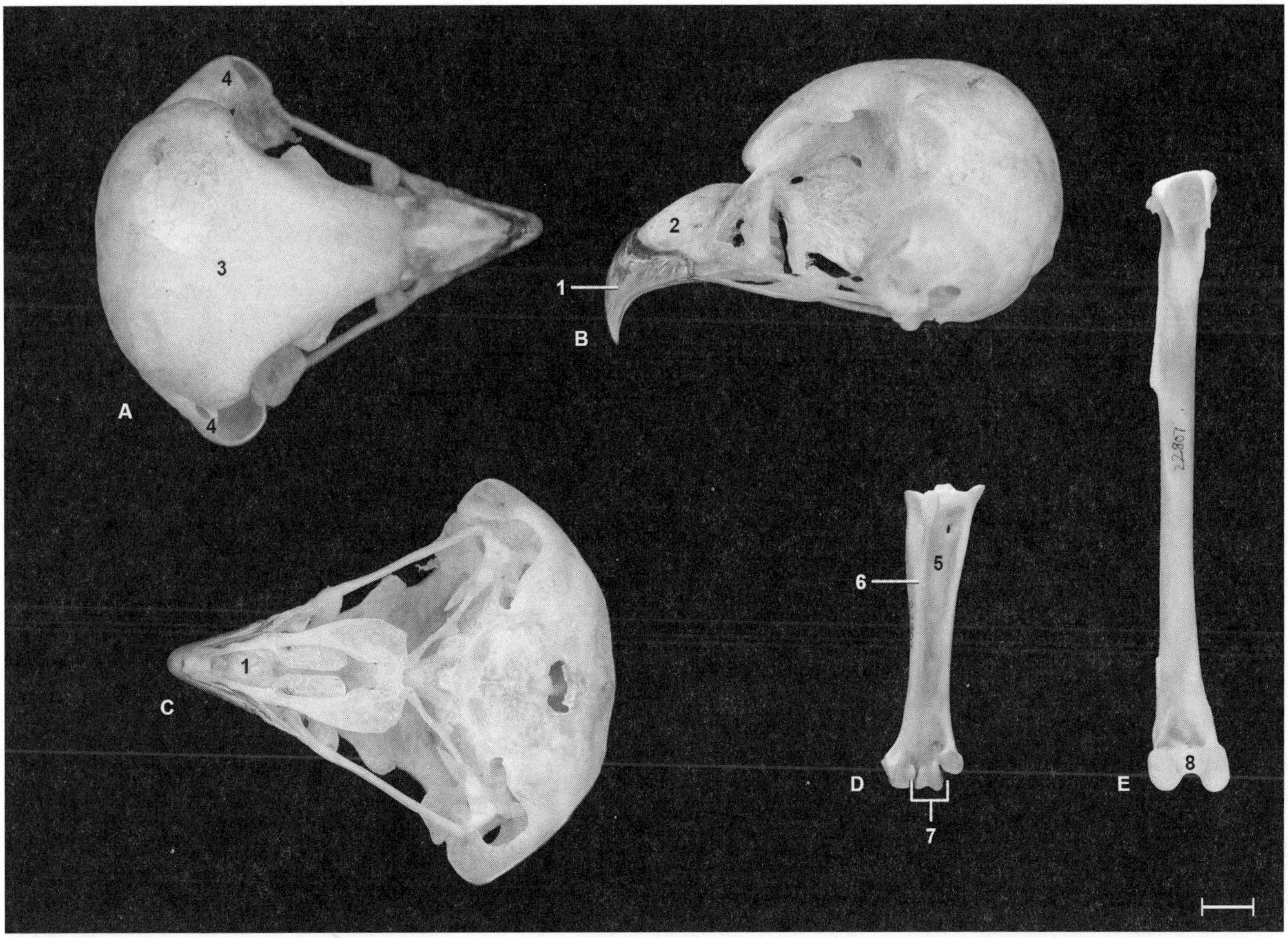

FIGURE 6.44. Saw-whet Owl (*Aegolius acadicus*) cranium, dorsal (A), lateral (B), and ventral (C) views; and Great Horned Owl (*Bubo virginianus*) r. tarsometatarsus, posterior view (D), and r. tibiotarsus, anterior view (E). Osteological features (Strigiformes): 1, premaxilla short, hooked, sharply pointed at the tip, with deep groove or pit at midline of ventral surface; 2, external nares oval, imperforate; 3, cranium, spongy, pneumatic—outline (dorsal aspect) of cranium is triangular, wider posteriorly relative to length (compared to Accipitriformes and Falconiformes); 4, positions of external auditory meati are asymmetrical in some taxa; tarsometatarsus with, 5, deep hypotarsal sulcus, 6, prominent medial hypotarsal crest and, 7, narrow gap posteriorly between 4th and 2nd trochlea; 8, distal end of tibiotarsus with deep groove between condyles.

at night and suck dry the teats of nanny goats, "which injures the udder and makes it perish, and the goats they have milked this way gradually go blind" (Jobling 2009:90). Three species have substantial ranges in the west: Common Nighthawk (*Chordeiles minor*; Figures 6.45–6.46), Lesser Nighthawk (*Chordeiles acutipennis*), and Common Poorwill (*Phalaenoptilus nuttallii*). In northern parts of their winter range, Common Poorwills are known to enter a hibernation-like state of deep and prolonged torpor where body temperatures drop from 94°F to below 50°F for periods up to several weeks. Male Common Nighthawks dive-bomb intruders, other males,

FIGURE 6.45. Common Nighthawk (*Chordeiles minor*).

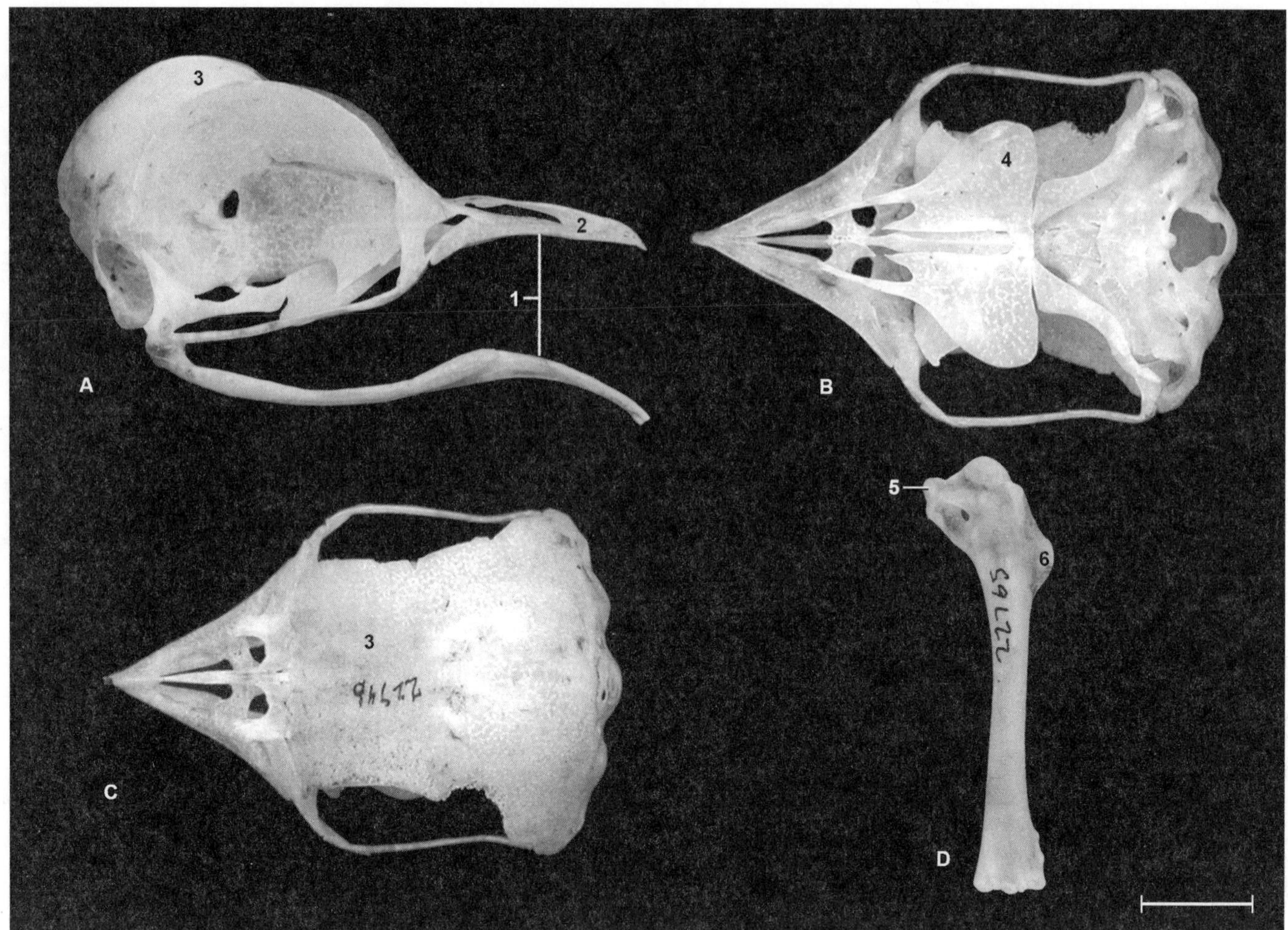

FIGURE 6.46. Common Nighthawk (*Chordeiles minor*) cranium, lateral (A), ventral (B), and dorsal (C) views; and r. humerus, posterior view (D). Osteological features (Caprimulgidae): 1, jaws open to form large gape; 2, bill short, flat, and wide; 3, cranium spongy, frontal region dished out, steeply ascending; 4, palatines enlarged; proximal humerus with, 5, proximo-ventral projection of internal tuberosity and, 6, dorsal extension of deltoid crest.

and potential mates as they produce a loud whooshing roar, referred to as "booming," caused by air rushing over the primary flight feathers.

ORDER Apodiformes (Hummingbirds and Swifts)

Order Apodiformes is represented by two families, Apodidae (swifts) and Trochilidae (hummingbirds) and 10 genera in the west. Both swifts and hummingbirds have very small legs and minute feet, hence the names Apodiformes and Apodidae (Gr., a = lacking, podos = foot). Swifts resemble the more familiar but unrelated swallows and are among the most adept fliers of all North American birds. They spend most of their lives on the wing, eating, drinking, bathing, copulating, and allegedly even sleeping in the air (the behavioral and electrophysiological parameters that define sleep have not, however, been measured in any wild bird). They feed on a wide variety of aerial arthropods and many species nest in cliff-based colonies (i.e., White-throated Swift [*Aeronautes saxatalis*]). Hummingbirds, the smallest of all birds, weigh only 2.8 to 8.5 gm. The Calliope Hummingbird (*Selasphorus calliope*) is the smallest North American bird. Their minute bodies, long, pointy bills, and the unique blur and buzz of their hovering wings, distinguish them from all other birds (Figures 6.47–6.48; Color Plate 45). Hummers feed on nectar but also insects and other arthropods. During courtship, males display with spectacular (species-specific) dives for perched females.

FIGURE 6.47. Broad-billed Hummingbird (*Cynanthus latirostris*).

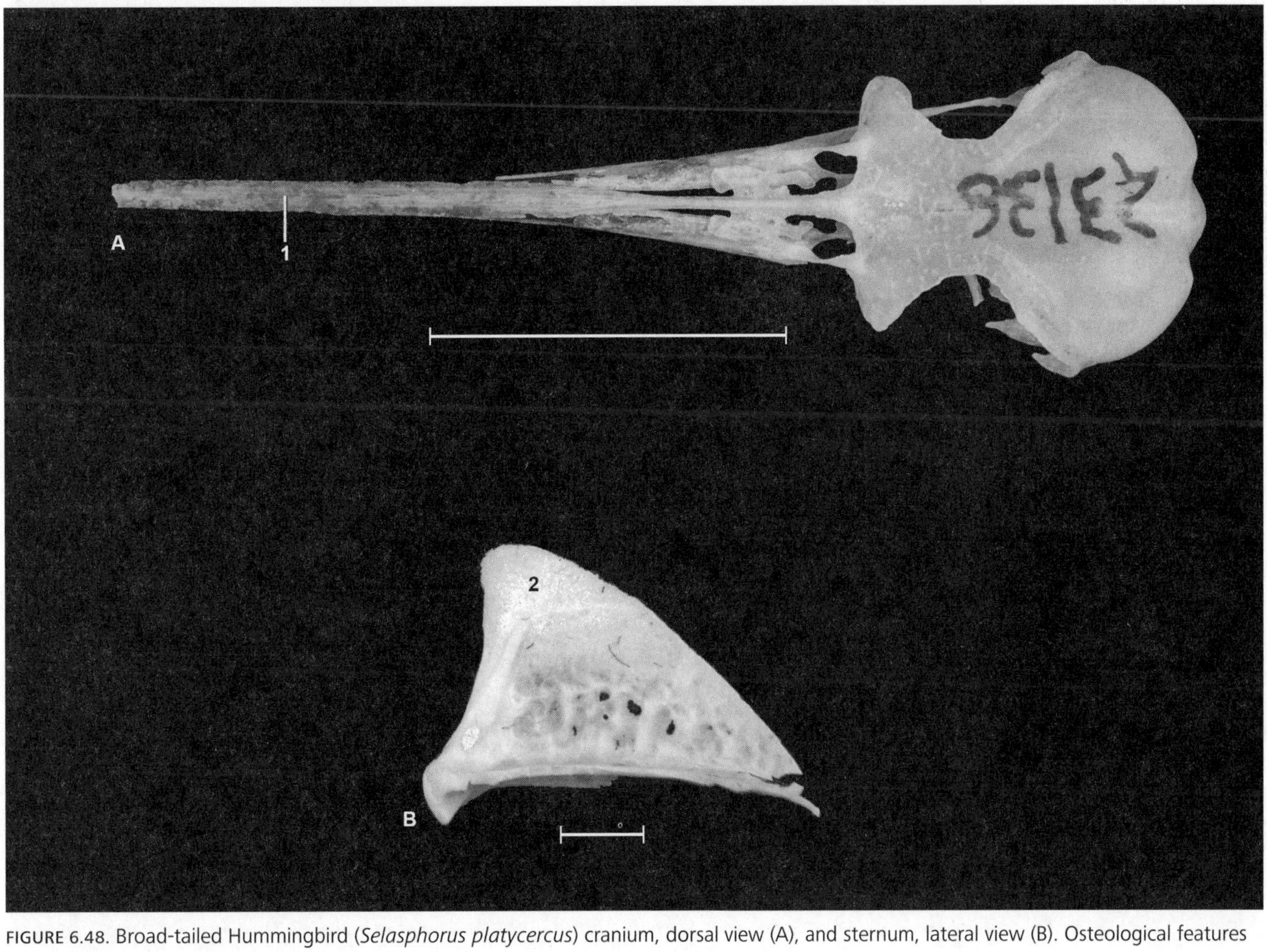

FIGURE 6.48. Broad-tailed Hummingbird (*Selasphorus platycercus*) cranium, dorsal view (A), and sternum, lateral view (B). Osteological features (Trochilidae): 1, bill long, pointed, delicate; 2, sternum with extremely large keel (carina).

FIGURE 6.49. Belted Kingfisher (*Megaceryle alcyon*).

ORDER Coraciiformes (Kingfishers, Rollers, Bee-eaters, and Allies)

FAMILY Alcedinidae (Kingfishers)

In western North America, the order Coraciiformes (L., coracia = type of crow) is represented by a single family (Alcedinidae) and a single species, the Belted Kingfisher (*Megaceryle alcyon*; Figures 6.49–6.50). Belted Kingfishers are stocky birds with short necks and tails, small feet, and big heads with shaggy crests and long, sturdy bills. Belted Kingfishers are found near perennial streams, rivers, freshwater ponds, and lakes where they hunt for fish and other aquatic animals by sight. They require clear water to locate and capture prey. These birds hover, or sit and wait from a prominent perch, before plunging headfirst into water to capture prey. They nest in deep burrows typically along dirt banks near water. The tunnels range from 0.3–2.4 m long and slope upward from the entrance to keep water from entering the nest.

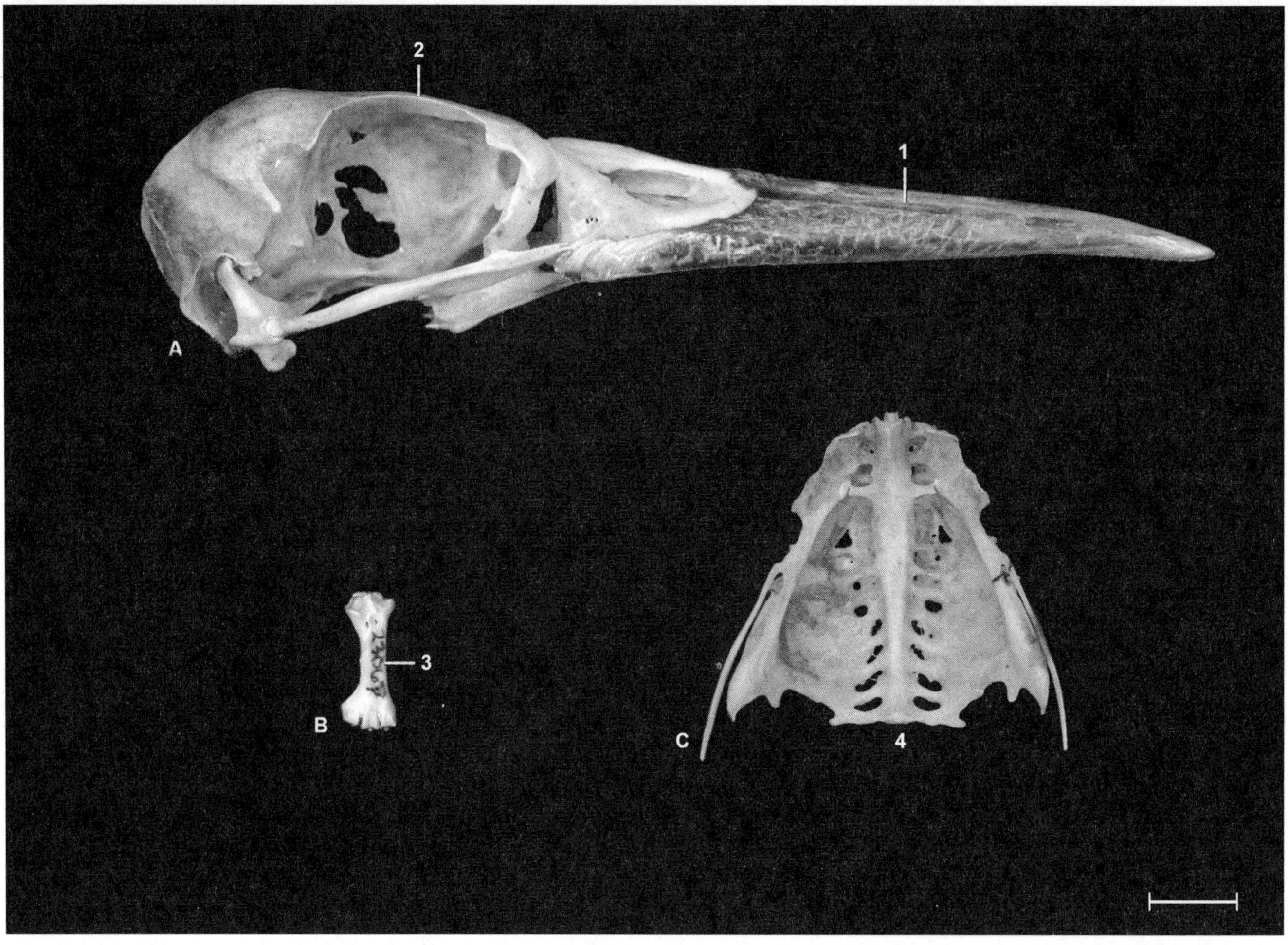

FIGURE 6.50. Belted Kingfisher (*Megaceryle alcyon*) cranium, lateral view (A); l. tarsometatarsus, anterior view (B); and os coxa and synsacrum, ventral view (C). Osteological features (Alcedinidae): 1, bill long, sturdy, imperforate external nares; 2, cranium long, flat, compared to Piciformes, Passeriformes; 3, leg elements (especially the tarsometatarsus) tiny, relative to body size; 4, broad posterior pelvis.

Kingfishers cast pellets primarily of fish bones and scales.

ORDER Piciformes (Woodpeckers and Allies)

FAMILY Picidae (Woodpeckers and Allies)

Piciformes (L., picus = woodpecker) are arboreal birds, typically with contrasting black-and-white plumage and a distinctive, vertical posture created by long, stiff tails that prop their bodies up against tree trunks. Five genera occur in the west (Figures 6.51–6.52; Color Plates 46–47). Woodpeckers use their chisel-shaped bills to excavate wood in search of insect prey and to make nest cavities; they also eat sap, seeds, and nuts. Found primarily in wooded areas, most species drum on wood for communication and territorial defense. Most are also monogamous and territorial; some breed cooperatively, most notably the Acorn Woodpecker (*Melanerpes formicivorus*), where multiple males

FIGURE 6.51. Acorn Woodpecker (*Melanerpes formicivorus*).

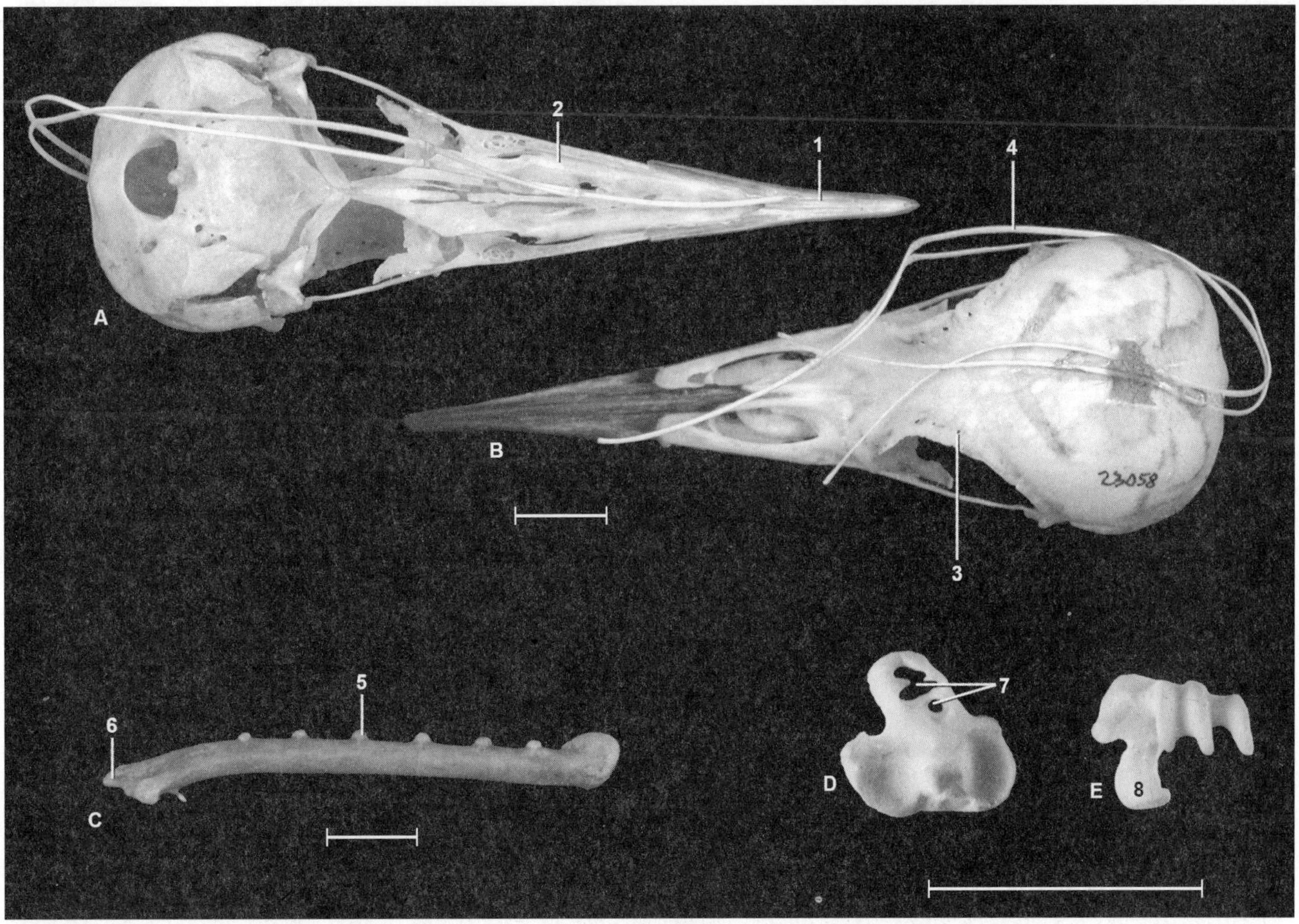

FIGURE 6.52. Northern Flicker (*Colaptes auratus*) cranium, ventral (A) and dorsal (B) views; l. ulna, dorsal view (C); and r. tarsometatarsus, proximal (D) and distal (E) views. Osteological features (Picidae): 1, bill strong, flat, wide base narrows to chisel-like tip; 2, palatines oriented ventral to premaxilla; 3, supraorbital margins (frontal) perforated; 4, greatly enlarged hyoid (hypobranchial) apparatus; ulna with, 5, pronounced papillae, 6, pointed proximal extension of olecranon; 7, proximal tarsometatarus (hypotarsus) with two closed tendinal canals; 8, distal end, with distinctive mesially hooked and thumblike 4th trochlea.

FIGURE 6.53. Peregrine Falcon (*Falco peregrinus*).

and females breed in a group. Many woodpeckers are nonmigratory. Northern Flickers (*Colaptes auratus*) spend more time on the ground than other woodpeckers; ants are their main food: they probe soil and rotten logs to locate them and use their long, barbed tongue to lap them up. Tongue morphology is distinctive: in some species (e.g., Northern Flicker) the tongue can be extended as far as 13 cm beyond the bill. This is enabled by a greatly elongated hyoid apparatus—a set of bones and muscles that controls tongue movements. The hyoid wraps clear around the back of the skull and anchors near the eye (Figure 6.52).

ORDER Falconiformes (Falcons and Caracaras)

FAMILY Falconidae (Falcons and Caracaras)

Only a single family of Falconiformes (L., falco = falcon), Falconidae, and two genera (*Caracara* and *Falco*) occur in the west. With long necks and legs and rounded wings, the shape and flight of the Northern Caracara (*Caracara cheriway*) is unlike that of any other raptor. This unique bird occurs in savanna or desert habitats and, in the west, reaches only extreme southern Arizona. All other local falconids are in the genus *Falco*, the true falcons (Figures 6.53–6.54). These birds are small- or medium-sized raptors, with thin, tapering wings that allow flight at very high speeds and rapid directional change. The Peregrine Falcon (*Falco peregrinus*; Figure 6.53) has been recorded diving at speeds over 200 mph (320km/h), making it the fastest moving animal on the planet. Most feed on vertebrate prey; smaller taxa however use insects and other arthropods. The American Kestrel (*Falco sparverius*) is the most common and familiar falcon in the west. It inhabits open areas with low-growing vegetation where it hunts using a "sit-and-wait" strategy, mostly using perches such as fence posts or utility wires but also by hovering when suitable perches are unavailable.

ORDER Passeriformes (Perching Birds)

The Passeriformes (L., passer = sparrow), or "passerines," is the largest and most diverse clade of birds. Of the roughly 10,000 or so extant species of birds, over half (~5,300) are perching birds (Figures 6.55–6.57; Color Plates 48–49). Represented by 27 families, this is easily the most diverse order of birds in the west; it also includes some of our most colorful and beloved species. Passerines are defined on the basis of the toe arrangement, which functions in grasping perches: all have

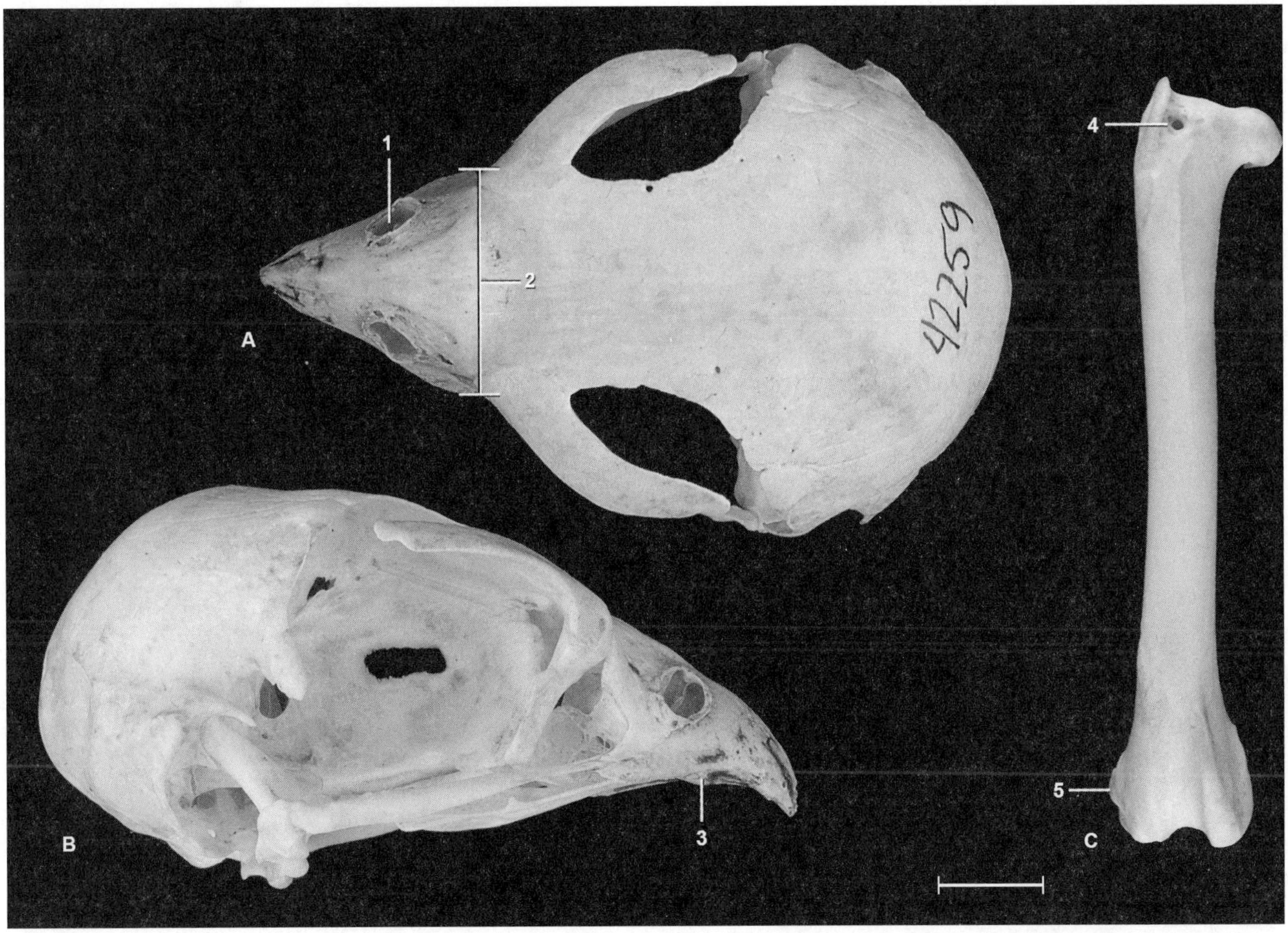

FIGURE 6.54. Prairie Falcon (*Falco mexicanus*) cranium, dorsal (A) and lateral (B) views; and Peregrine Falcon (*Falco peregrinus*), r. femur, anterior view (C). Osteological features (Falconiformes): 1, cranium similar to Accipitriformes but imperforate external nares smaller, rounded; 2, bill proportionately broader at base (posteriorly) compared to Accipitriformes; 3, premaxilla with prominent paired notches and "teeth" on ventro-lateral surfaces, posterior to the bill tip—lacking in Accipitriformes; femur with, 4, small paired peneumatic foramina near sharp-edged trochanter, and, 5, lacking prominent lateral epicondyle on distal end (present in Accipitriformes).

FIGURE 6.55. Black-capped Chickadee (*Poecile atricapillus*).

FIGURE 6.56. Yellow-headed Blackbird (*Xanthocephalus xanthocephalus*).

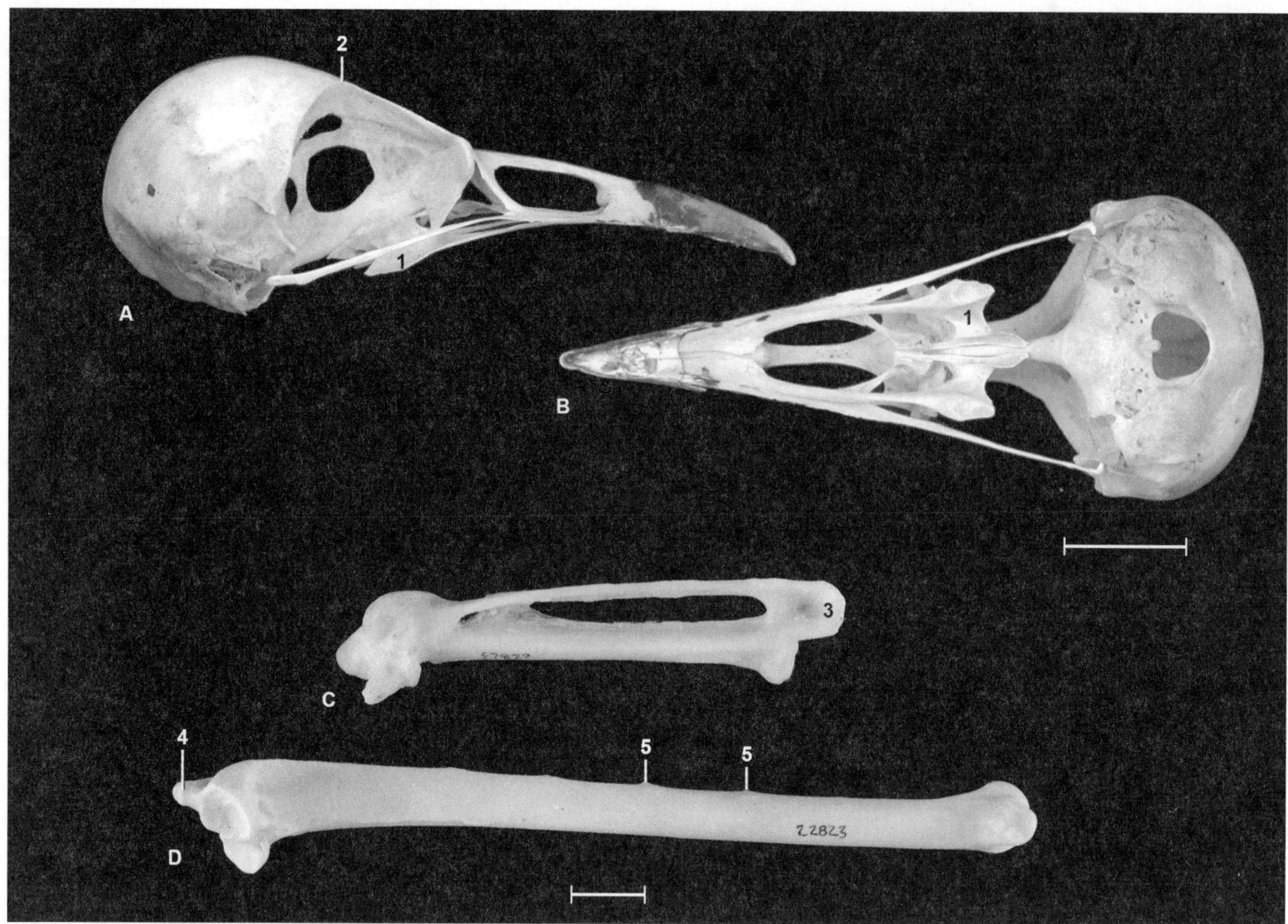

FIGURE 6.57. Pinyon Jay (*Gymnorhinus cyanocephalus*) cranium, lateral (A) and ventral (B) views; and Common Raven (*Corvus corax*), r. carpometacarpus, ventral view (C), and r. ulna (C), ventral view (D). Osteological features (Passeriformes): cranium distinguished from Piciformes by 1, posteriorly enlarged palatines, and 2, lack of supraorbital perforations; 3, carpometacarpus with prominent distal extension of process for digit III; ulna with, 4, pointed, proximal extension of olecranon process, but, 5, lacking prominent papillae present in Piciformes.

three toes pointed forward and one pointed backward, with all meeting the foot at the same level. They also have a distinctive sperm morphology, a unique bony palate, and distinctive fore- (wing) and hindlimb musculature. Body sizes of passerines vary from up to 1.4 kg in the Common Raven (*Corvus corax*) to less than 8.5 gm in some warblers (e.g., Parulidae). The family Corvidae (jays, magpies, crows, ravens) is represented by the largest sized species and is generally the best-represented family in archaeological deposits.

Notes

Our discussion on the characteristics, origins, and osteology of birds is based on Liem et al. (2001), Hildebrand (1995), and King and Custance (1982). Godefroit et al. (2014) present evidence for the presence of feathers in nontheropod dinosaurs. We primarily follow Howard's (1929) osteological nomenclature for birds, with some additional features from Baumel (1993). We follow the taxonomy for birds established by the AOU (2015). Jobling (2009) provides a dictionary of scientific bird names, and Olsen (1968, 1979) and Gilbert et al. (1981) provide osteological criteria to distinguish various bird taxa. Our presentation of the natural history, ecology, and behavior of birds is derived from Sibley (2014), Alderfer (2014), Elphick et al. (2001), and Poole and Gill (2015).

CHAPTER 7

Taphonomy and Bone Damage

The taxonomic composition of a fauna represents one of the most informative classes of zooarchaeological data. The previous chapters provide material that facilitates the collection of this information. We emphasize, however, that there are other types of data that can be obtained in analyses of archaeological and paleontological faunas. These include, but are not limited to, molecular data such as stable isotopes and ancient DNA, aspects of the age and size structure of animal populations, disease that is revealed in skeletal pathologies, and postmortem bone damage and modifications.

Of these data classes, bone damage is perhaps the most accessible and the most frequently gathered to address a wide range of issues in zooarchaeology, from reconstructing butchery and carcass processing activities to deciphering the bone accumulating agents responsible for an assemblage. The latter is a critical task if, for instance, human subsistence is of interest. It is also, however, a serious challenge, since in many settings, especially caves and rockshelters, a wide range of agents can accumulate bones. In these settings, bone damage such as burning or cut marks from stone tools, for instance, is often used to indicate human involvement, while carnivore chewing, rodent gnawing, and raptor damage would indicate material contributed or affected by other agents.

Studies designed to understand the depositional origin of faunal accumulations fall within the realm of taphonomy, which is the study of the processes involved in the transition of living communities of organisms into the archaeological or paleontological records. In this chapter, we examine and illustrate most of the more commonly encountered forms of bone-surface modifications that are routinely collected and incorporated in taphonomic analyses of archaeological faunas in western North America. A hand lens (at least 10×) or dissecting microscope and bright light is usually required to clearly observe many of these forms of bone damage and surface modification.

Human Damage

Stone Tool Cut Marks

Prehistoric peoples routinely skinned, dismembered, and filleted animal carcasses with stone tools to acquire nutritional resources (e.g., muscle, organs, blood) and raw material for tools and clothing (bone, sinew, fur, hide). Such activities may have occurred where the animal was harvested (the kill site) or the location to which a field-processed animal was returned. In many contexts, the bones themselves were subjected to intensive processing to extract the fat-rich marrow that occurs within their medullary cavities, or within the dense, compact, and trabecular bone itself where a smaller quantity of bone grease resides. These activities often result in repeated contact between stone tools (usually made from obsidian, chert, or basalt) and the underlying bone surface, producing cut marks. A host of variables influence the distribution and shape of those marks, depending on the specific task involved (skinning, dismembering, filleting, bone marrow processing, etc.), the size and shape of the tool, the force and approach taken by the butcher, and the nature of the overlying soft tissue (muscle tissue, periosteum). Still, most cut marks are elongate, linear striations, with V-shaped cross sections that have steeply intersecting walls at their bases, often terminating in sharp points (Figure 7.1). With low-power magnification (≤ 40×), multiple microstriae may be evident paralleling the sides of the V-shaped gouges. In most cases, cut marks appear in parallel groups of two or more, since a single cut is typically not sufficient to sever the overlying muscle or ligament. Cut marks are more commonly encountered on larger vertebrate taxa, relative to smaller animals that require less reduction prior to cooking and consumption. Finally, marks nearly identical to these attributes can be made by nonhuman taphonomic processes such as carnivore activity or the trampling of bones by large, hoofed mammals.

Percussion Impact Fractures

As noted, ancient peoples routinely processed animal bones for bone marrow and grease; such activities produce a range of damage characteristics. Access to within-bone nutrients minimally requires the fracture of long bones, readily accomplished with a blow from a simple hammerstone. In addition to producing a range of bone fragments, this activity creates percussion impact fractures that are highly variable in shape and morphology but typically leave smooth-edged, crescentic, or semicircular (chonchoidal) depressions or flake scars at the point of impact. Prominent impact notches are also often present directly adjacent to the flake scars (Figure 7.2). Depending on the nature of the blow and the shape of the bone, thin fractures not resulting in breaks are often produced. Fractures radiating through surrounding bone in a "starburst" pattern, or in a

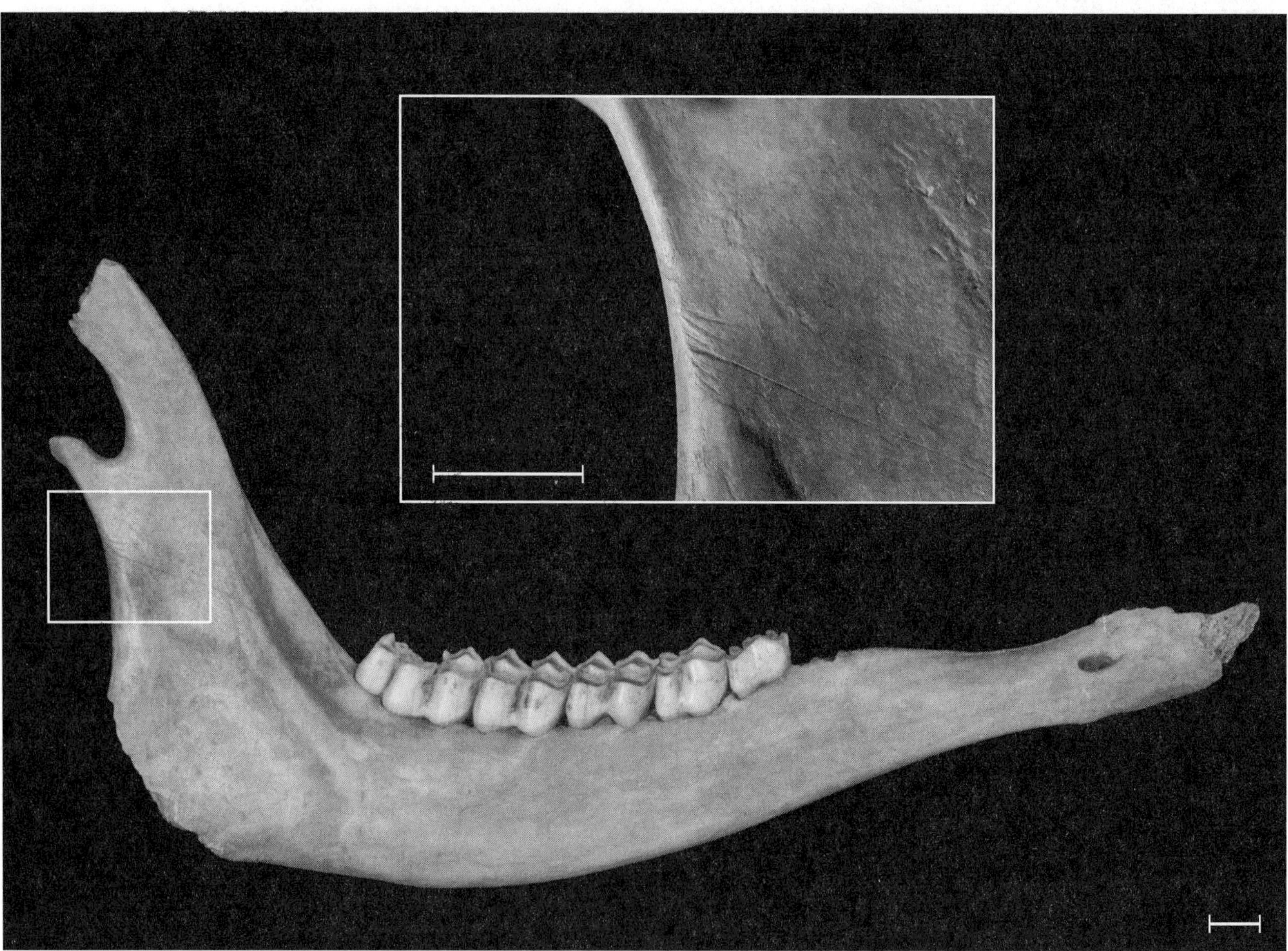

FIGURE 7.1. Mule Deer (*Odocoileus hemionus*) mandible with multiple cut marks on superior ramus.

series of concentric fracture rings encircling the impact, are both common.

Spiral Fractures

As described in chapter 1, bone is a composite material comprised of the mineral hydroxyapatite, which provides strength and rigidity but also protein (mostly collagen), which gives bone tensile strength and elasticity. This elasticity is reflected in the way fresh bone fractures: living bone or bone from recently deceased animals fractures with smooth, sharp edges in many ways similar to the sharp, feathered edges of a broken bottle. These breaks will often curve around the contour of a bone in a spiral fashion. As a result, such fresh bone fractures have long been referred to as "spiral fractures" (Figure 7.3). Spiral fractures are routinely produced by percussion impacts on longbones in the process of bone marrow and grease extraction. In fact, for some time it was thought that only human processing could produce them—and a long literature exists on the subject—but it is now known that many agents (carnivores, trampling, roof fall) can produce such damage in bone that still retains substantial collagen. A very different appearance is associated with bone fractures produced long after the collagen content has decayed (see Figure 7.12).

Burning

Since humans are the only organism to make and control fire and have long used it to cook animal resources, evidence of burning on bones is commonly recorded and used to indicate human involvement in bone deposition (Figure 7.4). Yet post-depositional burning can occur, for example, in dry cave contexts with substantial dry plant material where surface fires can burn and smolder

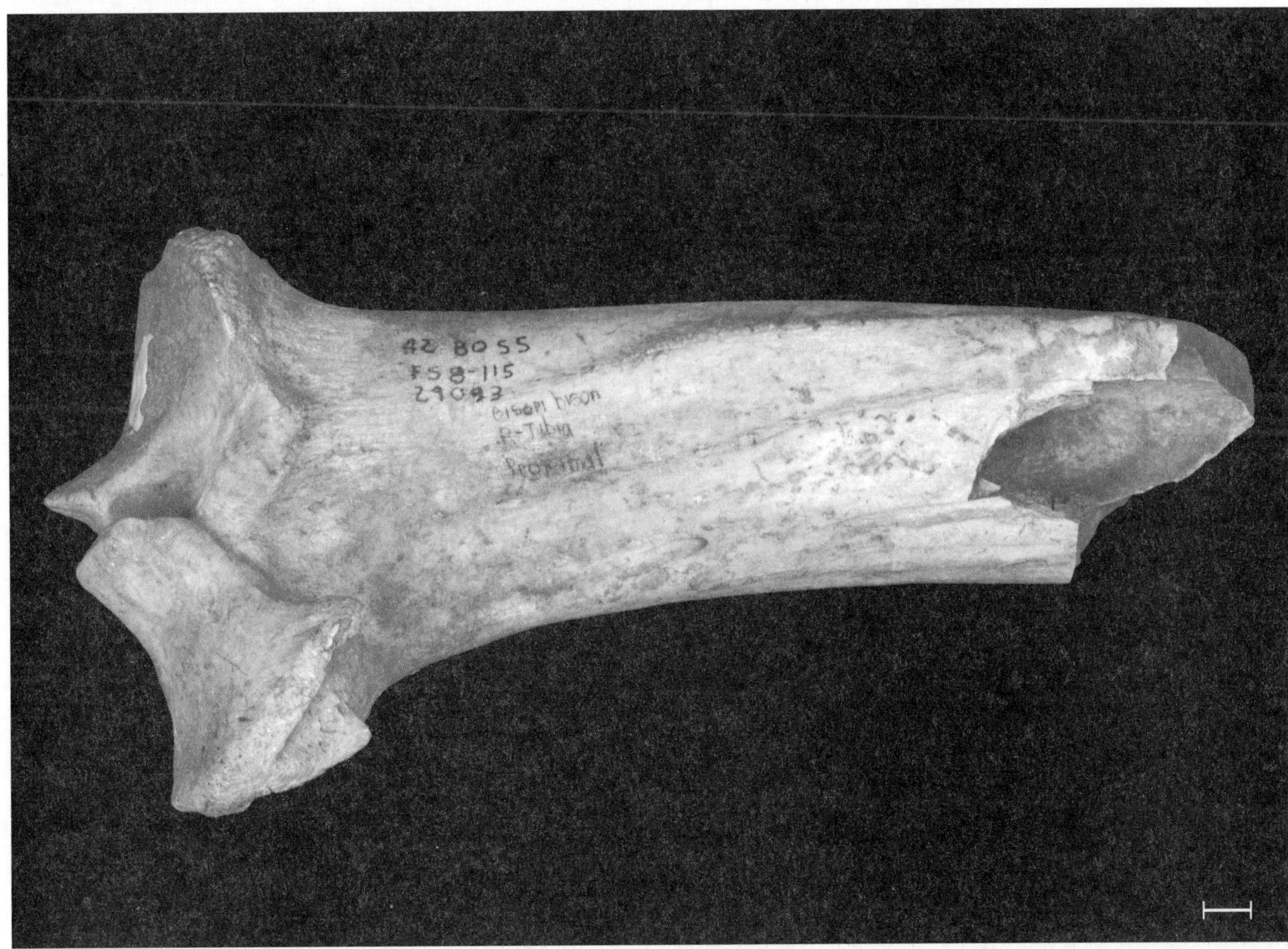

FIGURE 7.2. American Bison (*Bison bison*) proximal tibia with percussion impact fracture at midshaft.

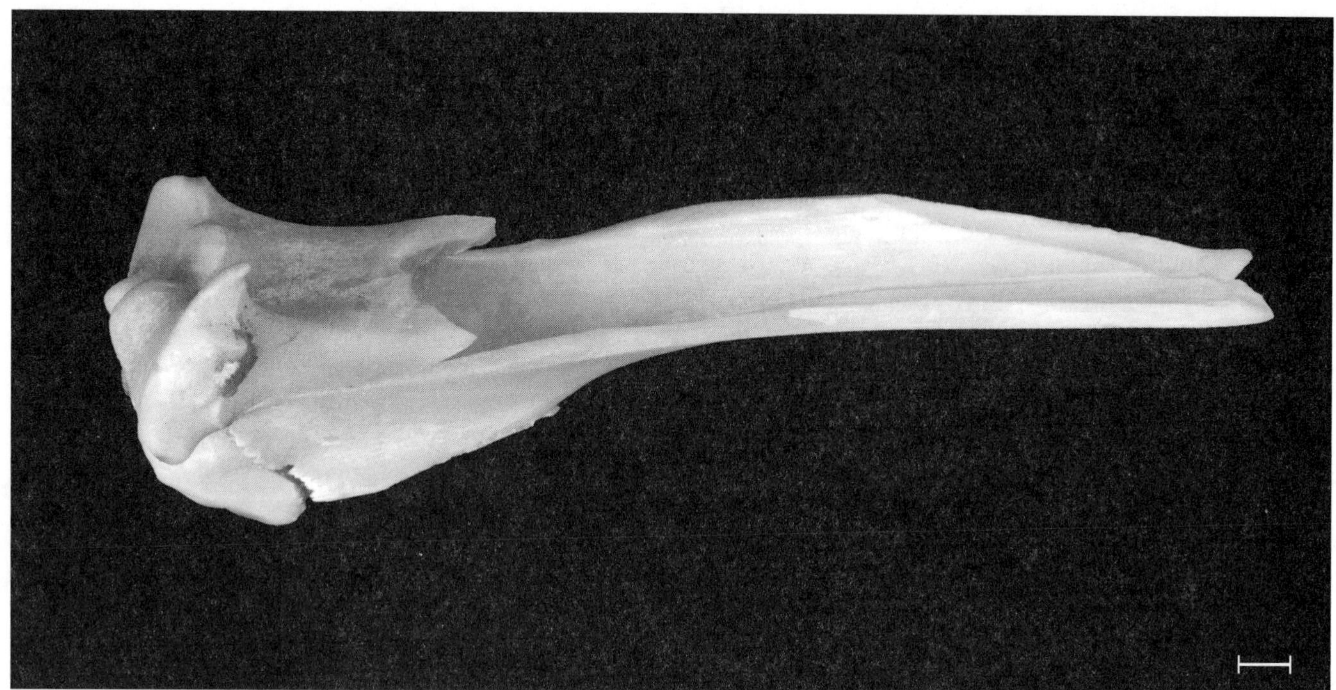

FIGURE 7.3. Pronghorn (*Antilocapra americana*) tibia with spiral fractures. Note sharp, feathered edges along breaks.

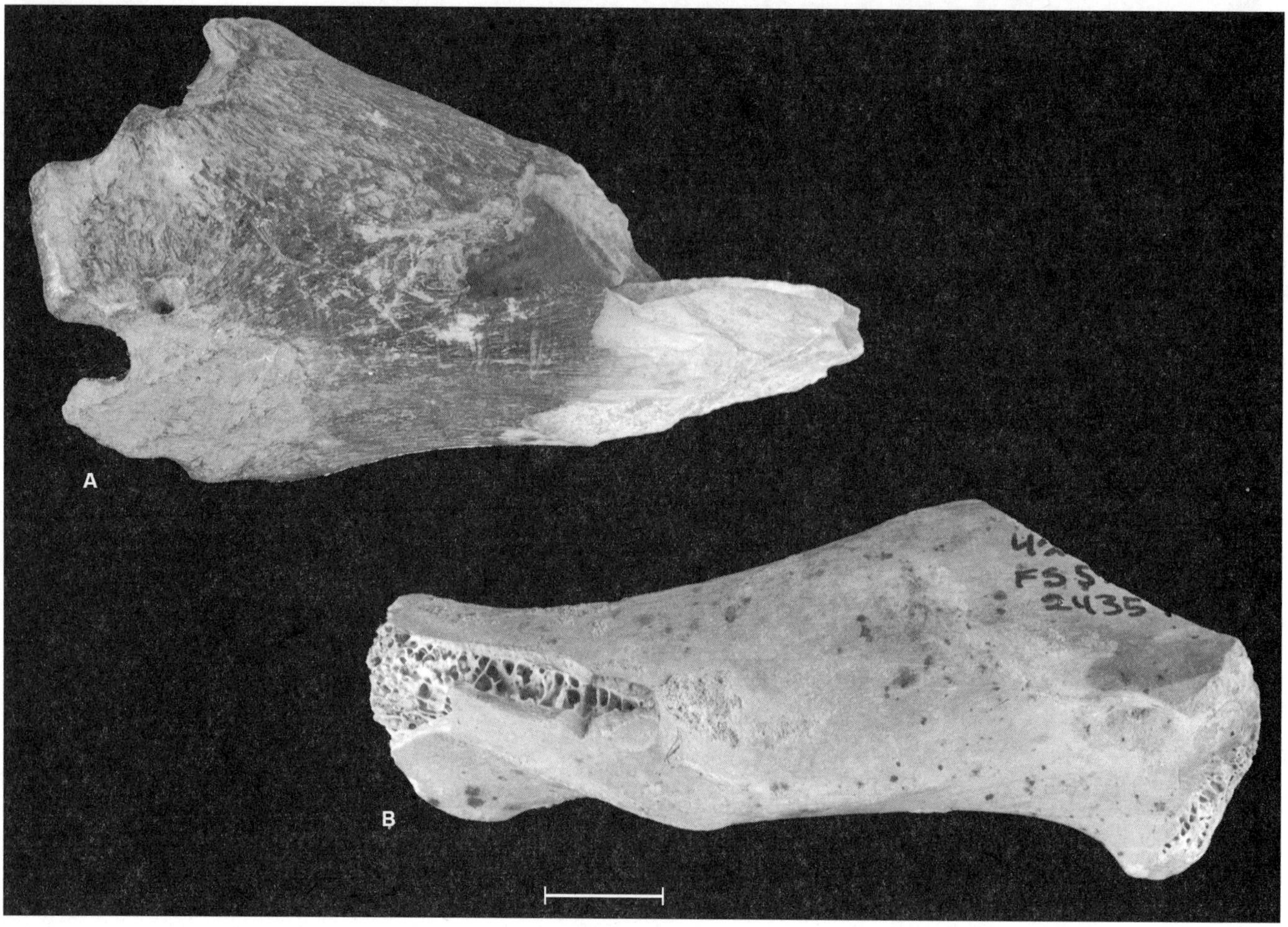

A

B

FIGURE 7.4. Medium artiodactyl burned distal tibia (A) and calcined os coxa (B).

deep through cave sediments. Patterns in the distribution of burning across elements (and element portions) can also be used to reconstruct cooking and disposal methods.

Although the weight, consistency, chemical composition, and other physical features of bone are altered when exposed to fire, the most obvious and commonly recorded damage involves color changes. The nature of bone color change is dependent on the temperature of the fire and the duration of time that a bone was exposed to it, but the general sequence of color change is from black, to gray, to white. Bones that have turned gray or white from prolonged fire exposure are commonly said to be "calcined"; bone specimens only blackened by fire are typically referred to as "burned" (Figure 7.4, A). Calcined bones that have lost much or all of their protein also have a noticeable porcelain or ceramic consistency (Figure 7.4, B). This can be revealed "by sound": when dropped onto the hard surface of a table, calcined bones produce a higher pitched "clinking" sound relative to bones only burned or those without fire damage.

Carnivore Damage

Mammals within the order Carnivora are among the most important bone-accumulating agents, and a wealth of literature exists regarding the various forms of damage that they impart to bone. And since many carnivore taxa occupy caves and rockshelters that were also favored by past peoples, zooarchaeologists are routinely faced with the question of whether bones recovered from these settings are of anthropogenic origin or result from the action of carnivores.

The most common carnivore taxa that accumulate and damage bone in the west are canids, including Domestic Dog (*Canis lupus familiaris*), Gray Wolf (*Canis lupus*), Coyote (*Canis latrans*), and foxes (*Urocyon, Vulpes*). The nature of carnivore bone damage is highly variable and depends on a wide range of variables including the unique dental morphology of the species involved, the types of teeth being used (e.g., canines versus carnassials), the action or goal of the carnivore (biting to subdue prey, versus chewing bone prior to swallowing), and the size and shape of the underlying bone element. Several general categories of damage include: 1) ragged-edged chewing, 2) punctures, 3) pitting, 4) striations or scoring, and 5) digestive corrosion. Ragged-edged chewing marks typically occur on the ends of thick, limb bone shafts and result from chips of bone being removed from broken edges—bone margins appear gnarled, ragged, or crenulated (Figure 7.5, A).

Punctures appear as circular or oval depressions of collapsed cortical bone that often contain bone flakes caving into the hole—textbook examples of carnivore damage, typically produced from their dagger-like canines (Figure 7.6). Punctures decrease in diameter with increasing depth into the bone, reflecting the conical nature of the canine teeth that produce them. Pitting appears as irregular depressions, often circular or oval in shape, that result from tooth-inflicted compression of dense bone that fails to puncture (Figure 7.5, B). Striations or scoring appear as elongate grooves, typically U-shaped in cross section, that result from prominent cusps (carnassials, canines) being dragged across bone surfaces (Figure 7.5, B).

Bones or bone fragments that have passed through the digestive tracts and have been deposited in feces or later regurgitated show substantial damage not only from chewing prior to ingestion but from the action of digestive acids and enzymes in the gut. Such bones appear corroded or melted, with multiple sieve-like perforations and feathered or thinned fracture edges (Figure 7.5, C). The entire carcasses of smaller vertebrates (e.g., birds, smaller fishes) are typically ingested by carnivores and when voided can produce whole elements that appear rounded, pitted, or deformed from digestive processes.

Raptor Damage

Hawks, eagles, and vultures (Accipitriformes), falcons (Falconiformes), and owls (Strigiformes), among other birds, rely extensively on vertebrate prey and can be important bone accumulating agents, especially in cave and rockshelter settings where many roost and nest. Again, since these places were also widely used by ancient humans, identifying raptor deposits is an important but challenging obstacle to reconstructing human subsistence on the basis of bones derived from such contexts.

Bone deposits produced by raptors are of two main sorts: 1) those derived from regurgitated pellets and, 2) uneaten parts of the skeleton that

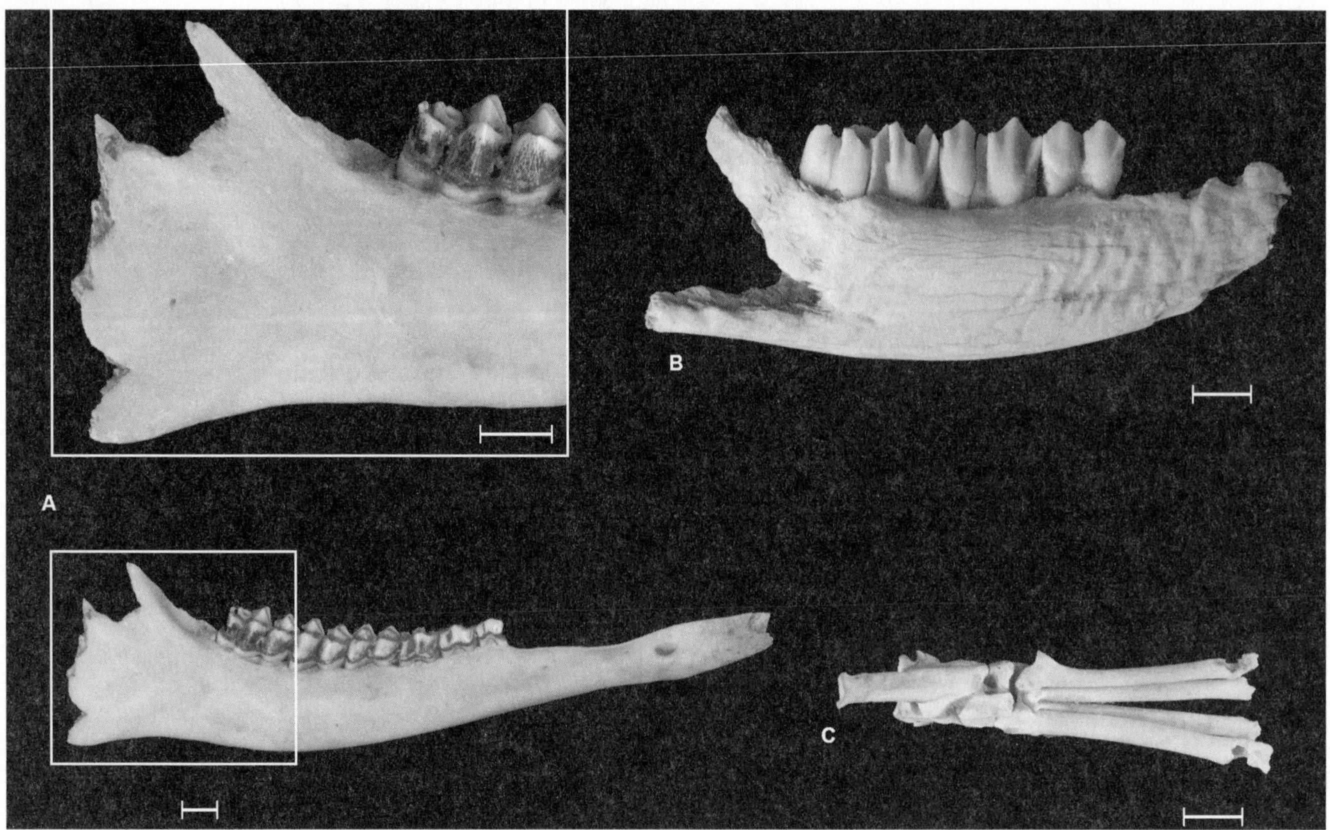

FIGURE 7.5. Carnivore damage: ragged-edged chewing on Mule Deer mandible (A), pitting and scoring on Mule Deer mandible (B), and digestive etching on Black-tailed Jackrabbit (*Lepus californicus*) hindfoot (C).

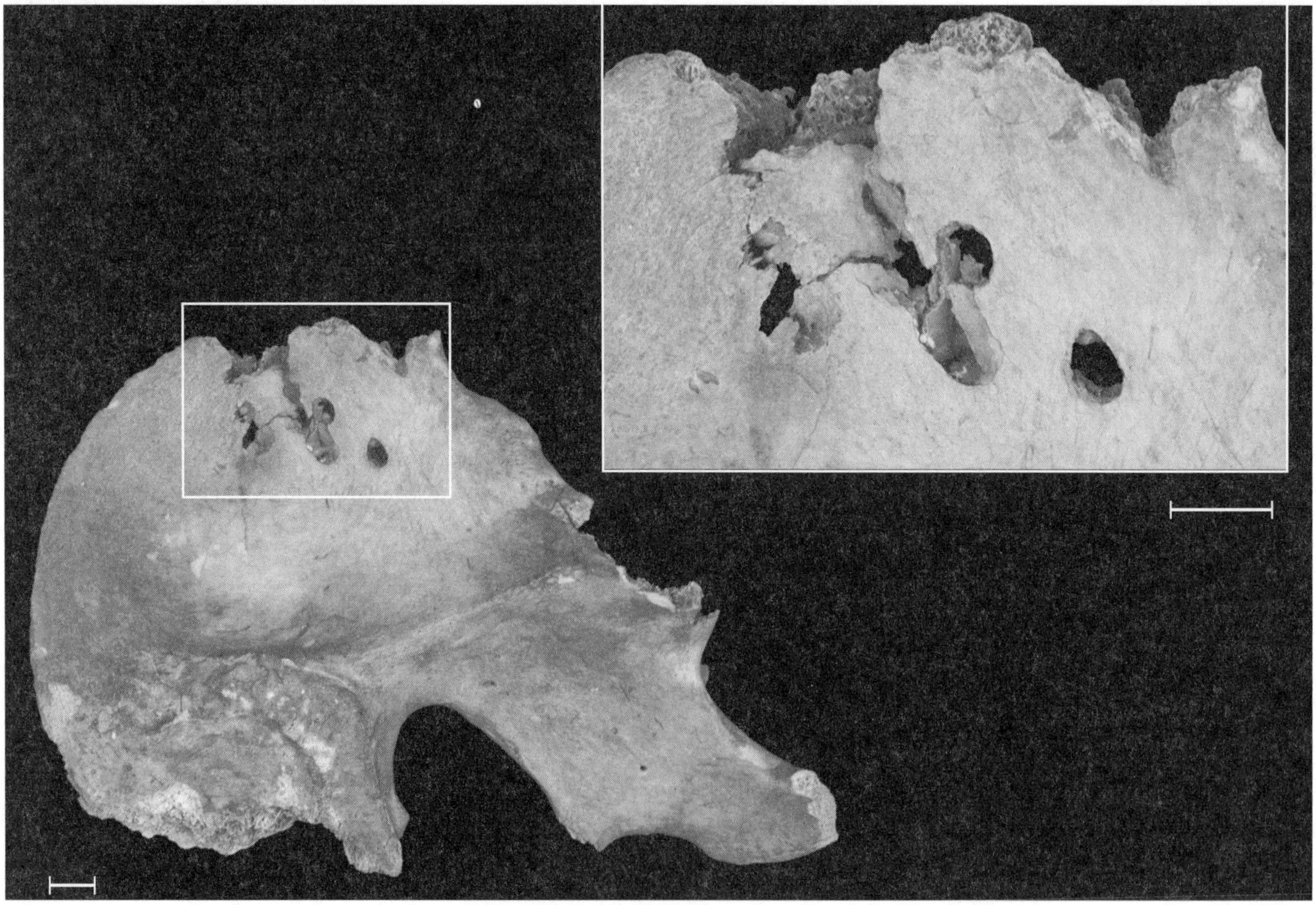

FIGURE 7.6. Carnivore damage: puncture marks on human os coxa.

are dropped from roosts or nests. Raptor pellets appear outwardly as oblong, cylindrical tubes of matted fur, but the insides reveal a range of indigestible portions of prey consumed whole or in large chunks and can include insect exoskeletons, fish scales, bones, and teeth (Figure 7.7). Depending on the size of the prey ingested and the raptor involved, damage to pellet-derived bone can be minimal or extensive. Barn owls, for example, typically ingest small rodents whole, and the pellets contain mostly unfragmented whole bone with minimal digestive damage. Bone derived from the pellets of other raptors such as Golden Eagles (*Aquila chrysaetos*) and Gyrfalcons (*Falco rusticolus*) is highly fragmented and exhibits extensive digestive corrosion, thinning and polish, virtually indistinguishable from the digestive damage described for carnivores above. There is also considerable variation in the nature of damage, fragmentation, and element representation exhibited in uneaten vertebrate bone, depending on the raptor taxon and context. Few specific damage types are clearly indicative of raptor activity. Punctures caused by talons or beaks are most noteworthy but still only occur in low (> 5%) frequencies in known, raptor produced deposits.

Rodent Gnawing

Rodents have evergrowing incisors to compensate for use wear. It is often suggested that rodents gnaw on bone to maintain the proper length and alignment of these chisel-shaped teeth that occur in pairs in both the premaxilla and mandible. Empirical observations, however, indicate that rodents gnaw on bone more commonly to access nutritional resources such as bone grease, marrow, and minerals (e.g., calcium phosphate). These studies have also documented that different rodent taxa appear to focus on different dietary qualities of bone. Brown Rats (*Rattus norvegicus*), for instance, preferentially target fat-rich bones, while Eastern Gray Squirrels (*Sciurus carolinensis*) focus on higher density elements that apparently produce greater returns of bone mineral.

In any case, rodent gnawing results in a unique pattern of bone damage characterized

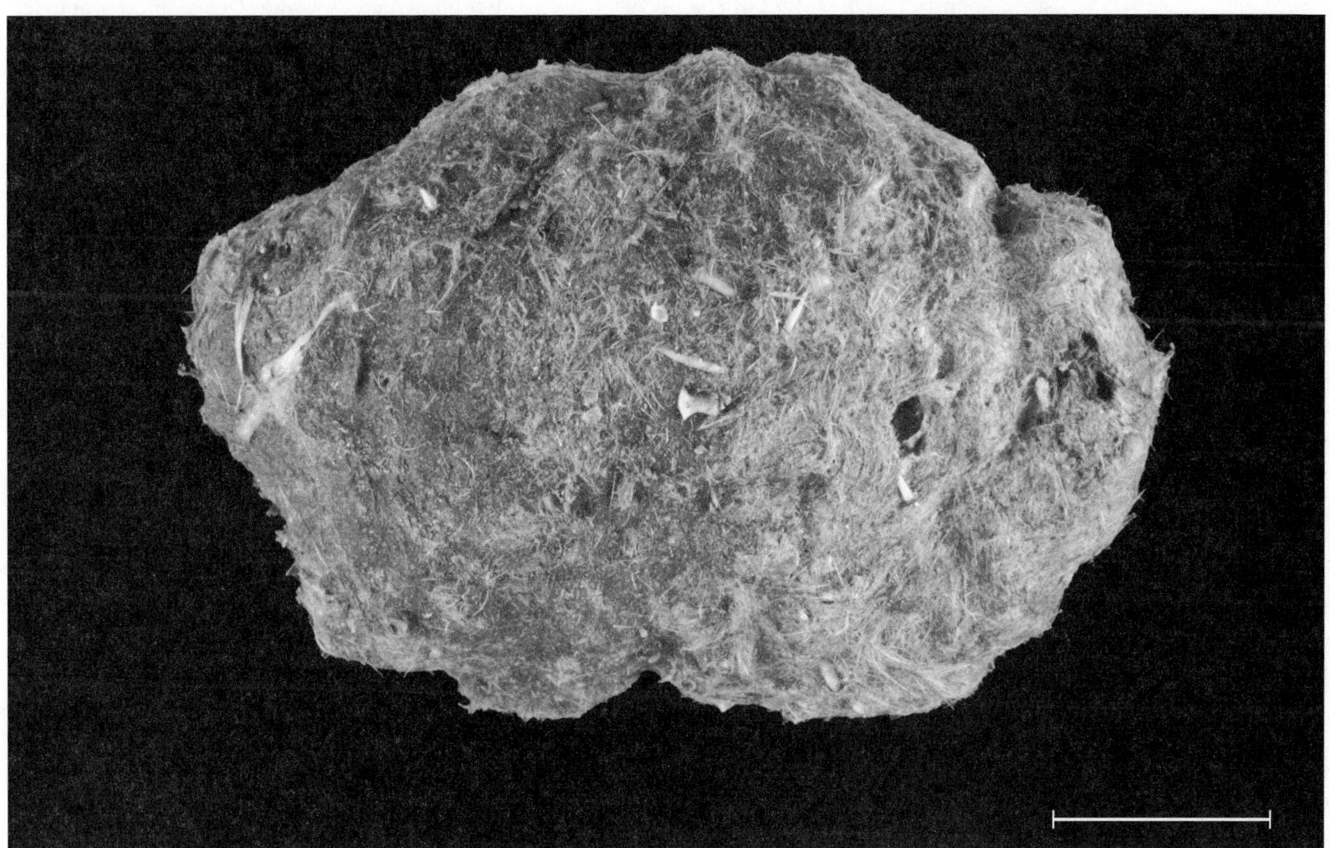

FIGURE 7.7. Barn owl (*Tyto alba*) pellet.

by extended, overlapping series of paired, broad, shallow, flat-bottomed grooves—bones heavily impacted by rodent gnawing have a pronounced "chiseled" or "whittled" appearance (Figure 7.8). The size of the grooves can also be used to narrow down the list of rodents that could have produced the damage. For example, the incisor impressions left by rats and squirrels are larger than those produced by mice, and beaver marks are larger than those left by porcupines. In western North America, woodrats (*Neotoma* spp.) are the most notorious bone gnawers and will often incorporate bones—typically gnawed—into their stick-pile nests. Tucked into protected rock crevices or caves, their own urine permeates and crystalizes the constituents into a substance known as amberat that, if kept dry, can last for tens of thousands of years (see chapter 5, Family Cricetidae (Cricetid Rats and Mice); Figure 7.9).

Root and Fungal Etching

Plant roots excrete humic acid; when they come into contact with bone, they can etch the cortical surfaces, producing a series of shallow grooves with U-shaped cross sections in a pattern that has been described as "dendritic," "wavy," "sinuous," and "sphagetti-like"(Figure 7.10). Nearly identical patterns of etching have been documented on bones derived from dry-cave contexts in western North America. Clearly, these caves could never have supported rooted vegetation, but their sediments contained plant remains, presumably introduced, still fresh, by prehistoric peoples. Etching in this context can occur when bones come into contact with plant stems containing acids produced in fungal decomposition. Root etching may indicate that the sedimentary context supported plant growth at least sometime during the past and may provide information on the taphonomic history of bone deposits. Stratigraphic variation in fungal etching, and possibly the presence of associated plant remains, may, similarly, have implications for paleoenvironmental or human dietary change. Etching can also hasten bone-surface degradation and otherwise obscure or obliterate the evidence for other forms of damage such as cut marks.

Surface Weathering

Long-term bone preservation normally requires burial in sediments. Indeed, archaeological faunas only exist at all because they were buried shortly after human processing and deposition. Bones left on the surface will be affected by solar radiation, freezing and thawing, wetting and drying, wind-carried sand abrasion, and many other factors that rapidly break down material. Various formal stages of bone weathering have been proposed to

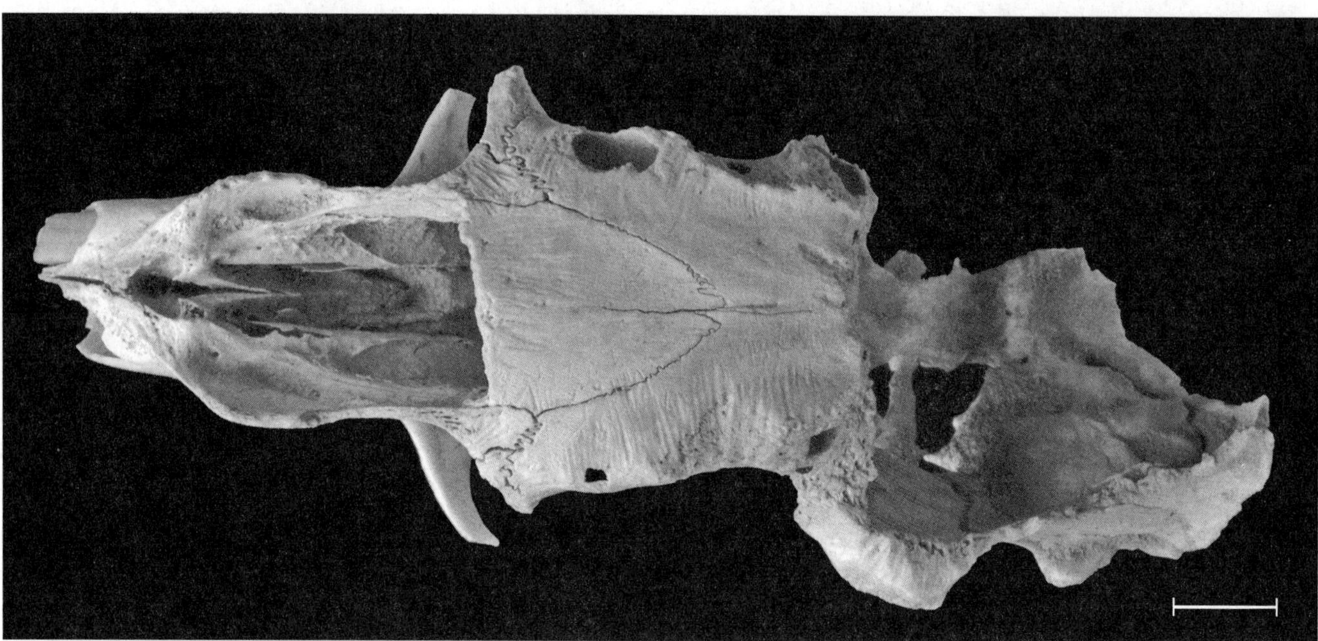

FIGURE 7.8. Rodent gnawing on dorsal cranium of North American Porcupine (*Erethizon dorsatum*).

FIGURE 7.9. Amberat collected from a woodrat midden.

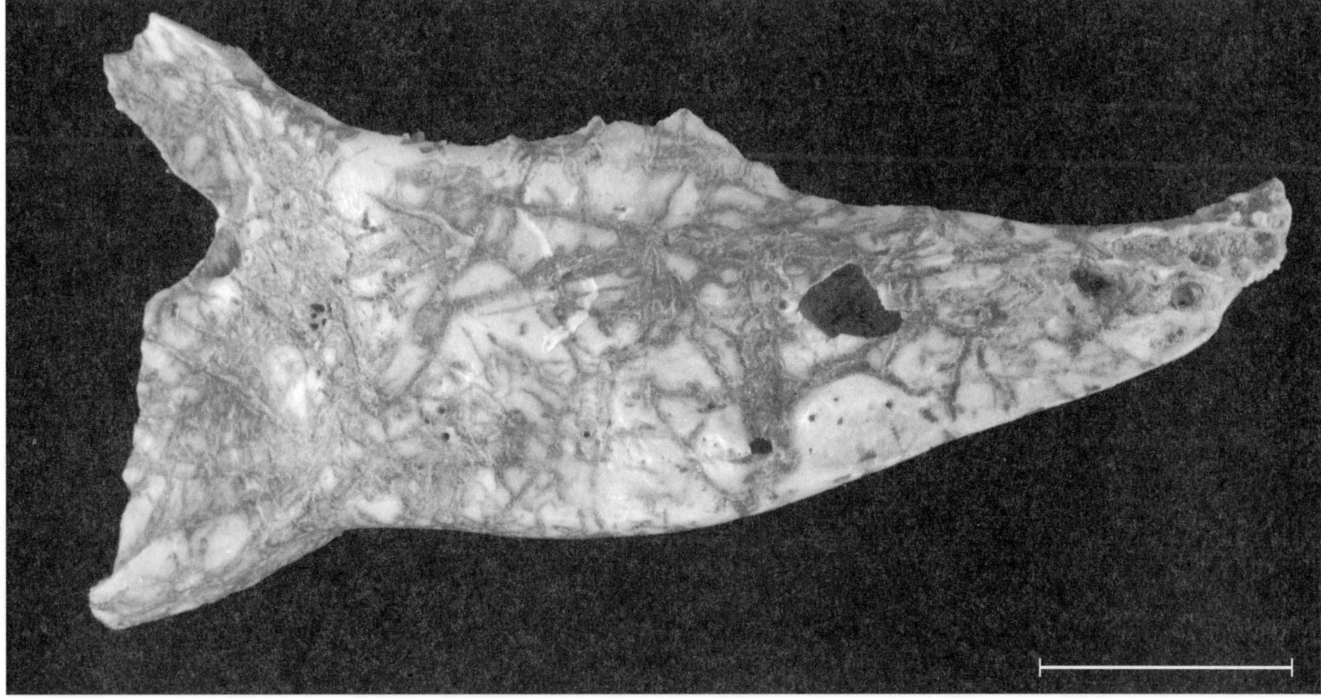

FIGURE 7.10. Root etching on Black-tailed Jackrabbit (*Lepus californicus*) mandible.

reflect increasing evidence of cracking, splitting, splintering, exfoliation, and disintegration (Figure 7.11). Bone fractures that occur due to weathering or other causes, long after collagen has been lost through decay, show jagged, angular breaks (Figure 7.12), very different from the sharp-edged, spiral fractures that occur from breaks in fresh bone. Bone weathering data can potentially give insight on bone accumulation processes and the length of time an assemblage was exposed on the surface prior to burial, but many factors influence the rate at which bone weathers. In addition, weathering needs to be carefully assessed in order to evaluate the degree to which other bone-surface modifications of interest (e.g., cut marks) may have been obscured or removed.

Bone Polish

Polished bones have surfaces that are smooth, rounded, and glossy resulting from the incremental removal of tiny bone particles through abrasion. Such abrasion can be produced by a variety of processes such as wind-borne sand, fluvial transport, licking by carnivores, extensive trampling in sandy sediments, or human processing of bones in ceramic or other containers—the latter producing the so-called "pot-polish." Evaluation of the depositional contexts can help determine a likely source for bone polish. The specimen in Figure 7.13 was collected from a gravel beach adjacent to a large Great Basin lake; the polish was caused by wave action and abrasion from sands and gravels.

Excavation Damage

Sharp-edged metal trowels, shovels, and other instruments used by archaeologists to loosen sediment and recover artifacts during excavations can impart scratches and striations on bone that may be confused with other similar types of damage such as cut marks and carnivore scoring. Such marks can, however, usually be distinguished

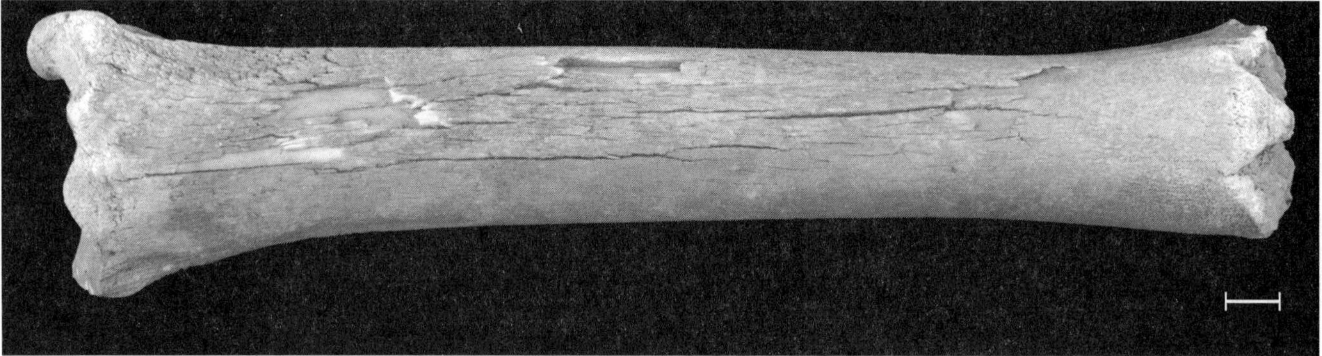

FIGURE 7.11. Bone weathering showing cracking, splitting, splintering, and exfoliation of Domestic Cattle (*Bos taurus*) metapodial. The top (anterior surface) was exposed to the elements, while the posterior surface was partially buried.

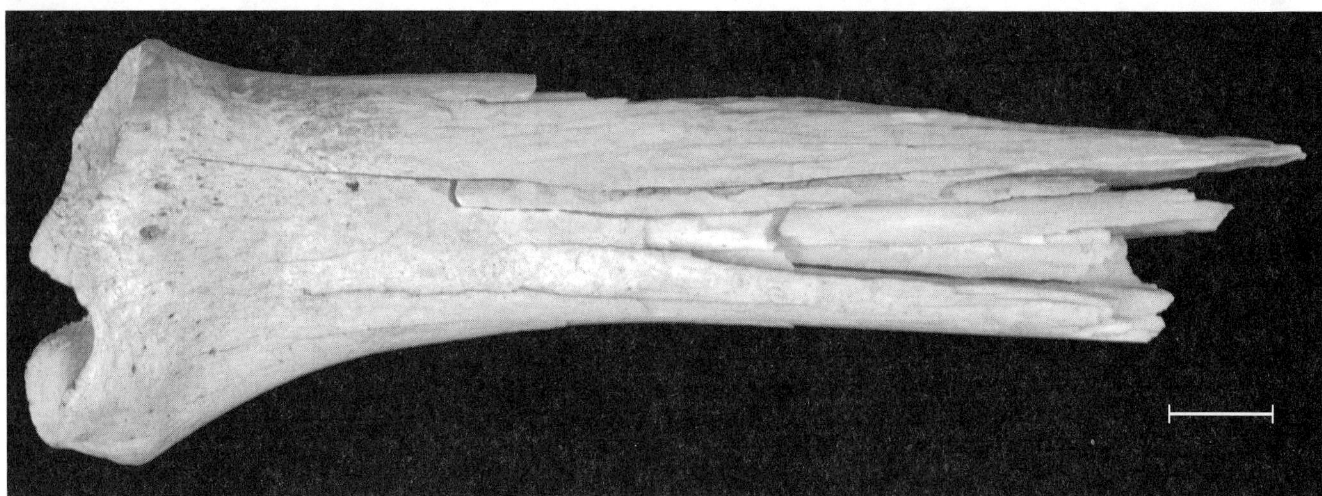

FIGURE 7.12. Mule Deer proximal radius showing dry bone fractures. Note jagged, angular, splintered break edges.

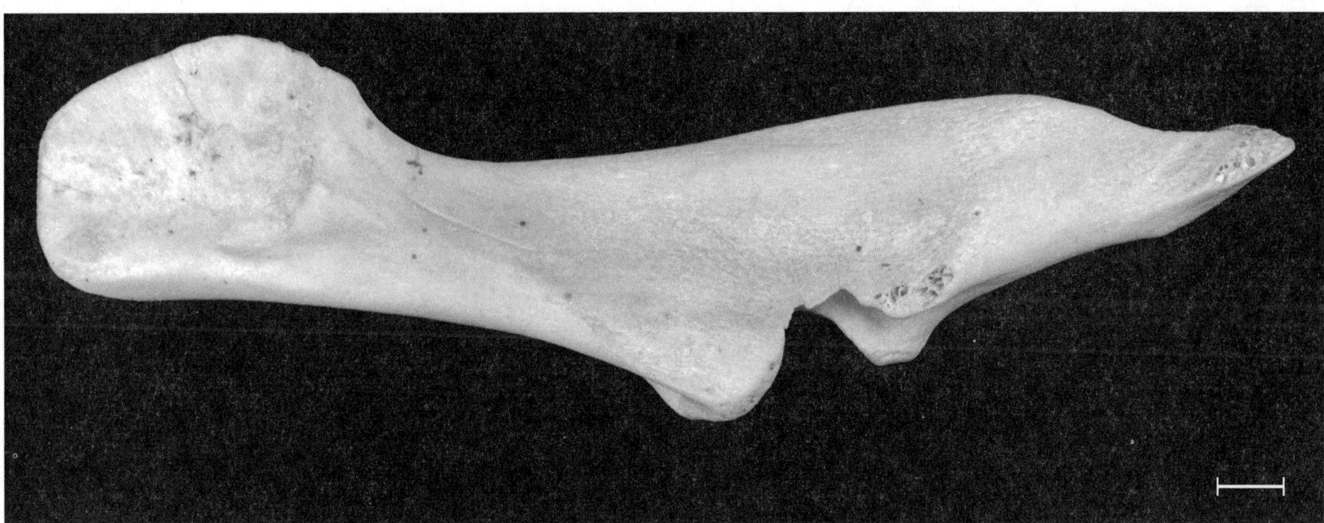

FIGURE 7.13. Mule Deer os coxa collected from a lake shoreline. The bone is heavily smoothed and polished from abrasion by waterborne sands and gravels.

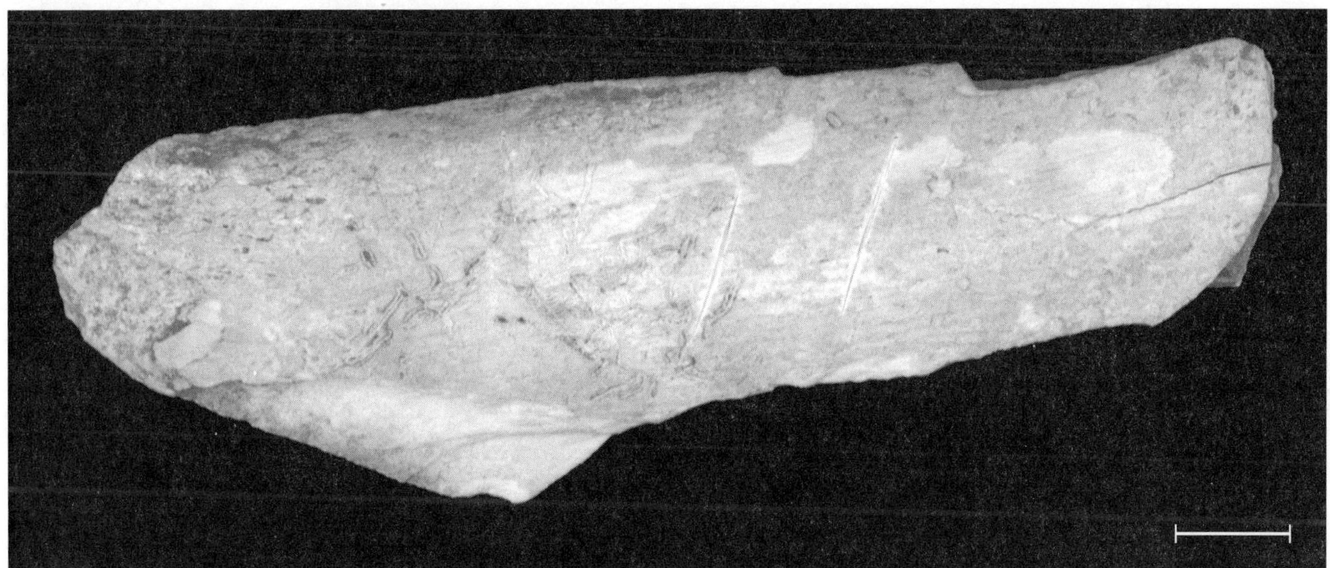

FIGURE 7.14. Trowel marks on large mammal longbone midshaft fragment. Note lighter color of the marks—these recent cuts exposed the unstained subsurface of the cortical bone.

by their lighter color relative to the surrounding bone. This is because bones are naturally white in color, but are typically stained in various shades of brown by the archaeological sedimentary matrix in which they have rested. Recent or modern scratches by trowels or shovels remove the stained surface and expose the lighter, natural bone color (Figure 7.14).

Notes

Our discussion of bone surface modifications is informed by descriptions and photographs of specimens in Binford (1981), Lyman (1994), Hockett and Jenkins (2013), Fisher (1995), and Dominguez-Rodrigo et al. (2010). Behrensmeyer (1978) provides a commonly cited sequence of bone-weathering stages. Bochenski and Tornberg's (2007) description of raptor damage on bird remains informs our discussion on this subject. Grayson (1988) provides an interesting discussion on interpreting burned bones from dry cave settings, and our discussion of rodent gnawing draws on the work of Klippel and Synstelien (2007).

References

Alderfer, J. (ed.)

2014 *National Geographic Complete Birds of North America*, 2nd edition. National Geographic, Washington D.C.,

American Ornithologists' Union

2015 *Checklist of North and Middle American Birds. Electronic document*, http://www.aou.org/checklist/north/print.php, accessed January 10, 2015.

Baumel, J. J. (ed.)

1993 *Handbook of Avian Anatomy: Nomina Anatomica Avium.* Publications of the Nuttall Orithological Club 23.

Behrensmeyer, A. K.

1978 Taphonomic and Ecologic Information from Bone Weathering. *Paleobiology* 4:150–162.

Bengston, P.

1988 Open Nomenclature. *Palaeontology* 31:223–227.

Benton, M. J.

1997 *Vertebrate Palaeontology*, 2nd ed. Chapman and Hall, London.

Bever, G. S., T. R. Lyson, D. J. Field, and B. A. S. Bhullar

2015 Evolutionary Origin of the Turtle Skull. *Nature* 525:239–242.

Binford, L. R.

1981 *Bones: Ancient Men and Modern Myths.* Academic Press, New York.

Bochenski, Z. M., and R. Tornberg

2007 Fragmentation and Preservation of Bird Bones in Uneaten Food Remains of the Gyrfalcon *Falco rusticolus*. *Journal of Archaeological Science* 30:1665–1671.

Bruzgul, J. E., W. Long, and E. A. Hadly

2005 Temporal Response of the Tiger Salamander (*Ambystoma tigrinum*) to 3,000 Years of Climatic Variation. *BMC Ecology* 5:7.

Collins, J. T, and T. W. Taggart

2015 *Standard Common and Current Scientific Names for North American Amphibians, Turtles, Reptiles, and Crocodilians*, 6th ed. Publication of the Center for North American Herpetology, Lawrence, KS. Electronic document, http://www.cnah.org, accessed January 4, 2015.

Dominguez-Rodrigo, M., T. R. Pickering, and H. T. Bunn

2010 Configurational Approach to Identifying the Earliest Hominin Butchers. *Proceedings of the National Academy of Sciences* 107:20929–20934.

Elbroch, M.

2006 *Animal Skulls: A Guide to North American Species*. Stackpole Books, Mechanicsburg, PA.

Elphick, C., J. B., Dunning, and D. A. Sibley

2001 *The Sibley Guide to Bird Life and Behavior.* Alfred Knopf, New York.

Fisher, J. W.

1995 Bone Surface Modifications in Zooarchaeology. *Journal of Archaeological Method and Theory* 2:7–68.

Froese, R., and D. Pauly (eds.)

2015 *Fishbase*. Electronic publication: www.fishbase.org

Gilbert, M. B.

1990 *Mammalian Osteology*. Missouri Archaeological Society, Columbia, MO.

Gilbert, B. M., L. D. Martin, and H. G. Savage

1981 *Avian Osteology*. B. M. Gilbert (Publisher), Laramie, WY.

Gilbert, S.

1975 *Pictorial Anatomy of the Cat.* University of Washington Press, Seattle.

Godefroit, P., S. M. Sinutsa, D. Dhouailly, Y. L. Bolotsky, A. V. Sizov, M. E. McNamara, M. J. Benton, and P. Spagna

2014 A Jurassic Ornithiscian Dinosaur from Siberia with Both Feathers and Scales. *Science* 345 (6195):451–455.

Grayson, D. K.

1988 *Danger Cave, Last Supper Cave, and Hanging Rock Shelter: The Faunas*. Anthropological Papers of the American Museum of Natural History 66.

Hildebrand, M.

1995 *Analysis of Vertebrate Structure.* John Wiley and Sons, Inc., New York, NY.

Hockett, B., and D. L. Jenkins

2013 Identifying Stone Tool Cut Marks and the Pre-Clovis Occupation of the Paisley Caves. *American Antiquity* 78:762–778.

Howard, H.

1929 *The Avifauna of the Emeryville Shellmound. University of California Publications in Zoology* 32.

Jameson, E. W., and H. J. Peeters

2004 *Mammals of California*. University of California Press, Berkeley.

Jobling, J. A.

2009 *Helm Dictionary of Scientific Bird Names*. Christopher Helm Publishers Ltd., London, UK.

Kardong, K. V.

2004 *Vertebrates: Comparative Anatomy, Function, Evolution*, 4th ed. McGraw Hill, New York.

Kays, R. W., and D. E. Wilson

2009 *Mammals of North America*, 2nd ed. Princeton University Press, Princeton, NJ.

King, G. M., and D. R. N. Custance

1982 *Colour Atlas of Vertebrate Anatomy: An Integrated Text and Dissection Guide*. Bolsover Press, London, UK.

Klippel, W. E., and J. A. Synstelien

2007 Rodents as Taphonomic Agents: Bone Gnawing by Brown Rats and Gray Squirrels. *Journal of Forensic Science* 52:765–773.

Lederer, R. J.

1984 *Ecology and Field Biology*. The Benjamin/Cummings Publishing Co, Menlo Park, CA.

Liem, K. F., W. E. Bemis, W. F. Walker, Jr., and L. Grande

2001 *Functional Anatomy of the Vertebrates: An Evolutionary Perspective*, 3rd ed. Brooks/Cole—Thomson Learning, Belmont, CA.

Lyman, R. Lee

1994 *Vertebrate Taphonomy*. Cambridge University Press, New York.

Moyle, P. B.

2002 *Inland Fishes of California: Revised and Expanded*. University of California Press, Berkeley.

Olsen, S. J.

1968 *Fish, Amphibian and Reptile Remains from Archaeological Sites: Part 1, Southeastern and Southwestern United States, Appendix: The Osteology of the Wild Turkey*. Papers of the Peabody Museum of Archaeology and Ethnology Vol. 56, No. 2. Harvard University, Cambridge.

1979 *Osteology for the Archaeologist. North American Birds: Skulls and Mandibles* (No. 4). *North American Birds: Postcranial Skeletons* (No. 5). Papers of the Peabody Museum of Archaeology and Ethnology Vol. 56, Nos. 4–5. Harvard University, Cambridge.

Poole, A., and F. Gill (eds.)

2015 *The Birds of North America*. The Academy of Natural Sciences, Philadelphia, and the American Ornithologists' Union, Washington, D.C., Electronic document, http://bna.birds.cornell.edu/bna/species, accessed 15 January 2015.

Reid, F.

2006 *Peterson Field Guides: Mammals of North America*. Houghton Mifflin, New York.

Sibley, D. A.

2014 *The Sibley Guide to Birds*, 2nd ed. Alfred A. Knopf, New York.

Sigler, W. F., and J. W. Sigler

1987 *Fishes of the Great Basin: A Natural History*. University of Nevada Press, Reno.

Smith R. E., and V. L. Butler

2008 Towards the Identification of Lamprey (*Lampetra* spp.) in Archaeological Contexts. *Journal of Northwest Anthropology* 42:131–142.

Stebbins, R. C.

2003 *A Field Guide to Western Reptiles and Amphibians*, 3rd ed. Houghton Mifflin, New York.

Walker, J. D., J. W. Geissman, S. A. Bowring, and L. E. Babcock.

2012 *Geologic Time Scale V. 4.0*. Geological Society of America. Electronic document: http://www.geosociety.org/science/timescale/timescl.pdf, accessed January 15, 2015.

Wheeler, A. and A. K. G. Jones

1989 *Fishes*. Cambridge University Press, Cambridge, UK.

Whitaker, J. O.

1980 *The Audubon Society Field Guide to North American Mammals*. Alfred A. Knopf, New York.

White, T. D., and P. A. Folkens

2005 *The Human Bone Manual*. Elsevier Academic Press, Oxford, UK.

Wilson, D. E., and D. M. Reeder

2015 *Mammal Species of the World. A Taxonomic and Geographic Reference*, 3rd ed. Electronic resource, http://www.vertebrates.si.edu/msw/mswcfapp/msw/, accessed January 23, 2015.

Wilson, D. E., and S. Ruff

1999 *The Smithsonian Book of North American Mammals*. UBC Press, Vancouver, B.C.

Wydoski, R. S. and R. R. Whitney

2003 *Inland Fishes of Washington*, 2nd ed. University of Washington Press, Seattle.

Suggestions for Further Reading

Chapter 1: Introduction

Bilezikian, J. P., L. G. Raisz, G. A. Rodan
1996 *Principles of Bone Biology*. Academic Press, San Diego.

Cooper, J. E., and M. E. Cooper
2013 *Wildlife Forensic Investigation. Principles and Practice*. CRC Press, Taylor and Francis Group, London, UK.

Deng, H. and Y. Liu
2005 *Current Topics in Bone Biology*. World Scientific Hackensack, NJ.

Driver, J. D.
2011 Identification, Classification, and Zooarchaeology. *Ethnobiology Letters* 2:19–39.

Eisen, J. A.
2007 Phylogenetic Reconstruction. In *Evolution*, edited by N. H. Barton, D. E. G. Briggs, J. A. Eisen, D. B. Goldstein, and N. H. Patel. Cold Spring Harbor Laboratory Press, Cold Spring Harbor, NY. Online chapter 27:1–55.

C. R. Gilbert, R. N. Lea, and J. D. Williams
2004 *Common and Scientific Names of Fishes from the United States, Canada, and Mexico*. American Fisheries Society, Special Publication 29, Bethesda, MD.

Hamilton, A.
2014 *The Evolution of Phylogenetic Systematics*. University of California Press, Berkeley.

Hancox, N. M.
1972 *Biology of Bone*. Cambridge University Press, Cambridge, UK.

Hennig, W.
1999 *Phylogenetic Systematics*. University of Illinois Press, Urbana.

Harland, W. B.
1990 *A Geologic Time Scale*. Cambridge University Press, Cambridge.

Hillson, S.
2005 *Teeth*, 2nd ed. Cambridge University Press, Cambridge.

Mayne, K. I, R. S. J. Lambert, and D. York
1959 The Geological Time-Scale. *Nature* 183:212–214.

Mayr, E.
1981 Biological Classification: Toward a Synthesis of Opposing Methodologies. *Science* 214 (4520): 510–516.

Ogg, G., J. G. Ogg, and F. M. Gradstein
2008 *The Concise Geologic Time Scale*, 1st ed. Cambridge University Press, Cambridge, UK.

Pourquie, O. (ed.)
2009 *The Skeletal System*. Cold Spring Harbor Laboratory Press, Cold Spring Harbor, NY.

Schuh R. T.
2000 *Biological Systematics: Principles and Applications*. Cornell University Press, Ithaca, NY.

Simpson, G. G.
1961 *Principles of Animal Taxonomy*. Columbia University Press, New York.

Virtual Zooarchaeology of the Arctic Project
2014 Electronic document, vzap.iri.isu.edu/, accessed October 22, 2014.

Waldon, T.
2009 *Palaeopathology*. Cambridge University Press, Cambridge, UK.

White, T. D., M. T. Black, and P. A. Folkens
2011 *Human Osteology*, 3rd ed. Academic Press, San Diego.

Williams, D. M., and S. Knapp
2010 *Beyond Cladistics: The Branching of a Paradigm*. University of California Press, Berkeley.

Chapter 2: Fishes

Branson, B. A.
1962 Comparative Cephalic and Appendicular Osteology of the Fish Family Catostomidae. Part I, *Cycleptus elongates* (Lesueur). *Southwestern Naturalist* 7:81–153.

Brinkhuizen, D. C., and A. T. Clason
1986 *Fish and Archaeology: Studies in Osteometry, Taphonomy, Seasonality and Fishing Methods*. BAR International Series 294.

Broughton, J. M.
2000 Terminal Pleistocene Fish Remains from Homestead Cave, Utah, and Implications for

Fish Biogeography in the Bonneville Basin. *Copeia* 2000:645–656.

Broughton, J. M., D. B. Madsen, and J. Quade
2000 Fish Remains from Homestead Cave and Lake Levels of the Past 13,000 Years in the Bonneville Basin. *Quaternary Research* 53:392–401.

Broughton, J. M., E. Martin, B. McEneaney, T. Wake, and D. D. Simons
2015 Late Holocene Anthropogenic Depression of Sturgeon in San Francisco Bay, California. *Journal of California and Great Basin Anthropology* 35:3–27.

Broughton, R. E., R. Betancur-R, C. Li, G. Arratia, and G. Orti
2013 Multi-locus Phylogenetic Analysis Reveals the Pattern and Tempo of Bony Fish Evolution. *PLOS Currents: Tree of Life*. 2013 Apr 16 [last modified: 2013 Apr 17]. Edition 1. doi: 10.1371/currents.tol.2ca8041495ffafd0c92756e75247483e.

Butler, V. L.
1996 Tui Chub Taphonomy and the Importance of Marsh Resources in the Western Great Basin of North America. *American Antiquity* 61(4): 699–717.

Butler, V. L. and J. E. O'Connor
2004 9,000 Years of Fishing on the Columbia River. *Quaternary Research* 62 (1):1–8.

Cannon, D. Y.
1987 *Marine Fish Osteology: A Manual for Archaeologists*. Archaeology Press, Simon Fraser University, Burnaby, BC.

Casteel, R.W.
1976 *Fish Remains in Archaeology and Paleoenvironmental Studies*. Academic Press, London.

Colley, S. M.
1990 The Analysis and Interpretation of Archaeological Fish Remains. *Archaeological Method and Theory* 2:207–253.

Follett, W.
1980 Fish Remains from the Karlo Site (CA-Las-7), Lassen County, California. *Journal of California and Great Basin Anthropology* 2:114–122.

Gobalet, K. W., P. D. Schulz, T. Wake, and N. Siefkin
2004 Archaeological Perspectives on Native American Fisheries of California, With Emphasis on Steelhead and Salmon. *Transactions of the American Fisheries Society* 133:801–833.

Gregory, W. K.
1933 Fish Skulls: A Study of the Evolution of Natural Mechanisms. *Transactions of the American Philosophical Society* 23:2.

Groot, C. and L. Margolis (eds.)
1991 *Pacific Salmon Life Histories*. UBC Press, Vancouver.

Miller, J.
2012 Lamprey "Eels" in the Greater Northwest: A Survey of Tribal Sources, Experiences, and Sciences. *Journal of Northwest Anthropology* 46:65–84.

Minckley, W. L.
1973 *Fishes of Arizona*. Arizona Game and Fish Dept.

Minckley, W. L., and P. C. Marsh
2009 *Inland Fishes of the Greater Southwest: Chronicle of a Vanishing Biota*. University of Arizona Press, Tucson.

Moss, M. L and A. Cannon (eds.)
2011 *The Archaeology of North Pacific Fisheries*. University of Alaska Press, Fairbanks.

Mundell, T. R.
1975 *An Illustrated Osteology of the Channel Catfish (Ictalurus punctatus)*. National Park Service, Midwest Archaeological Center, Lincoln, NE.

Nelson, J. S., E. J. Crossman, H. Espinosa-Perez, L. T. Findley, C. R. Gilbert, R. N. Lea, and J. D. Williams
2004 *Common and Scientific Names of Fishes from the United States, Canada, and Mexico*. American Fisheries Society, Special Publication 29, Bethesda, MD.

Norden, C. R.
1961 Comparative Osteology of Representative Salmonid Fishes, with Particular Reference to the Grayling (*Thymallus arcticus*) and Its Phylogeny. *Journal of the Fisheries Research Board of Canada* 18:679–791.

Page, L. M. and B. M. Burr
2011 *Peterson Field Guide to Freshwater Fishes of North America, North of Mexico*, 2nd ed. Houghton Mifflin Harcourt, New York, NY.

Roberts, T. R.
1974 Osteology and Classification of the Neotropical Characoid Fishes of the Families Hemiodontidae (including Anodontinae) and Parodontidae. *Bulletin of the Museum of Comparative Zoology* 146:411–472.

Rojo, A. L.
1991 *Dictionary of Evolutionary Fish Osteology*. CRC Press, Boca Raton, FL.

Ross, S. T.
2013 *Ecology of North American Freshwater Fishes*. University of California Press, Berkeley.

Smith, G. R., T. E. Dowling, K.W. Gobalet, T. Lugaski, D. K. Shiozawa, and R. P. Evans
2002 Biogeography and Timing of Evolutionary Events among Great Basin Fishes. In *Great Basin Aquatic Systems History*, R. Hershler, D. B. Madsen, and D. R Currey (editors). Smithsonian Contributions to the Earth Sciences 33:175–234.

Sublette, J. E., M. Hatch, and M. F Sublette
1990 *The Fishes of New Mexico*. University of New Mexico Press, Albuquerque.

Weisel, G. F.
1960 The Osteocranium of the Catostomid Fish,

Catostomus macrocheilus: A study in Adaptation and Natural Relationship. *Journal of Morphology* 106:109-2-129.

Chapter 3: Amphibians

Bartlett, R. D. and P. P. Bartlett
2013 *New Mexico's Reptiles and Amphibians: A Field Guide*. University of New Mexico Press, Albuquerque.
Bever, G. S.
2005 Variation in the Ilium of North American *Bufo* (Lissamphibia: Anura) and its Implications for Species-Level Identification of Fragmentary Anuran Fossils. *Journal of Vertebrate Paleontology* 25:548–560.
Buckley, D., M. H. Wake, and D. B. Wake
2009 Comparative Skull Osteology of *Karsenia koreana* (Amphibia, Cuadata, Plethodontidae). *Journal of Morphology* 271:533–558.
Chantel, C. J.
1968 The Osteology of *Pseudacris* (Amphibia: Hylidae). *American Midland Naturalist* 80:381–391.
Corkran, C. C., and C. Thoms
1996 *Amphibians of Oregon, Washington, and British Columbia*. Lone Pine Publishing, Edmonton, AB.
Degenhardt, W. G., C. W. Painter, and A. H. Price
1996 *Amphibians and Reptiles of New Mexico*. University of New Mexico Press, Albuquerque.
Duellman, W. E., and L. Trueb
1994 *Biology of Amphibians*. Johns Hopkins University Press, Baltimore.
Gaudin, A. J.
1974 An Osteological Analysis of Holarctic Tree Frogs, Family Hylidae. *Journal of Herpetology* 8:141–152.
Green, D. M., L. Veir, G. Casper, and M. Lannoo
2014 *North American Amphibians: Distribution and Diversity*. University of California Press, Berkeley.
Holman, J. A.
1995 *Pleistocene Amphibians and Reptiles of North America*. Oxford University Press, New York.
Hoyos, J. M., M. R. Sanchez-Villagra, A. A. Carlini, and C. Mitgutsch
2012 Skeletal Development and Adult osteology of *Hypsiboas pulchellus* (Anura: Hylidae). *Acta Herpetologica* 7:119–138.
Jones, L. L. C., W. P. Leonard, and D. H. Olson
2005 *Amphibians of the Pacific Northwest*. Seattle Audubon Society, Seattle, WA.
Heatwole, H.
2003 *Amphibian Biology (Volume 5), Osteology*. Surrey Beatty and Sons, Baulkham Hills BC, NSW, Australia.
Kysely, R.
2008 Frogs as a Part of the Eneolithic diet. Archaeozoological Records from the Czech Republic (Kutná Hora-Denemark Site, Řivnáč Culture). *Journal of Archaeological Science* 35:143–157.
Maglia, A. M. and L. A. Pugener
1998 Skeletal Development and Adult Osteology of *Bombina orientalis* (Anura: Bombinatoridae). *Herpetologia* 54:344–363.
Mead, J. I. and C. J. Bell
1994 Late Pleistocene and Holocene Herpetofaunas of the Great Basin and Colorado Plateau. In *The Natural History of the Colorado Plateau and Great Basin*, K. Harper, L. St. Clair, K. Thorne, and W. Hess (editors), pp. 225–275. University Press of Colorado, Niwot.
Olsen, S. J.
1971 *Zooarchaeology: Animal Bones in Archaeology and Their Interpretation, Issue* 2, 1971, pp. 1–30. Addison-Wesley Publishing Co., New York.
Ponssa, M. L.
2008 Cladistic Analysis and Osteological Descriptions of the Frog Species in the *Leptodactylus fuscus* Species Group (Anura, Leptodactylidae). *Journal of Zoological Systematics and Evolutionary Research* 46:249–266.
Ponssa, M. L., F. Brusquetti, and F. L. Souza
2011 Osteology and Intraspecific Variation of *Leptodactylus podicipinus* (Anura: Leptodactylidae), with Comments on the Relationship Between Osteology and Reproductive Modes. *Journal of Herpetology* 45:79–93.
Sampson, C. G.
2003 Amphibians from the Acheulean Site at Duinefontein 2 (Western Cape, South Africa). *Journal of Archaeological Science* 30:547–557.
Stebbins, R. C. and S. McGinnis
2012 *Field Guide to Amphibians and Reptiles of California*. University of California Press, Berkeley.
Velez-Rodriguez, C. M.
2005 Osteology of *Bufo sternosignatus* Gunther, 1858 (Anura:Bufonidae) with Comments on Phylogenetic Implications. *Journal of Herpetology* 39:299–303.
Wake, D. B.
1963 Comparative Osteology of the Plethodontid Salamander Genus *Aneides*. *Journal of Morphology* 113:77–118.
Wells, K. D.
2007 *The Ecology and Behavior of Amphibians*. University of Chicago Press, Chicago.
Werner, J. K., B. A Maxwell, P. Hendricks, and D. Flath
2004 *Amphibians and Reptiles of Montana*. Mountain Press Publishing, Missoula.
Young, M. T.
2011 *The Guide to Colorado Reptiles and Amphibians*. Fulcrum Publishing, Golden Colorado.

Chapter 4: Reptiles

Bartlett, R. D., and P. P. Bartlett
2013 *New Mexico's Reptiles and Amphibians: A Field Guide*. University of New Mexico Press, Albuquerque.

Blasco, R.
2008 Human Consumption of Tortoises at Level IV of Bolomor Cave (Valencia, Spain). *Journal of Archaeological Science* 35:2839–2848.

Brennan, T. C.
2015 *Online Field Guide to The Reptiles and Amphibians of Arizona*. Electronic document, http://www.reptilesofaz.org, accessed January 20, 2015.

Degenhardt, W. G., C. W. Painter, and A. H. Price
1996 *Amphibians and Reptiles of New Mexico*. University of New Mexico Press, Albuquerque.

de Queiroz, K.
1987 Phylogenetic Systematics of the Iguanine Lizards: A Comparative Osteological Study. *University of California Publications in Zoology* 118.

Gans, C., A. S. Gaunt, and K. Adler
2008 *Biology of the Reptilia: Volume 20, Morphology H: The Skull of Lepidosauria*. Society for the Study of Amphibians and Reptiles, Salt Lake City, UT.

Hollenshead, M.
2004 Mineral Hill Cave Reptiles. In *Paleontological Investigations at Mineral Hill Cave*, B. Hockett and E. Dillingham (editors), pp.72–135. Contribution to the Study of Cultural Resources Technical Report No. 18. U.S. Department of Interior, BLM, Elko, NV.

Khanna, D. R., and P. R. Yadav
2004 *Biology of Reptiles*. Discovery Publishing House, New Delhi, India.

Lee, M. S. Y, and J. D. Scanlon
2002 Snake Phylogeny Based on Osteology, Soft Anatomy and Ecology. *Biological Review* 77: 333–401.

Lewis, D.
2011 *A Field Guide to the Amphibians and Reptiles of Wyoming*. The Wyoming Naturalist, Douglas, WY.

Mead, J.
1983 Paleontology of Hidden Cave: Amphibians and Reptiles. In *The Archaeology of Hidden Cave, Nevada*. Anthropological Papers of the American Museum of Natural History 61:162–170.

Mead, J. I., and C. J. Bell.
1994 Late Pleistocene and Holocene Herpetofaunas of the Great Basin and Colorado Plateau. In *The Natural History of the Colorado Plateau and Great Basin*, K. Harper, L. St. Clair, K. Thorne, and W. Hess (editors), pp. 225–275. University Press of Colorado, Niwot.

Perez, J. V. M. and A. S. Serra
2009 The Quaternary Fossil Record of the Genus *Testudo* in the Iberian Peninsula. Archaeological Implications and Diachronic Distribution in the Western Mediterranean. *Journal of Archaeological Science* 36:1152–1162.

Romer, A. S.
1997 *Osteology of the Reptiles*. Krieger Publishing, Malabar, FL.

Rybczynski, N., D. Gifford-Gonzalez, and K. M. Stewart
1996 The Ethnoarchaeology of Reptile Remains at a Lake Turkana Occupation Site, Kenya. *Journal of Archaeological Science* 23:863–867.

Sampson, C. G.
2000 The Taphonomy of Turtles by Birds and Bushmen. *Journal of Archaeological Science* 27:779–788.

Schneider, J. S., and G. D. Dickson.
1989 The Desert Tortoise (*Xerobates agassizii*) in the Prehistory of the Southwestern Great Basin and Adjacent Areas. *Journal of California and Great Basin Anthropology* 11:175–202.

Sobolik, K. D., and D. G. Steele
1996 *A Turtle Atlas to Facilitate Archaeological Identifications*. Mammoth Site of Hot Springs, South Dakota.

Stahl, P. W.
1996 The Recovery and Interpretation of Microvertebrate Bone Assemblages from Archaeological Contexts. *Journal of Archaeological Method and Theory* 3:31–75.

Stiner, M. C., N. D. Munro, and T. A. Surovell
2000 The Tortoise and the Hare: Small Game Use, the Broad-Spectrum Revolution, and Paleolithic Demography. *Current Anthropology* 41:39–73.

St. John, A.
2002 *Reptiles of the Northwest: California to Alaska; Rockies to the Coast*. Lone Pine Publishing, Edmonton, AB.

Stoetzel, E., C. Denys, S. Bailon, M. A. El Hajraoui, and R. Nespoulet
2012 Taphonomic Analysis of Amphibian and Squamate Remains from El Harhoura 2 (Rabat-Témara, Morocco): Contributions to Palaeoecological and Archaeological Interpretations. *International Journal of Osteoarchaeology* 22:616–635.

Van Devender, T. R.
2002 *The Sonoran Desert Tortoise: Natural History, Biology, and Conservation*. University of Arizona Press, Tucson.

Werner, J. K., B. A Maxwell, P. Hendricks, and D. Flath
2004 *Amphibians and Reptiles of Montana*. Mountain Press Publishing, Missoula.

Chapter 5: Mammals

American Society of Mammalogists

2015 *Mammalian Species*. Electronic document, www.science.smith.edu/msi/, accessed January 14, 2015.

Adams, B., and P. Crabtree

2011 *Comparative Osteology: A Laboratory and Field Guide to Common North American Animals*. Elsevier, Waltham, MA.

Bailey, V.

1936 *The Mammals and Life Zones of Oregon*. North American Fauna No. 55. U.S. Fish and Wildlife Service, Washington, D.C.

Balkwill, D. M., and S. L. Cumbaa

1992 A Guide to the Identification of Postcranial Bones of *Bos taurus* and *Bison bison*. *Syllogeus* 71. Canadian Museum of Nature, Ottawa.

Beisaw A. M.

2013 *Identifying and Interpreting Animal Bones: A Manual*. Texas A&M University Press, College Station.

Brown, C. L., and C. E. Gustafson

1979 *Key to Postcranial Skeletal Remains of Cattle/Bison, Elk, and Horse*. Washington State University Laboratory of Anthropology Reports on Investigations No. 57. Pullman.

Dalquest, W. W.

1948 *Mammals of Washington*. University of Kansas Publications, Museum of Natural History, Vol. 2., Lawrence, KS.

Durrant, S. D.

1952 *Mammals of Utah: Taxonomy and Distribution*. University of Kansas Publications, Museum of Natural History, Vol. 6, Lawrence, KS.

Feldhamer, G. A., L. C. Drickamer, S. H. Vessey, J. F. Merritt, and C. Krajewski

2007 *Mammalogy: Adaptation, Diversity, Ecology*, 3rd ed. Johns Hopkins University Press, Baltimore.

Feldhamer, G. A., B. C. Thompson, and J. A. Chapman

2003 *Wild Mammals of North America: Biology, Management, and Conservation*, 2nd ed. Johns Hopkins University Press, Baltimore.

Findley, J. S., A. H. Harris, D. E. Wilson, and C. Jones

1975 *Mammals of New Mexico*. University of New Mexico Press, Albuquerque.

Ford, P. J.

1990 Antelope, Deer, Bighorn Sheep, and Mountain Goats: A Guide to the Carpals. *Journal of Ethnobiology* 10:169–181.

Foresman, K. R., and A. V. Badyaev

2012 *Mammals of Montana*, 2nd ed. Mountain Press, Missoula.

France, D.

2008 *Human and Non-human Bone Identification: A Color Atlas*. CRC Press, Boca Raton, FL.

Gilbert, S.

1975 *Pictorial Anatomy of the Cat*. University of Washington Press, Seattle.

Glass, B. P.

1973 *A Key to the Skulls of North American Mammals*, 2nd ed. Privately printed, Department of Zoology, Oklahoma State University, Stillwater.

Grayson, D. K.

1984 *Quantitative Zooarchaeology*. Academic Press, Orlando, FL.

2006 The Late Quaternary Biogeographic Histories of some Great Basin Mammals (Western USA). *Quaternary Science Reviews* 25: 2964–2991.

Hall, E. R.

1946 *Mammals of Nevada*. University of California Press, Berkeley.

1981 *The Mammals of North America*, 2nd ed. John Wiley and Sons, New York.

Hoffmeister, D. F.

1986 *Mammals of Arizona*. University of Arizona Press, Tucson.

Ingles, L. G.

1965 *Mammals of the Pacific States: California, Oregon, and Washington*. Stanford University Press, Stanford.

Jacobson, J. A.

2004 Determining Human Ecology on the Plains Through the Identification of Mule Deer (*Odocoileus hemionus*) and White-tailed Deer (*Odocoileus virginianus*) Postcranial Remains. Unpublished PhD dissertation, Department of Anthropology, University of Tennessee, Knoxville.

Jones, J. K., and R. W. Manning

1992 *Illustrated Key to Skulls of North American Land Mammals*. Texas Tech University Press, Lubbock.

Larrison, E. J.

1967 *Guide to Idaho Mammals*. Journal of the Idaho Academy of Science 7.

Lawrence, B.

1951 *Post-cranial Skeletal Characters of Deer, Pronghorn, and Sheep-goat with notes on Bos and Bison*. Papers of the Peabody Museum of American and Ethnology, Harvard University 35:9–43.

Lyman, R. Lee

2008 *Quantitative Paleozoology*. Cambridge University Press, Cambridge.

Maser, C.

1998 *Mammals of the Pacific Northwest: From the Coast to the High Cascades*. Oregon State University Press, Corvallis.

Mullican, T. R., and L. N. Carraway

1990 Shrew Remains from Moonshiner and Middle Butte Caves, Idaho. *Journal of Mammalogy* 71:351–356.

Olsen, S. J.
1960 *Post-cranial Skeletal Characters of Bison and Bos*. Papers of the Peabody Museum of Archaeology and Ethnology, Harvard University 35(4).
Olsen, S. J.
1973 *Mammal Remains from Archaeological Sites: Part I—Southeastern and Southwestern United States*. Papers of the Peabody Museum of Archaeology and Ethnology, Harvard University, Vol. 56 (1):1–162.
Sargis, E. J., and M. Dagosto (eds.)
2008 *Mammalian Evolutionary Morphology: A Tribute to Frederick S. Szalay*. Springer, Dordrecht, the Netherlands.
Searfoss, G.
1995 *Skulls and Bones: A Guide to Skeletal Structures and Behavior of North American Mammals*. Stackpole Books, Mechanicsburg, PA.
Vaughan, T. A., J. M. Ryan, and N. J. Czaplewski
2011 *Mammalogy*, 5th ed. Jones and Bartlett Publishers, Sudbury, MA.
Verts, B. J., and L. N. Carraway
1998 *Land Mammals of Oregon*. University of California Press, Berkeley.
Wolverton, S., and R. L. Lyman
2012 *Conservation Biology and Applied Zooarchaeology*. University of Arizona Press, Tucson.
Zeveloff, S. I., and F. R. Collett
1988 *Mammals of the Intermountain West*. University of Utah Press, Salt Lake City.

Chapter 6: Birds

Alcorn, J. R.
1988 *The Birds of Nevada*. Fairview West Publishing, Fallon, NV.
Beedy, E. C., E. R. Pandolfino, and K. Hansen
2013 *Birds of the Sierra Nevada: Their Natural History, Status, and Distribution*. University of California Press, Berkeley.
Bochenski, Z. M.
2008 Identification of Skeletal Remains of Closely Related Species: The Pitfalls and Solutions. *Journal of Archaeological Science* 35:1247–1250.
Bochenski, Z. M., and T. Tomek
1997 Preservation of Bird bones: Erosion Versus Digestion by Owls. *International Journal of Osteoarchaeology* 7:372–387.
Bovy, K. M.
2012 Why So Many Wings? A Re-examination of Avian Skeletal Part Representation in the South-central Northwest Coast, USA. *Journal of Archaeological Science* 39:2049–2059.
Broughton, J. M.
2004 *Prehistoric Human Impacts on California Birds: Evidence from the Emeryville Shellmound Avifauna*. Ornithological Monographs 56. American Ornithologists' Union, Washington, D.C.
Broughton, J. M., D. Mullins, and T. Ekker
2007 Avian Resource Depression or Intertaxonomic Variation in Bone Density? A Test with San Francisco Bay Avifaunas. *Journal of Archaeological Science*, 34, 374–391.
Cohen, A., and D. Serjeantsen
1995 *A Manual for the Identification of Bird Bones from Archaeological Sites*. Archetype Books, London.
Dacke, C. G., S. Arkle, D. J. Cook, I. M. Wormstone, S. Jones, M. Zaidi, and Z. A. Bascal.
1993 Medullary Bone and Avian Calcium Regulation. *Journal of Experimental Biology* 184:63–88.
Driver, J. C., and K. A. Hobson
1992 A 10,500-year Sequence of Bird Remains from the Southern Boreal Forest Region of Western Canada. *Arctic* 45:105–110.
Dunn, J. L., and J. Alderfer
2011 *National Geographic Society Field Guide to the Birds of North America*, 6th ed. National Geographic Society, Washington, D.C.
Feducia, A.
1999 *The Origin and Evolution of Birds*. Yale University Press, New Haven, CT.
Floyd, T., and B. Small
2014 *The American Birding Association Field Guide to the Birds of Colorado*. Scott and Nix, New York, NY.
Gabrielson, I., and S. G. Jewett
1940 *Birds of Oregon*. Oregon State College, Studies in Zoology 2, Corvallis.
Gill, F.
2006 *Ornithology*, 3rd ed. W. H. Freeman, New York.
Grayson, D. K.
1973 The Avian and Mammalian Remains from Nightfire Island. Unpublished PhD dissertation, Department of Anthropology, University of Oregon, Eugene.
Larrison, E. J., and K. G. Sonnenberg
1968 *Washington Birds: Location, and Identification*. Seattle Audubon Society, Seattle.
Larrison, E. J., J. L. Tucker, and M. T. Jollie
1967 *Guide to Idaho Birds*. Journal of Idaho Academy of Science 5.
Livezey, B. C.
1993 Morphology of Flightlessness in *Chendytes*, Fossil Seaducks (Anatidae: Mergini) of Coastal California. *Journal of Vertebrate Paleontology* 13:185–199.
Livingston, S.D.
2000 The Homestead Cave Avifauna. In *Late Qua-*

ternary Paleoecology in the Bonneville Basin, D. Madsen (ed.), pp. 91–102. Utah Geological Survey Bulletin 130.

Nuechterlein, G.L., and R. W. Storer.
1982 The Pair-formation Displays of the Western Grebe. *The Condor* 84:350–369.

Phillips, A., J. Marshall, G. Monson, G. M. Sutton, and E. Porter
1964 *The Birds of Arizona*. University of Arizona Press, Tucson.

Pimm, S., P. Raven, A. Peterson, C. H. Sekercioglu, and P. R. Erlich
2006 Human Impacts on the Rates of Recent, Present, and Future Bird Extinctions. *PNAS* 103:10941-10946.

Proctor, N. S., and P. J. Lynch
1998 *Manual of Ornithology: Avian Structure and Function*. Yale University Press, New Haven.

Prummel, W., J. T. Zeiler, and D. C. Brinkhuizen
2010 *Birds in Archaeology. Proceedings of the 6th Meeting of the ICAZ Bird Working Group in Groningen (23.8–27.8. 2008)*. Groningen University Library, Gronigen, the Netherlands.

Russell, N.
2011 *Social Zooarchaeology: Humans and Animals in Prehistory*. Cambridge University Press, Cambridge, UK.

Ryser, F. A.
1985 *Birds of the Great Basin: A Natural History*. University of Nevada Press, Reno.

Shufeldt, R. W.
1886 The Skeleton in *Geococcyx*. *Journal of Anatomy and Physiology* 20:241–266.
1909 *Osteology of Birds*. New York State Museum, Museum Bulletin 130.

Sibley, D. A.
2014 *The Sibley Guide to Birds*, 2nd ed. Alfred A. Knopf, New York.

Serjeantson, D.
2009 *Birds*. Cambridge University Press, Cambridge, UK.

Steadman, D. W.
1995 Prehistoric Extinctions of Pacific Island Birds: Biodiversity Meets Zooarchaeology. *Science* 24:1123–1131.

Woolfenden, Glen
1961 *Postcranial Osteology of the Waterfowl*. Bulletin of the Florida State Museum 6.

Chapter 7: Taphonomy and Bone Damage

Andrews, P.
1990 *Owls, Caves, and Fossils*. University of Chicago Press, Chicago.

Behrensmeyer, A. K., and A. P. Hill
1980 *Fossils in the Making: Vertebrate Taphonomy and Paleoecology*. University of Chicago Press, Chicago.

Blumenschine, R. J., C. W. Marean, and S. Capaldo
1996 Blind Tests of Inter-analyst Correspondence and Accuracy in the Identification of Cut Marks, Percussion Marks, and Carnivore Tooth Marks on Bone Surfaces. *Journal of Archaeological Science* 23:493–507.

Bobe, R., A. Zeresenay, and A. K. Behrensmeyer
2005 *Hominin Environments in the East African Pliocene: and Assessment of the Faunal Evidence*. Springer, Dordrecht.

Bonnichsen, R. and M. Sorg (eds.)
1989 *Bone Modification*. University of Maine Center for the Study of the First Americans, Orono.

Brain C. K.
1981 *The Hunters or the Hunted? An Introduction to African Cave Taphonomy*. University of Chicago Press, Chicago.

Broughton, J. M., V. I. Cannon, S. Arnold, R. J. Bogiatto, and K. Dalton
2006 The Taphonomy of Owl-deposited Fish Remains and the Origin of the Homestead Cave Ichthyofauna. *Journal of Taphonomy* 4:69–95.

Butler, V. L., and R. A. Schroeder
1998 Do Digestive Processes Leave Diagnostic Traces on Fish Bones? *Journal of Archaeological Science* 25:957–971.

Domínguez-Rodrigo, M.
2012 *Stone Tools and Fossil Bones: Debates in the Archaeology of Human Origins*. Cambridge University Press, New York.

Domínguez-Rodrigo M., S. de Juana, A. B. Galán, and M. Rodríguez
2009 A New Protocol to Differentiate Trampling Marks from Butchery Cut Marks. *Journal of Archaeological Science* 36:2643–2654.

Galán, A. B., M. Rodríguez, S. de Juana, and M. Domínguez-Rodrigo
2009 A New Experimental Study on Percussion Marks and Notches and their Bearing on the Interpretation of Hammerstone-broken Faunal Assemblages. *Journal of Archaeological Science* 36:776–784.

Hockett, B.
1996 Corroded, Thinned, and Polished Bones Created by Golden Eagles (*Aquila chrysaetos*): Taphonomic Implications for Archaeological Interpretations. *Journal of Archaeological Science* 23:587–591.

Lloveras, L., M. Moreno-Garcia, and J. Nadal
2008 Taphonomic Study of Leporid Remains Accumulated by the Spanish Imperial Eagle (*Aquila adalberti*). *Geobios* 41:91–100.

Lupo, K. D., J. M. Fancher, and D. N. Schmitt
2013 The Taphonomy of Resource Intensification: Zooarchaeological Implications of Resource Scarcity among Bofi and Aka Forest Foragers. *Journal of Archaeological Method and Theory* 20:420–447.

Lupo, K. D., and J. F. O'Connell
2002 Cut and Tooth Mark Distributions on Large Mammal Bones: Ethnoarchaeological Data from the Hadza and Their Implications for Current Ideas about Early Human Carnivory. *Journal of Archaeological Science* 29:85–109.

Lyman, R. L.
1987 Archaeofaunas and Butchery Studies: A Taphonomic Perspective. In *Advances in Archaeological Method and Theory*, M. B. Schiffer (ed.), Vol. 10, pp. 249–337. Academic Press, San Diego.
2005 Analyzing Cut Marks: Lessons from Artiodactyl Remains in the Northwestern United States. *Journal of Archaeological Science* 32:1722–1732.

O'Conner, T. P. (ed.)
2002 *Biosphere to Lithosphere: New Studies in Vertebrate Taphonomy*. International Council for Archaeozoology Conference 2002. Durham England. Oxbow Books, Oakville, CT.

Schmitt, D. N.
1995 The Taphonomy of Golden Eagle Prey Accumulations at Great Basin Roosts. *Journal of Ethnobiology* 15:237–256.

Schmitt, D. N., and K. E. Juell
1994 Toward the Identification of Coyote Scatological Faunal Accumulations in Archaeological Context. *Journal of Archaeological Science* 21:249–262.

Shipman, P.
1981 *Life History of a Fossil: An Introduction to Taphonomy and Paleoecology*. Harvard University Press, Cambridge.

Index

Note: Entries in *italics* refer to figures. Entries in italics and beginning with a *C* refer to illustrations in the colored insert in the center of the book.